# SELENIUM
*Its Molecular Biology and Role in Human Health*

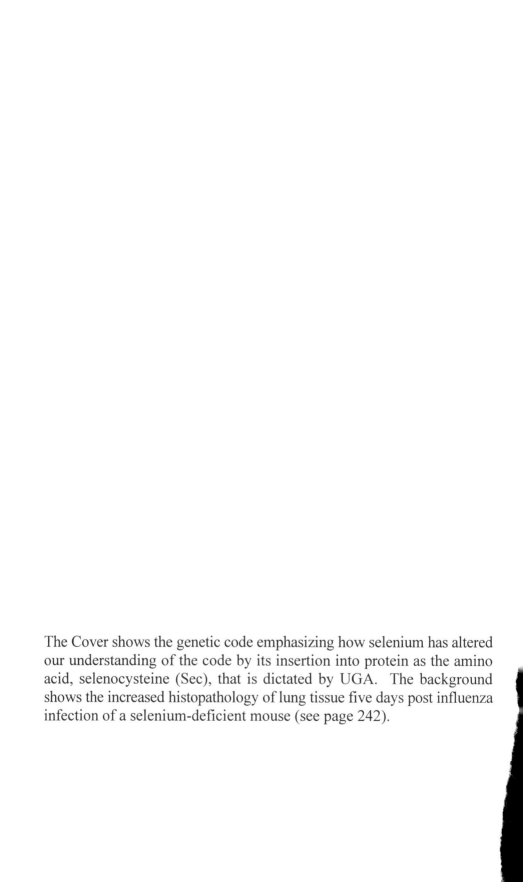

The Cover shows the genetic code emphasizing how selenium has altered our understanding of the code by its insertion into protein as the amino acid, selenocysteine (Sec), that is dictated by UGA. The background shows the increased histopathology of lung tissue five days post influenza infection of a selenium-deficient mouse (see page 242).

# SELENIUM
## *Its Molecular Biology and Role in Human Health*

*edited by*

**Dolph L. Hatfield**
*National Cancer Institute, U.S.A.*

KLUWER ACADEMIC PUBLISHERS
Boston / Dordrecht / London

**Distributors for North, Central and South America:**
Kluwer Academic Publishers
101 Philip Drive
Assinippi Park
Norwell, Massachusetts 02061 USA
Telephone (781) 871-6600
Fax (781) 681-9045
E-Mail <kluwer@wkap.com>

**Distributors for all other countries:**
Kluwer Academic Publishers Group
Distribution Centre
Post Office Box 322
3300 AH Dordrecht, THE NETHERLANDS
Telephone 31 78 6392 392
Fax 31 78 6546 474
E-Mail <services@wkap.nl>

 Electronic Services <http://www.wkap.nl>

**Library of Congress Cataloging-in-Publication Data**

Selenium : its molecular biology and role in human health / edited by Dolph L. Hatfield.
    p. cm.
    Includes bibliographical references and index.
    ISBN 0-7923-7335-9 (alk paper)
    1. Selenium—Health aspects. 2. Selenium—Physiological effect. 3. Selenium in human nutrition. I. Hatfield, Dolph L.

    QP535.S5 S445 2001
    612'.01524—dc21

                                                                          2001020364

**Copyright** © 2001 by Kluwer Academic Publishers

All rights reserved. No part of this publication may be reproduced, stored in a retrieval system or transmitted in any form or by any means, mechanical, photo-copying, recording, or otherwise, without the prior written permission of the publisher, Kluwer Academic Publishers, 101 Philip Drive, Assinippi Park, Norwell, Massachusetts 02061

*Printed on acid-free paper.*

Printed in the United States of America

***The Publisher offers discounts on this book for course use and bulk purchases. For further information, send email to <joanne.tracy@wkap.com>.***

**DEDICATION**

to my wife, Mary Wilson Hatfield

to my grandchildren, Amber Nicole Hatfield, Logan Alexa Hatfield and Tristan Lee Clubb.

# TABLE OF CONTENTS

**List of Contributors** ..................................................................................xi

**Foreword** ....................................................................................................xv
Raymond F. Burk

**Preface** ......................................................................................................xvii
Dolph L. Hatfield

**Acknowledgements** ..................................................................................xix

Chapter 1
**Introduction** ..................................................................................................1
Dolph L. Hatfield

## Part I. Biosynthesis of selenocysteine and its incorporation into protein

Chapter 2
**Selenium metabolism in bacteria** ..................................................................7
August Böck

Chapter 3
**Mammalian selenocysteine tRNA** ..............................................................23
Bradley A. Carlson, F. Javier Martin-Romero, Easwari
Kumaraswamy, Mohamed E. Moustafa, Huijun Zhi,
Dolph L. Hatfield and Byeong Jae Lee

Chapter 4
**Selenophosphate – selenium donor for protein and
 tRNA** ..........................................................................................................33
Gerard M. Lacourciere

Chapter 5
**SECIS elements** ..........................................................................................45
Glover W. Martin, III and Marla J. Berry

Chapter 6
**SECIS binding proteins** .................................................................55
Paul R. Copeland and Donna M. Driscoll

Chapter 7
**Towards a mechanism for selenocysteine incorporation in eukaryotes** ................................................................................69
John B. Mansell and Marla J. Berry

Chapter 8
**Regulation of selenoprotein expression** .....................................81
Roger A. Sunde

## Part II. Selenium-containing proteins

Chapter 9
**Identity, evolution and function of selenoproteins and selenoprotein genes** ...................................................................99
Vadim N. Gladyshev

Chapter 10
**Bacterial selenoenzymes and mechanisms of action** ................115
Thressa C. Stadtman

Chapter 11
**Selenoprotein P** ..........................................................................123
Kristina E. Hill and Raymond F. Burk

Chapter 12
**Selenoprotein W: A muscle protein in search of a function** ....................................................................................137
L. Walt Ream, William R. Vorachek and Phillip D. Whanger

Chapter 13
**The 15 kDa selenoprotein (Sep15): functional studies and a role in cancer etiology** .................................................147
Vadim N. Gladyshev, Alan M. Diamond and Dolph L. Hatfield

*Table of Contents*

Chapter 14
**Selenoproteins of the glutathione system** ...............................157
*Leopold Flohé and Regina Brigelius-Flohé*

Chapter 15
**Selenoproteins of the thioredoxin system** ..............................179
*Arne Holmgren*

Chapter 16
**Selenium, deiodinases and endocrine function** .........................189
*Donald L. St. Germain*

## Part III. Selenium and human health

Chapter 17
**Selenium as a cancer preventive agent** ..................................205
*Gerald F. Combs, Jr. and Junxuan Lü*

Chapter 18
**Selenium deficiency and human disease** ................................219
*Ruth J. Coppinger and Alan M. Diamond*

Chapter 19
**Selenium as an antiviral agent** ...........................................235
*Melinda A. Beck*

Chapter 20
**Role of selenium in HIV/AIDS** ............................................247
*Marianna K. Baum, Adriana Campa, Maria José Miguez-Burbano, Ximena Burbano and Gail Shor-Posner*

Chapter 21
**Effects of selenium on immunity and aging** ............................257
*Roderick C. McKenzie, Teresa S. Rafferty, Geoffrey J. Beckett and John R. Arthur*

## Chapter 22
**Selenium and male reproduction** ................................................273
*Leopold Flohé, Regina Brigelius-Flohé, Matilde Maiorino,
Antonella Roveri, Josef Wissing and Fulvio Ursini*

## Chapter 23
**Role of low molecular weight, selenium-
   containing compounds in human health** ................................283
*Henry J. Thompson*

## Chapter 24
**Evolution of human dietary standards for
   selenium** ................................................................................299
*Orville A. Levander*

## Chapter 25
**Selenium in biology and human health: controversies
   and perspectives** ...................................................................313
*Vadim N. Gladyshev*

**Index** ...........................................................................................319

# CONTRIBUTORS

**John R. Arthur**
Division of Cell Integrity
Rowett Research Institute,
  Bucksburn
Aberdeen, Scotland AB21 9SB,
  UK

**Mariana K. Baum**
Division of Metabolism and
  Disease Prevention
Department of Psychiatry and
  Behavioral Sciences
University of Miami School of
  Medicine
Miami, FL 33136, USA

**Melinda A. Beck**
Department of Pediatrics and
  Nutrition
University of North Carolina at
  Chapel Hill
Chapel Hill, NC 27599, USA

**Geoffrey J. Beckett**
Department of Clinical
  Biochemistry
University of Edinburgh
Edinburgh, Scotland, EH3 9YW,
  UK

**Marla J. Berry**
Thyroid Division
Brigham and Women's Hospital
  and the Department of
  Biological Chemistry and
  Molecular Pharmacology
Harvard Medical School
Boston, MA 02115, USA

**August Böck**
Lehrstuhl für Mikrobiologie der
  Universität Müchen
D-80638 Munich, Germany

**Regina Brigelius-Flohé**
Department of Vitamins and
  Atherosclerosis
German Institute for Research of
  Nutrition (DIfE)
Arthur-Scheunert-Allee 114-116
D-14558 Potsdam, Germany

**Ximena Burbano**
Division of Metabolism and
  Disease Prevention
Department of Psychiatry and
  Behavioral Sciences
University of Miami School of
  Medicine
Miami, FL 33136, USA

**Raymond F. Burk**
Department of Medicine
Vanderbilt University School of
  Medicine
Nashville, TN 37232, USA

**Adriana Campa**
Division of Metabolism and
  Disease Prevention
Department of Psychiatry and
  Behavioral Sciences
University of Miami School of
  Medicine
Miami, FL 33136, USA

**Bradley A. Carlson**
Section on the Molecular Biology
  of Selenium
Basic Research Laboratory
National Cancer Institute
National Institutes of Health
Bethesda, MD 20892 USA

**Gerald F. Combs, Jr.**
Division of Nutritional Sciences
Cornell University
Ithaca, NY 14853, USA

**Paul R. Copeland**
Department of Cell Biology
Lerner Research Institute
Cleveland Clinic Foundation
Cleveland, OH 44195, USA

**Ruth J. Coppinger**
Department of Human Nutrition
University of Illinois at Chicago
Chicago, IL 60612, USA

**Alan M. Diamond**
Department of Human Nutrition
University of Illinois at Chicago
Chicago, IL 60612, USA

**Donna M. Driscoll**
Department of Cell Biology
Lerner Research Institute
Cleveland Clinic Foundation
Cleveland, OH 44195, USA

**Leopold Flohé**
Department of Biochemistry
Technical University of
  Braunschweig
Mascheroder Weg 1
D-38124 Braunschweig, Germany

**Vadim N. Gladyshev**
Department of Biochemistry
University of Nebraska
Lincoln, NE 68588, USA

**Dolph L. Hatfield**
Section on the Molecular Biology
  of Selenium
Basic Research Laboratory
National Cancer Institute
National Institutes of Health
Bethesda, MD 20892, USA

**Kristina E. Hill**
Department of Medicine
Vanderbilt University School of
  Medicine
Nashville, TN 37232, USA

**Arne Holmgren**
Medical Nobel Institute for
  Biochemistry
Department of Medical
  Biochemistry and Biophysics
Karolinska Institute
SE-171 77 Stockhom, Sweden

**Easwari Kumaraswamy**
Section on the Molecular Biology
  of Selenium
Basic Research Laboratory
National Cancer Institute
National Institutes of Health
Bethesda, MD 20892, USA

**Gerard M. Lacourciere**
Laboratory of Biochemistry
National Heart, Lung, and Blood
  Institute
National Institutes of Health
Bethesda, MD 20892, USA

## Contributors

**Byeong Jae Lee**
Laboratory of Molecular Genetics
Institute for Molecular Biology
  and Genetics
Seoul National University
Seoul 151-742, Korea

**Orville A. Levander**
Beltsville Human Nutrition
  Research Center
U.S. Department of Agriculture
Agricultural Research Service
Beltsville, MD 20705, USA

**Junxuan Lü**
AMC Cancer Center
1600 Pierce Street
Denver, CO 80214, USA

**Matilde Maiorino**
Dipartimento di Chimica
  Biologica
Università di Padova
Viale G. Colombo 3
I-35121, Padova, Italy

**John B. Mansell**
Thyroid Division
Brigham and Women's Hospital
  and the Department of
  Biological Chemistry and
  Molecular Pharmacology
Harvard Medical School
Boston, MA 02115, USA

**Glover W. Martin, III**
Harvard/MIT Division of Health
  Sciences and Technology,
  and the Department of
  Microbiology and Molecular
  Genetics
Harvard Medical School
Cambridge, MA, 02139, USA

**F. Javier Martin-Romero**
Section on the Molecular Biology
  of Selenium
Basic Research Laboratory
National Cancer Institute
National Institutes of Health
Bethesda, MD 20892 USA

**Roderick C. McKenzie**
Department of Medical and
  Radiological Sciences
University of Edinburgh
Edinburgh, Scotland, EH3 9YW,
  UK

**Maria José Miguez-Burbano**
Division of Metabolism and
  Disease Prevention
Department of Psychiatry and
  Behavioral Sciences
University of Miami School of
  Medicine
Miami, FL 33136, USA

**Mohamed E. Moustafa**
Section on the Molecular Biology
  of Selenium
Basic Research Laboratory
National Cancer Institute
National Institutes of Health
Bethesda, MD 20892, USA

**Teresa S. Rafferty**
Department of Medical and
  Radiological Sciences
University of Edinburgh
Edinburgh, Scotland, EH3 9YW,
  UK

**L. Walt Ream**
Departments of Microbiology and
 Environmental and
 Molecular Toxicology
Oregon State University
Corvallis, OR 97331, USA

**Antonella Roveri**
Dipartimento di Chimica
 Biologica
Università di Padova
Viale G. Colombo 3
I-35121, Padova, Italy

**Gail Shor-Posner**
Division of Metabolism and
 Disease Prevention
Department of Psychiatry and
 Behavioral Sciences
University of Miami School of
 Medicine
Miami, FL 33136, USA

**Donald L. St. Germain**
Departments of Medicine and of
 Physiology
Dartmouth Medical School
Lebanon, NH 03756, USA

**Thressa C. Stadtman**
National Heart, Lung, and Blood
 Institute
National Institutes of Health
Bethesda, MD 20892, USA

**Roger A. Sunde**
F21C Nutrition Cluster Leader
Professor of Nutritional Sciences
 and of Biochemistry
University of Missouri
Columbia, MO 65211, USA

**Henry J. Thompson**
Center for Nutrition in the
 Prevention of Disease
AMC Cancer Research Center
Denver, CO 80214, USA

**Fulvio Ursini**
Dipartimento di Chimica
 Biologica
Università di Padova
Viale G. Colombo 3
I-35121, Padova, Italy

**William R. Vorachek**
Departments of Microbiology and
 Environmental and
 Molecular Toxicology
Oregon State University
Corvallis, OR 97331, USA

**Philip D. Whanger**
Departments of Microbiology and
 Environmental and
 Molecular Toxicology
Oregon State University
Corvallis, OR 97331, USA

**Josef Wissing**
Department of Biochemistry
Technical University of
 Braunschweig
Mascheroder Weg 1
D-38124 Braunschweig, Germany

**Huijun Zhi**
Section on the Molecular Biology
 of Selenium
Basic Research Laboratory
National Cancer Institute
National Institutes of Health
Bethesda, MD 20892, USA

# FOREWORD

From molecular biology to human health, selenium research is booming. It has become clear that most physiologic effects of selenium are exerted by selenoproteins. Therefore, most of the basic research on the element examines selenoprotein expression and regulation. Synthesis of a selenoprotein requires a complex process with several elements that are unique. The details of this process are emerging from the work of a number of groups.

The significance of selenium to human health is a hot topic. There is little or no selenium deficiency in the USA, but large numbers of people elsewhere in the world are selenium deficient. Animal studies have shown that selenium deficiency in the host can modify the course of viral infections and cause mutations in the virus. Whether this occurs in human beings will be important to determine. Such selenium deficiency induced mutations could possibly be responsible for emergence of new viral strains.

Evidence has been presented from studies in animals that selenium has anticancer effects when given in pharmacologic amounts. An intervention study in human beings has supported this idea. A number of intervention studies are currently being pursued in this important, but difficult area.

The selenium field has not always been so active. Selenium came to the notice of biologists in the 1930s when it was discovered to be the dietary substance responsible for hair and hoof loss in animals grazing certain areas of the American Great Plains. This recognition of its toxicity led the USDA to map the selenium content of forage in the USA. The map that was produced demonstrated that the selenium content of plants varied tremendously according to where they were grown. Most areas of the country produced plants that contained a moderate (non-toxic) concentration of selenium. However, a few locales were distinctly different. Some produced plants that had very high selenium contents (toxic) and some produced plants that had very low selenium contents. Thus, the effort to locate high-selenium areas in order to prevent selenium toxicity in animals led also to the identification of low-selenium areas.

After selenium had been shown to be an essential nutrient in the mid-1950s, some veterinary disease endemic to areas that had been identified as low in selenium on the USDA selenium map were investigated. White muscle disease in sheep responded to selenium and thus was classed as a selenium deficiency disease. Several other pathological conditions that were vitamin E responsive, including mulberry heart disease of pigs, also responded to selenium. The association of the two nutrients in these conditions linked vitamin E with selenium and suggested that selenium might function as an oxidant defense because vitamin E was known to be an antioxidant.

In the early 1970s, Chinese workers investigating the cause of a childhood cardiomyopathy called Keshan disease took note of the veterinary literature of selenium deficiency diseases. Keshan disease only occurred in certain areas of China and the investigators noted that animals in those areas had clinical illnesses that were similar to those caused by selenium deficiency. They determined that hair selenium was low in the children with Keshan disease. Then they carried out a placebo-controlled study of selenium administration to children at risk for Keshan disease. Children that received selenium were protected from Keshan disease. Thus, a manifestation of selenium deficiency in human beings is a susceptibility to cardiomyopathy in children. A second (and unknown) factor is needed to cause the disease as not all selenium deficient children develop it. It is not known whether selenium deficiency has other adverse effects on human health but this is an important topic for further research.

Efforts in the 1930s and 1940s to understand selenium biochemistry and metabolism focused on mechanisms of its toxicity and excretion. However, the realization in the 1950s that selenium was an essential nutrient led to investigations of its role in normal biochemistry. Selenium deficiency was induced in experimental animals by feeding them a low selenium diet. Then biochemical systems were studied in selenium deficient animals and in controls. Using this approach, a group at the University of Wisconsin identified glutathione peroxidase as the first animal selenoenzyme in 1973. This solidified the nutritional essentiality of selenium and supported the idea that it provided defense against oxidative injury.

After almost a decade as the only know animal selenoprotein, glutathione peroxidase was joined by selenoprotein P in 1982 and by other selenoproteins soon thereafter. Presently, approximately 15 animal selenoproteins are known and that number is rising as new ones are discovered.

Studies in bacteria yielded more selenoproteins and led to the pioneering work by the Munich group in the late 1980s that elucidated the mechanism of selenoprotein synthesis in bacteria. Their work served as the starting point of the work on selenoprotein synthesis in animals.

This book is an up-to-date collection of research summaries and reviews written by active selenium researchers. Its coverage is broad with detailed basic science reviews as well as health related chapters. This book will provide an in depth summary of the field of selenium as well as a guide for future research.

Raymond F. Burk

# PREFACE

Research over the last 30 years involving selenium biology has yielded new and surprising insights into biochemical, molecular and genetic aspects of this fascinating element. During this same time period, data have also been obtained from both human and animal nutritional studies suggesting a vital role for selenium in disease incidence and severity. These two disciplines of study are beginning to merge and the interrelationships between the different observations are becoming apparent as is discussed in this book.

An active area of selenium research has been the functional characterization of selenoproteins. While several chapters in this book deal with this significant topic, there is, in fact, considerable debate in the selenium field whether selenoproteins or low molecular weight selenium compounds are responsible for the many reported beneficial effects of dietary selenium. Both viewpoints are expressed herein, so that the reader will have an opportunity to see how different leaders in the field view the topic of selenium and human health, and, at the same time, obtain a more comprehensive appreciation of the present status of the thinking on these subjects.

Diet plays an extremely important role in health and disease. For example, an increasing number of health professionals contend that environmental factors such as infection, unbalanced diets, smoking, excessive alcohol consumption and insufficient exercise contribute more to chronic disease in man than genetic disposition. These maladies include cancer, cardiovascular disease, diabetes and liver dysfunction. The origins of these diseases may be attributed, at least in part, to disruption of redox homeostasis, leading to the presence of excess reactive oxygen species and oxidative damage of essential cellular components, including nucleic acids and proteins. As dietary selenium is viewed as a potent regulator of cellular redox homeostatis, selenium may be an important dietary contributor to reducing the incidence of many debilitating disorders. Although the specific biological mechanisms of selenium that are responsible for promoting better health, and the extent to which this element is involved in promoting better health, remain to be established, the available data suggest that the benefits of dietary selenium are highly significant. Caution must be exercised, however, in touting the advantages of this element as too much selenium in the diet can have toxic consequences and the range between too little and too much selenium is not much more than an order of magnitude. The guidelines for the recommended daily allowance in humans have recently been revised and are discussed in this book.

Combs and Combs wrote in 1986 that "One of the most important discoveries in nutrition in the last 30 years has been the recognition of the essentiality of the element selenium....." (GF Combs, SB Combs 1986 *The Role of Selenium in Nutrition* Academic Press, Inc New York). This statement was motivated by evidence that adequate dietary selenium has important health benefits that were primarily observed in laboratory animals and livestock. These findings followed the initial observations in the mid 1950s that selenium had a role in bacterial metabolism and in preventing liver necrosis in rats. Combs and Combs also noted in 1986 that the livestock industry had prevented losses estimated in the 100s of millions of dollars by supplementing the diet of livestock with selenium or with selenium and vitamin E. Today, these savings are in the billions of dollars.

In the last 15 years, many of the molecular mysteries surrounding selenium have been solved. We now know that selenium is incorporated into protein as selenocysteine, the $21^{st}$ naturally occurring amino acid in protein. Unlike any of the other 20 amino acids that are present in protein, the biosynthesis of selenocysteine occurs on its transfer RNA. The donation of selenocysteine to the growing selenopeptide in protein synthesis requires a novel mRNA binding protein and a selenocysteine-tRNA specific elongation factor in archaea and eukaryotes, while a single protein carries out both these functions in eubacteria. Selenoproteins have clearly evolved to take advantage of the fact that the pKa of selenocysteine is lower than the physiological pH and, therefore, this amino acid residue is ideally suited to participate in redox reactions. This and other chemical properties of selenium result in the unique redox characteristics of selenocysteine and its use in antioxidant enzymes.

As discussed in the chapters in this book, a large body of evidence indicates that selenium is a cancer chemopreventive agent. Further evidence points to a role of this element in reducing viral expression, in preventing heart disease, and other cardiovascular and muscle disorders, and in delaying the progression of AIDS in HIV infected patients. Selenium may also have a role in mammalian development, in male fertility, in immune function and in slowing the aging process. If dietary selenium does indeed influence these diseases and cellular processes, then the significance of including adequate amounts of this element in the diet may be regarded as one of the more important discoveries in nutrition in the last century.

Dolph L. Hatfield

# ACKNOWLEDGMENTS

The support and generous help of Vadim N. Gladyshev and Bradley A. Carlson throughout the preparation of this book are gratefully acknowledged.

# Chapter 1. Introduction

Dolph L. Hatfield

*Section on the Molecular Biology of Selenium, Basic Research Laboratory, National Cancer Institute, National Institutes of Health, Bethesda, MD 20892, USA*

Historically, the image of selenium has changed dramatically over the last century. Initially, selenium was widely considered to be a toxic agent in mammals. In the 1930s, this element was found to be responsible for severe illnesses that had been documented, primarily in livestock, many years earlier [reviewed in 1]. Thus, it was quite surprising when Schwarz and Foltz reported in 1957 [2] that selenium prevented liver necrosis in rats. This finding and a report that selenium was important in anaerobic growth of *Escherichi coli* when grown on glucose [3] provided evidence that selenium was an important micronutrient for mammals and for certain bacteria.

In the 1960s and 70s, selenium was reported to have anticarcinogenic properties [reviewed in 1], to have a role in preventing heart disease [1, 4-9 and references therein] and in other muscle disorders [reviewed in 1], as well as a role in male fertility [reviewed in 1]. In the 80s and 90s, this element was found to have a role in immune function [reviewed in 10 and 11], a role in viral suppression [reviewed in 10 and 12], a role in AIDS [13-19 and references therein], was implicated in delaying the aging process [1, 20 and references therein] and was further substantiated to be a chemopreventive agent in cancer [21-26 and references therein]. Most of these observations were based on studies involving laboratory animals and livestock.

The molecular biology of selenium has developed at a rapid pace over the last 10-12 years. Several important discoveries in the selenium field occurred in the 1970s and 1980s that provided the foundations for this rapid development. Perhaps the first of these occurred in 1973, when two independent groups found that selenium is an essential component in mammalian glutathione peroxidase [27-28]. Subsequently, the selenium component in selenoprotein A of the glycine reductase complex in bacteria was identified as selenocysteine [29]. Bovine glutathione peroxidase was sequenced in the mid-1980s and the location of the selenocysteine residue established [30]. Finally, in 1986, the gene sequences for two selenoproteins, mammalian glutathione peroxidase [31] and bacterial formate dehydrogenase [32], were reported and the sequences aligned with the corresponding protein sequences. In both cases, the selenocysteine moiety in the protein coincided with a TGA codon in the corresponding gene. The latter studies demonstrated that a codon, previously recognized only as a termination codeword for the cessation of protein synthesis, had a dual function of dictating stop and selenocysteine. Each of the above studies served as a milestone for providing the foundations for unraveling much of the molecular biology of selenium. Major, recent discoveries that defined the field of selenium molecular biology are covered in subsequent chapters of this book.

The wide range of beneficial effects of selenium on health has been recognized for several years, as noted above, largely through studies involving laboratory animals and livestock. However, only in the past decade has the focus of this element in human health begun to receive substantial attention.

The purpose of this book is to bring together our current knowledge of the molecular biology of selenium and the role that this element plays in health. Although the book's emphasis is on our understanding of selenium metabolism in mammals and the role of this element in human health, such a book would be incomplete without the bacterial work. Clearly, much of the elucidation of the molecular biology of selenium in mammals is based on what was initially found in bacteria and this work is primarily described in Chapter 2. Additional major findings that were first reported in bacteria are presented in Chapters 4 and 10.

The book is divided into three parts. The chapters in Part I, which is entitled "Biosynthesis of selenocysteine and its incorporation into protein," define selenocysteine as the $21^{st}$ naturally occurring amino acid in protein and describe how it is inserted into protein in both prokaryotes and eukaryotes. Addition of selenocysteine to the genetic code as its $21^{st}$ amino acid marks the first addition to the code (see Figure 1) since it was deciphered in the mid-1960s by Nirenberg and collaborators [33] and by Khorana and collaborators [34]. In addition, our current understanding of how selenoprotein expression is regulated in mammals is covered in Part I.

Part II, which is entitled "Selenium-containing proteins," discusses our current understanding of selenocysteine-containing proteins, primarily in higher eukaryotes. Also, the mechanism of action of selenoenzymes is described in this section using bacterial selenoenzymes as a model and a new technique for identifying selenoprotein genes in eukaryotes is presented.

Part III, which is entitled "Selenium and human health," cover our current understanding of the role of selenium in various diseases, including cancer and heart disease, in HIV infection and AIDS, in male reproduction, and as an antiviral agent. The role of small molecular weight, selenium-containing compounds in human health and the dietary selenium requirements for humans are also discussed.

# Introduction

This book provides an up-to-date review of much of the on-going research in the selenium field and it is the first compilation of such research since the book, *Selenium in Biology and Human Health*, edited by R.F. Burk in 1994 [35]. It is hoped that this book will provide an important resource for scientists working in the selenium field, as well as for physicians, other scientists and students who wish to learn more about this fascinating micronutrient.

| Middle Base  5' Base | U | C | A | G | Middle Base  3' Base |
|---|---|---|---|---|---|
| U | Phenylalanine | Serine | Tyrosine | Cysteine | U |
|   | Phenylalanine | Serine | Tyrosine | Cysteine | C |
|   | Leucine | Serine | *Terminator* | **Selenocysteine** / *Terminator* | A |
|   | Leucine | Serine | *Terminator* | Tryptophan | G |
| C | Leucine | Proline | Histidine | Arginine | U |
|   | Leucine | Proline | Histidine | Arginine | C |
|   | Leucine | Proline | Glutamine | Arginine | A |
|   | Leucine | Proline | Glutamine | Arginine | G |
| A | Isoleucine | Threonine | Asparagine | Serine | U |
|   | Isoleucine | Threonine | Asparagine | Serine | C |
|   | Isoleucine | Threonine | Lysine | Arginine | A |
|   | Methionine / *Initiator* | Threonine | Lysine | Arginine | G |
| G | Valine | Alanine | Aspartic Acid | Glycine | U |
|   | Valine | Alanine | Aspartic Acid | Glycine | C |
|   | Valine | Alanine | Glutamic Acid | Glycine | A |
|   | Valine | Alanine | Glutamic Acid | Glycine | G |

**Figure 1.** The genetic code showing selenocysteine as the 21$^{st}$ naturally occurring amino acid.

## References

1. GF Combs Jr, SB Combs 1986 *The Role of Selenium in Nutrition* Academic Press, Inc New York
2. K Schwarz, CM Foltz 1957 *J Am Chem Soc* 79:3292
3. J Pinsent 1954 *Biochem J* 57:10

4. GQ Yang, YM Xia 1995 *Biomed Environ Sci* 8:187
5. JK Huttunen 1997 *Biomed Environ Sci* 10:220
6. GL Wang, SC Wang, BQ Gu, YX Yang, HB Song, WL Xue, WS Liang, PY Zhang 1997 *Biomed Environ Sci* 10:316
7. C Russo, O Olivieri, D Girelli, G Faccini, ML Zenari, R Corrocher 1998 *J Hypertens* 16:1267
8. WM Tong, F Wang 1998 *Metabolism* 47:415
9. M Navarro-Alarcon, H Lopez-Garcia, V Perez-Valero, C Lopez-Martinez 1999 *Ann Nutr Metab* 43:30
10. MA Beck, OA Levander 1998 *Annu Rev Nutr* 18:93
11. RC McKenzie, TS Rafferty, GJ Beckett 1998 *Immunol Today* 19:342
12. MA Beck 2000 *Am J Clin Nutr* 71:1676S
13. J Constans, JL Pellegrin, C. Sargeant, M Simonoff, I Pellegrin, H Fleury, B Leng, C Conri 1995 *J Acquir Immune Defic Syndr Hum Retrovirol* 10:392
14. P Chariot, ML Dubreuil-Lemaire, JY Zhou, B Lamia, L Dume, B Larcher, I Monnet, Y Levy, A Astier, R. Gherardi 1997 *Muscle Nerve* 20:386
15. MP Look, JK Rockstroh, GS Rao, KA Kreuzer, S Barton, H Lemoch, T Sudhop, J Hoch, K Stockinger, U Sprengler, T Sauerbruch 1997 *Eur J Clin Nutr* 51:266
16. BM Dworkin 1994 *Chem Biol Interact* 91:181
17. MK Baum, G Shor-Posner, S Lai, G Zhang, H Lai, MA Fletcher, H Sauberlich, JB Page 1997 *J Acquir Immune Defic Syndr Hum Retrovirol* 15:370
18. K Hori, DL Hatfield, F Maldarelli, BJ Lee, KA Clouse 1997 *AIDS Res Human Retrovir* 13: 1325
19. VN Gladyshev, TC Stadtman, DL Hatfield, K-T Jeang 1999 *Proc Natl Acad Sci USA* 96:835
20. O Olivieri, AM Stanzial, et al 1994 *Am J Clin Nutr* 61: 1174
21. GF Combs Jr 1997 *Antioxidants and Disease Prevention* HS Garewell (Ed) CRC Press New York p97
22. LC Clark, GF Clark Jr, et al. 1996 *JAMA* 276:1957
23. VA Knizhnikov, NK Shandala, VA Komleva, T Knyazhev, VA Tutelyan 1996 *Nutr Res* 16:505
24. SY Yu, YJ Zhu, WG Li 1997 *Biol Trace Elem Res* 56:117
25. P Knekt, J Marniemi, L Teppo, M Heliovaara, A Aromaa 1998 *Am J Epidemiol* 148:975
26. K Yoshizawa, WC Willett, SJ Morris, MJ Stampfer, D Spieglelman, EB Rimm, E Giovannucci 1998 *J Natl Can Insti* 90:1219
27. L Flohè, WA Gunzler, HH Schock 1973 *FEBS Lett* 32:132
28. JT Rotruck, AL Pope, HE Ganther, AB Swanson, DG Hafeman, WG Hoekstra 1973 *Science* 179:588
29. JE Cone, RM Del Rio, JM Davis, TC Stadtman 1976 *Proc Natl Acad Sci USA* 73:2659
30. WA Gunzler, GJ Steffens, A Grossmann, S Kim, F Otting, A Wendel, L Flohé 1984 *Hoppe Seylers Z Physiol Chem* 365:195
31. I Chambers, J Frampton, P Goldfarb, N Affara, W McBain, PR Harrison 1986 *EMBO J* 5:1221
32. F Zinoni, A Birkmann, TC Stadtman, A Böck 1986 *Proc Natl Acad Sci USA* 83:4650
33. M Nirenberg, T Caskey, R Marshall, R Brimacombe, D Kellog, B Doctor, D Hatfield, J Levin, F Rottman S Pestka, M Wilcox and F Anderson 1966 *Cold Spr Harbor Symp Qunat Biol* 3:11
34. GH Khorana, H Büchi, H Ghosh, N Gupta, TM Jacob, H Kössel, R Morgan, SA Narang, E Ohtuska, RD Wells *Cold Spr Harbor Symp Qunat Biol* 3:39
35. RF Burk (Ed) 1994 *Selenium in Biology and Human Health* Springer-Verlag, New York, pp 221

# — Part I —

Biosynthesis of selenocysteine and its incorporation into protein

# Chapter 2. Selenium metabolism in bacteria

August Böck

*Lehrstuhl für Mikrobiologie der Universität München, D-80638 Munich, Germany*

**Summary:** The biosynthesis and specific incorporation of selenocysteine into protein in bacteria requires the function of two *cis* and four *trans* elements. The cis elements are a UGA codon determining the position of selenocysteine insertion into the nascent polypeptide and a secondary/tertiary structure within the mRNA, designated the SECIS element, following the UGA at its 3´side. The *trans* elements consist of a tRNA species, tRNA$^{Sec}$, which is charged by the cellular seryl-tRNA synthetase and serves as an adaptor for the conversion of the seryl moiety into the selenocysteyl product, catalysed by the biosynthetic enzyme selenocysteine synthase. Monoselenophosphate, provided by selenophosphate synthetase, is the selenium donor species. Selenocysteyl-tRNA$^{Sec}$ is bound by the special translation factor SelB which in its N-terminal part displays sequence similarity to elongation factor Tu. With its C-terminal 17 kDa domain, SelB binds to the apical part of the SECIS stem-loop structure thus forming a quaternay complex with the two nucleic acids and the guanosine nucleotide. Besides tethering the charged tRNA to the ribosomal A-site, binding to the SECIS confers to SelB a conformation competent for interaction with the ribosome. SECIS binding also stabilizes the interaction of SelB with charged tRNA. Selenocysteine insertion is competing with termination at the UGA codon whereby competition is influenced by codon context, the two amino acids preceding selenocysteine in the polypeptide and the balance of the components of the insertion machinery. Selenocysteine insertion in archaea resembles the eucaryal situation in that the SECIS element is localized in the 3´untranslated region of the mRNA. The archaeal homolog of SelB is unable to recognize the SECIS motif and may require the function of a second protein, as has been shown to be the case for the mammalian counterpart.

## Introduction

When bacteria are challenged with low molecular weight selenium compounds in the medium, they can process selenium in a nonspecific or a specific manner. The nonspecific metabolism rests on the chemical similarity between selenium and its neighbor element in the periodic system, sulfur. When present above a critical concentration in *Escherichia coli*, i.e., at selenite concentrations higher than 1 µM, selenium intrudes the sulfur pathways and is metabolized along the routes of sulfur metabolism [1,2] (Figure 1). Thus, selenium in the form of selenate is taken up by the sulfate transport system and reduced to selenide via the assimilatory sulfate reduction system. When offered as selenite, reduction appears to proceed chemically by interaction with thiol compounds like glutathione [see 3 for review].

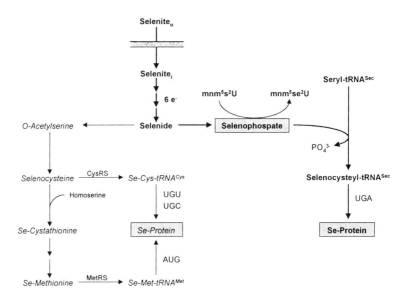

**Figure 1.** Specific and nonspecific metabolism and incorporation of selenium into macromolecules. The specific pathway is highlighted in bold. mnm$^5$s$^2$U is the abbreviation for 5-methylamino-methyl-2-thiouridine and mnm$^5$se$^2$U for 5-methyl-aminomethyl-2-selenouridine.

The first organic selenium compound formed is free selenocysteine which can be converted to selenocystathionine and eventually to selenomethionine. On the other hand, selenocysteine has been shown to be a substrate for cysteyl-tRNA ligase which forms selenocysteyl-tRNA$^{Cys}$ and in this way incorporates selenocysteine at cysteine positions in proteins [4-6]. The decision, whether selenium is incorporated nonspecifically as either selenocysteine or selenomethionine, therefore, should be dependent on the relative catalytic efficiencies of cysteyl-tRNA synthetase and cystathionine synthetase for the substrate cysteine and its analog selenocysteine. Nonspecific incorporation into macromolecules is drastically reduced when the cysteine biosynthetic pathway is interrupted by mutations or when it is fully repressed [6].

When selenomethionine is provided in the medium, it is almost indiscriminately incorporated in place of methionine. This replacement is frequently used in x-ray analysis of protein crystals by multiwavelength anomalous diffraction [7] or in NMR spectroscopy, because of the large chemical shift of the $^{77}$Se isotope [8]. Selenomethionine as the major selenium compound has also been detected when bacteria were grown on excessive amounts of selenite [9,10]. Free selenocysteine, on the other hand, is highly toxic and therefore growth inhibitory. Its incorporation in place of cysteine requires an overexpression system like the promoter-polymerase system of phage T7 to circumvent toxicity [11].

The specific incorporation of selenocysteine, on the other hand, is effective at much lower concentrations of selenite in the medium. With the aid of a *fdhF-lacZ* fusion reporter gene, in which readthrough into *lacZ* is dependent on the availability of selenium (see below), saturation has already occurred by 0.1 µM selenite [12]. Specific incorporation does not involve free, low molecular weight selenocysteine since the biosynthesis of the molecule takes place from a precursor amino acid esterified with tRNA. It should be emphasized that the capacity to synthesize selenoproteins by the specific pathway is not ubiquitous. Actually, it is absent in the majority of microorganisms. In this chapter we will first discuss the specific incorporation of selenocysteine by bacteria, mainly *E. coli*, and then by members of archaea.

Identification of the components involved in selenocysteine biosynthesis and specific insertion rests to a considerable degree on the early work of several groups studying the anaerobic formate metabolism of *E. coli* [13-18]. Genes had been analyzed which, when mutated, abolished the ability of *E. coli* to synthesize active isoenzymes of formate dehydrogenase known as formate dehydrogenase N and formate dehydrogenase H which couple formate oxidation to the reduction of nitrate or protons, respectively. Thus, some mechanism must have been affected in the mutants that is required for generating activity of both enzymes. The genes had been mapped on the chromosome of *E. coli* and some of them (*fdhA fdhB and fdhC*) turned out to be involved in selenium metabolism [19]. Merits also go to two technical developments, namely the establishment of a plate overlay technique for screening large numbers of colonies for formate dehydrogenase activity [15] and the set-up of a precedure for specific incorporation of radioactive selenium into selenopolypeptides [20]. With the aid of this technique, it was easy to differentiate between specific and nonspecific incorporation (see Figure 1).

**Specific incorporation of selenocysteine by bacteria**
The first genes discovered to contain an in-frame UGA codon directing selenocysteine insertion were *gpx*, coding for glutathione peroxidase from mouse [21], and *fdhF* from *E. coli*, coding for the selenopolypeptide of formate dehydrogenase H [22]. Whereas an amino acid sequence was available for glutathione peroxidase showing complete colinearity between the UGA in the mRNA and selenocysteine in the protein, this was not the case for the bacterial enzyme. Evidence was obtained, however, by leading truncations from the 3´end into the gene and showing that removal of the segment containing the UGA also abolished selenium incorporation into the truncated gene product. Definite proof for the cotranslational insertion was then provided by fusion of the *lacZ* reporter gene upstream and downstream of the UGA in *fdhF* and the demonstration that readthrough of the UGA required the presence of selenium in the medium [12]. Analysis of mutations which affected readthrough led to the identification of the genetic elements involved in selenium metabolism in *E. coli* [19].

**tRNA$^{Sec}$**
The key element was identified as the product of the *fdhC* gene, now designated as the *selC* gene [23]. It codes for a tRNA with an unusual sequence and structural properties (Figure 2A).
  With 95 nucleotides, tRNA$^{Sec}$ is the largest tRNA in *E. coli* mainly because of an aminoacyl acceptor stem of eight possible base pairs and a 22 nucleotide long extra arm. There are also a number of deviations from the consensus structure characteristic of canonical elongator tRNAs, namely a G at position 8, an A at position 14, a Y-R pair at the 10-25 sites and an R-Y base pair at positions 11-24. Moreover, the R-Y Levitt pair between the positions 15-48 is missing. As expected, extensive enzymatic and chemical probing of the solution structure of tRNA$^{Sec}$ from *E. coli*, compared with that of canonical tRNA$^{Ser}$, showed that these deviations, plus the fact that the D stem is closed to a six base pair helix minimizing the D loop to four nucleotides, also restrict the types of tertiary interactions within the molecule [24]. Whereas the canonical G19-C56 interaction is still present, there are new interactions between C16 of the D loop and C59 of the T loop and the canonical A21-(U8-A14) triple pair is substituted by a G8-(A21-U14) triple interaction. The extra arm is closed by a G45-A48 pair and connected to the anticodon coaxial helix only by interaction of A44 with U26. All these unusual sequence and structural properties are conserved in the sequences of other bacterial tRNA$^{Sec}$ species [25]. In view of the still open discussion on the structure of the eucaryal (see Chapter 3) and archaeal (Figure 2B) counterparts and of the lack of an x-ray structure, the conclusions can be concentrated on three characteristic features: (i) the acceptor-T stem stacked helix is extended to 13 base pairs, (ii) the closure of the D stem and the

deviations from the sequence in canonical positions restrict the possibilities for tertiary interactions within the molecule, and (iii) the extra arm appears to be less well fixed to the body of the molecule than in classical elongator tRNAs.

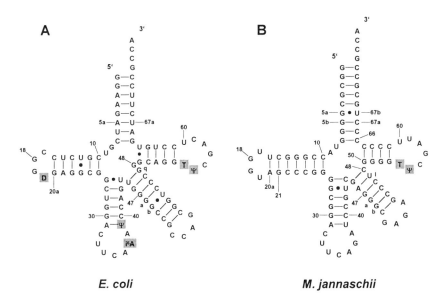

**Figure 2.** Cloverleaf models of the *Escherichia coli* (A) and *Methanococcus jannaschii* (B) tRNA$^{Sec}$ species. The modified bases are shaded. The bacterial tRNA$^{Sec}$ is drawn in the 8 + 5 arrangement of the acceptor stem/T-stem helices and the archaeal one in the 9 + 4 structure. See also discussion in Chapter 3.

## Biosynthesis of selenocysteine

Formally, the biosynthesis of selenocysteine takes place in the following three steps:

(a) L-Serine + ATP + tRNA$^{Sec}$ → L-Seryl-tRNA$^{Sec}$ + AMP + PP$_i$
(b) L-Seryl-tRNA$^{Sec}$ + SS → Dehydroalanyl-tRNA$^{Sec}$ + SS + H$_2$O
(c) Dehydroalanyl-tRNA$^{Sec}$ + SePO$_3^{3-}$ → Selenocysteyl-RNA$^{Sec}$ + PO$_4^{3-}$

tRNA$^{Sec}$ is charged with L-serine by seryl-tRNA synthetase (equation a) [23] which is in accordance with the presence of the serine identity elements [24]. The overall catalytic efficiency of charging, however, is only about 1% of that measured for a cognate serine acceptor which reflects the limited requirement of serine carbon flux into the minor pathway [26]. Each of the structural properties differentiating tRNA$^{Sec}$ from cognate serine acceptors could be responsible for the reduced acceptor activity.

The conversion of seryl-tRNA$^{Sec}$ into selenocysteyl-tRNA$^{Sec}$ is catalysed by selenocysteine synthase (SS), which is the *selA* gene product. Selenocysteine synthase is a decameric protein made up of 50 kDa subunits which contain pyridoxal-5´phosphate as prosthetic group [27]. The amino group of serine forms an aldimine linkage with the carbonyl of pyridoxal phosphate and a water molecule is eliminated yielding dehydroalanyl-tRNA$^{Sec}$ (equation b). Chemical proof for this intermediate was brought about via reduction by potassium borohydride which yields alanyl-tRNA$^{Sec}$ [28]. Nucleophilic addition of selenide then gives rise to selenocysteyl-RNA$^{Sec}$ (equation c). The source of selenide is monoselenophosphate [29], whose formation via selenophosphate synthetase is discussed in detail in Chapter 4 of this book. Since elevated levels of selenide can substitute for monoselenophosphate in the reaction, it has been speculated why free selenide is not used as the natural substrate. A possibility considered is that the phosphate serves as a specificity "handle" to discriminate selenide from the highly similar sulfide. Indeed, substituting selenophosphate by thiophosphate gave rise to cysteyl-tRNA$^{Sec}$ [30].

Biochemical and high resolution electron microscopic analysis demonstrated that two subunits of the decameric enzyme bind one seryl-tRNA$^{Sec}$ molecule [27,31]. The fully loaded enzyme thus contains five molecules of charged tRNAs bound to it. Binding seems uncooperative, only depending on the stoichiometry between the protein and the substrate [31]. It appears that once the tRNA is charged with serine it is immediately bound to selenocysteine synthase and stays in the activated state until selenophosphate is available as the substrate molecule. The cellular numbers of tRNA$^{Sec}$ molecules (about 250) [32] and selenocysteine synthase decamers (about 150, which can bind five tRNA$^{Sec}$ molecules simultaneously) [27,28,31] favor the assumption of the enzyme functioning as a sink for capturing the charged tRNA which may be an efficient way to optimize the efficiency of utilization of the trace element.

**Translation factor SelB**
To participate in the decoding process, the 20 classical aminoacyl-tRNAs each have to enter a ternary complex with elongation factor Tu and GTP. When the affinity of EF-Tu to selenocysteyl-tRNA$^{Sec}$ was determined, it was found that it is about 200-fold lower than that of the standard aminoacyl-tRNAs [33]. Under competitive conditions, therefore, it cannot serve as a substrate for EF-Tu. This role is taken over by the specialized translation factor SelB [34]. SelB (encoded by the *selB* gene, previously *fdhA*) is 69 kDa in size, and in its N-terminal part, it displays significant sequence similarity to EF-Tu (Figure 3). Since the three-dimensional structure of EF-Tu was available, the sequence of SelB which was homologous was modeled into the coordinates of elongation factor Tu [35]. An excellent match could be

observed in that the carbon backbone was almost perfectly superimposable, with the exception of that side of the molecule which in EF-Tu interacts with the guanosine nucleotide release factor EF-Ts [36].

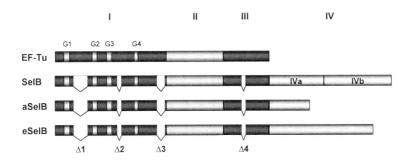

Figure 3. Domain structures of the bacterial SelB protein and its archaeal (aSelB) and eucaryal (eSelB) homologs, in comparison to the three structural domains of elongation factor Tu (EF-Tu). The G motifs involved in binding of the guanosine nucleotides are indicated by G1 to G4. Deletions within the SelB sequences relative to the EF-Tu primary structure are indicated.

Lack of this domain correlated well with the finding that the affinity of SelB to GTP is around 10-fold higher than to GDP which obviates the necessity for a guanosine nucleotide release factor (see below). Structural modeling suggested a homologous function and this was corroborated by the finding that SelB indeed tightly binds selenocysteyl-tRNA$^{Sec}$ [34]. Intriguingly, its precursor seryl-tRNA$^{Sec}$ was not recognized, which may be the reason why it does not function in protein synthesis. SelB, in contrast to EF-Tu, therefore, can discriminate between the aminoacyl residues of its tRNA substrate. The structural basis for this ability is still unresolved.

**Domain structure of SelB and the interaction with the SECIS element** A property of SelB crucial for its function is that it can bind to a secondary/tertiary structure of the mRNA, the SECIS element [37] (selenocysteine insertion sequence). (The designation SECIS was originally coined for the eucaryal mRNA element directing UGA decoding with selenocysteine by Berry and collaborators [38,39]. It is adopted for the bacterial in-frame element.) Bacterial SECIS is an approximately 40 nucleotide long stem-loop structure that follows the UGA codon at the immediate 3' side (Figure 4A). It is conserved by structure rather than by sequence in different organisms. Exceptions are organisms like *Clostridium sticklandii* [40] or *Eubacterium acidaminophilum* [41] in which no such structures can be formed within the reading frames of mRNAs coding for selenopolypeptides.

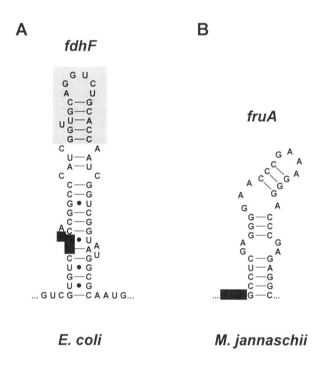

**Figure 4.** SECIS elements of the *Escherichia coli fdhF* mRNA (A) and of the *Methanococcus jannaschii fruA* mRNA (B). The bacterial SECIS structure is located within the reading frame immediately downstream of the UGA. The 17 nucleotide minihelix which is the minimal structure required for binding to SelB and the UGA codon are shaded. The archaeal SECIS element (B) is located in the 3'-untranslated region immediately downstream of the UAA stop codon (shaded) of the *fruA* gene [64].

On the basis of chemical probing results, the solution structures of the SECIS elements of the *E. coli fdhF* and *fdnG* mRNAs, were derived by computer modelling [42]. The results point to unique features within the loop regions indicating the existence of a compact tertiary structure. Despite differences in the sequence, identical bases within the upper part of the loop and stem of both hairpins interacted specifically with SelB indicative of identical tertiary structures.

The binding of SelB to the *E. coli* SECIS element is mediated via the 17 kDa penultimate C-terminal domain which is not present in EF-Tu [43]. This domain, designated 4b, is separated by a 14.5 kDa linker from the part of the molecule which is homologous to EF-Tu. Binding capacity was maintained when the domain 4b was separately expressed from an appropriately truncated *selB* gene. By gel retardation experiments it was proven that SelB can form a complex simultaneously with selenocysteyl-tRNA$^{Sec}$ and the SECIS element. The unusual feature that SelB interacts with two nucleic acids was also demonstrated in vivo by overexpression of *selB*. Overproduction of the protein resulted in a complete cessation of selenocysteine incorporation since the probability of interaction of one SelB molecule with each of the two RNA ligands was statistically impaired. It could be restored by overproduction of tRNA$^{Sec}$ in the same cell which increased the statistics of forming the quaternary complex between SelB, charged tRNA, the SECIS element and GTP [44].

The specificity of the interaction between SelB and the SECIS motif was analyzed via several techniques. The first employed the introduction of mutations into the SECIS element and examined their consequences on binding to SelB and on readthrough of the UGA in a *fdhF-lacZ* fusion reporter gene [45]. Base changes in the apical loop had the most prominent effects. They abolished readthrough of UGA located about 20 nucleotides upstream. Shortening of the lower helical stem by three bases on each side of the helix also caused a break-down of readthrough. On the other hand, insertion of a single triplet on each side at the lower part of the helix only quantitatively reduced readthrough, whereas the insertion of two triplets on each side was detrimental. In a later study, these results were confirmed and also extended by showing that the lower helical part of the SECIS element tolerated mutations which impaired or abolished base pairing whereas the upper helix was an essential element for function [46]. The relevance of the upper part of the stem-loop structure was also shown by in vitro synthesis of minihelices and by the capacity of SelB to bind them [43]. It was found that the smallest RNA structure still capable of interacting with either SelB or domain 4b was a 17 nucleotide minihelix. Mutant minihelices with base changes which did not promote readthrough with the intact mRNA were unable to bind indicating that the binding was still specific.

The second method with which interaction between SelB and the SECIS element was studied was by SELEX, i.e., selected *de novo* synthesis of aptamers with the capacity to bind to the SelB protein [47]. About 50% of the aptamers had the apical part of the stem-loop structure identical with the 17 base minihelix mentioned above. With this technique, the importance of the bulged-out U residue could be highlighted, which later was also demonstrated by a mutational approach [48]. Neither the lower helical part nor the UGA codon were selected amongst the aptamers. Even more interesting was the sequence of the aptamers which differed from the sequence and/or structure of the cognate SECIS element. Binders with higher affinity than the wild type structure were found as well as others with no structural similarity to the cognate SECIS element at all. The results showed that there is more than one solution for the formation of an RNA structure able to fit into the given binding pocket of a protein. However, these novel structures were unable to function as SECIS elements in decoding UGA with selenocysteine [47].

The third approach entailed selection of intergenic suppressor mutations of SelB which counteracted the detrimental effect of a mutation in the SECIS element [49]. Allele-specific and allele non-specific mutations could be selected. Most of the mutations led to an amino acid exchange within a 25 amino acid stretch in domain 4b of SelB and at least some of them appeared to be residues involved in contact formation between the protein and the RNA ligand. A mutation in domain 4a, E437K, was obtained several times; it caused allele non-speific readthrough at a low level and may be involved in the intramolecular communication between the N-terminal part of SelB binding the charged tRNA and the C-terminal domain binding the SECIS element.

**Readthrough versus termination**
The first evidence that readthrough of the UGA codon, i.e., the insertion of selenocysteine, competes with termination came from an experiment in which the *fdhF* gene was overexpressed from a plasmid in an otherwise wild-type genetic background. A truncated gene product was formed whose size correlated with the assumption that the specific UGA was not decoded but used as a termination signal [19]. The truncated polypeptide was not observed in a mutant in which the UGA was exchanged for a sense codon. In such mutants, the efficiency of product formation was increased dramatically both in the bacterial and the eucaryal systems [50-52]. The competition of termination with readthrough was also confirmed with genetic constructs in which the UGA plus the SECIS element were sandwiched in reporter gene constructs between two genes whose amounts and activities could be easily measured.

An intruiging demonstration of the competition between termination and UGA readthrough was recently provided by the expression of the thioredoxin reductase gene (TR) from mammalia in *E. coli* [53]. In TR mRNA, UGA is located next to last codon, a fact that prompted the fusion of the bacterial SECIS element 3´ to the indigenous TR stop codon. TR could be synthesized in significant amounts in *E.coli*, but in a mixture with a predominant variant lacking the last two amino acids. The conclusion was that in the synthesis of this variant UGA served as a stop codon. When the components of the selenocysteine insertion machinery were overproduced concomitantly in the same cells, the amount of full length gene product could be increased dramatically [53]. Altogether these results indicate that readthrough depends on a balanced supply of mRNA and the components of selenocysteine biosynthesis and incorporation. A second important conclusion drawn from the expression of the mammalian TR gene in *E. coli* was that the function of the SECIS element resides solely in the binding of the SelB protein; its sequence does not need to be translated.

Elements favoring termination at the cost of readthrough at the UGA directing selenocysteine insertion were recently elucidated [54]. It was shown that the rules of quadruplet stop codon recognition established by Tate and collaborators [55] also hold for the read-out of the selenocysteine specific UGA codon. A or C at the fourth position, which is used predominantly in selenoprotein mRNAs antagonizes termination. In the same communication, it was pointed out that the two amino acids preceding selenocysteine in the selenopolypeptides also greatly influence termination. The rules followed are essentially those elucidated by Isaksson and coworkers for premature termination of protein synthesis at prokaryotic ribosomes [56]. Therefore, the counteraction of termination of polypeptide synthesis at UGA does not reflect a system specific for selenocysteine insertion, but is subject to the same mechanisms as observed with ordinary sense and stop codons. A limitation of selenium supply signals the ribosome that the cognate aminoacyl-tRNA is unavailable and induces the same response via the empty A site as is observed upon withdrawal of one of the 20 classical amino acids [see 57 for review].

Termination at the selenocysteine-specific UGA is certainly connected with the fact that there is a considerable pause in translation once the UGA is in the ribosomal A site. Its duration is about 8 seconds, the time in which *E. coli* ribosomes (at $37^0$ C) insert about 100 of the standard amino acids [52]. The pause was not connected with the process of melting out of the SECIS structure, since it was not observed in the absence of SelB when the UGA was replaced by a sense codon and the anticodon of tRNA$^{Sec}$ was altered to match that particular sense codon. In the presence of SclB, ribosomal pausing took place irrespective of whether the mRNA contained the UGA or a classical sense codon, or whether the anticodon of tRNA$^{Sec}$ was complementary. Pausing, therefore, must be mechanistically connected with the resolution of the complex between the SECIS structure and SelB.

## Mechanism of decoding of the UGA with selenocysteine
The situation that one and the same codon can have two different meanings in information transfer in the same cell is unique and therefore attracted considerable attention. Mechanistically, the fact that termination efficiency is reduced must be separated from the issue why that particular UGA and not any other ordinary UGA stop codon is recognised as such a signal. A model originally proposed implicated that binding of the translation factor carrying the charged tRNA in the immediate proximity of the codon leads to the tethering of the particular tRNA to the ribosomal A site [45]. Several lines of

experimental evidence gathered later, however, indicated that this cannot be the sole explanation.

First, no selenocysteine insertion can be detected in the absence of the SECIS structure, even when SelB and tRNA$^{Sec}$ are overproduced. Thus, binding of SelB must have some other effect beyond increasing the local concentration of the tRNA in the proximity of the ribosomal A site. Secondly, in the SELEX approach, a considerable number of RNA aptamers were obtained which bound to SelB, some even with higher affinity than the wild type structure, but which were biologically inactive. Thus, binding can be separated functionally from the biological activity [47]. Insight into the nature of the biological effect then came from two sides. First, in the gel retardation assays it was observed that SelB formed a stronger complex with selenocysteyl-tRNA$^{Sec}$ when the SECIS element was also present as a ligand [37]. The conclusion was that the two RNA binding sites of SelB must communicate with each other. Secondly, when using GTP hydrolysis catalysed by SelB as an indicator for the activity state of the protein, it was found that the activity could be greatly stimulated by ribosomes, but only, when the SECIS element was present as a ligand [58]. Intriguingly, the 17 nucleotide minihelix constituting the minimal SECIS structure could suffice in stimulation. The only possible conclusion was that binding to the SECIS element led to a conformational switch that rendered SelB compatible for productive interaction with the ribosome. This interaction is required for the induction of GTP hydrolysis [58].

A rather detailed insight in the decoding process could be gained by rapid kinetic analysis of the interaction of *E. coli* translation factor SelB with guanosine nucleotides and the two RNA ligands via the stop-flow technique [59]. Intrinsic fluorescence of tryptophane residues, the fluorescence of methylanthraniloyl derivatives of guanosine nucleotides and fluorescence resonance energy transfer from tryptophane to the methylanthraniloyl group were used as signals. It was found that the affinities of SelB to GTP and GDP were considerably lower than those reported for other translation factors. The rate constant of the release of GDP from its complex with SelB was several orders of magnitude larger than that displayed by elongation factor EF-Tu which explains why no guanosine nucleotide release factor is required for the function of SelB.

On the other hand, the rate constant for the release of GTP is two orders of magnitude lower and in the same range as that measured for elongation factor EF-Tu. When the interaction of SelB with the 17 nucleotide minihelix of the *fdhF* SECIS element which carried a fluorescent group was assessed, an affinity of 1 nM was observed which was even increased when selenocysteyl-tRNA$^{Sec}$ was present. Binding of the charged tRNA, therefore, maximizes the affinity of SelB for the RNA ligand; dissociation of the tRNA decreases the affinity, on the other hand, which leads to dissociation of SelB from the mRNA, a necessary requirement for the translation of codons downstream of the UGA.

In conclusion, the following scenario can be visualized for decoding UGA as selenocysteine on the basis of the information presently available:

(i) A quaternary complex between SelB, selenocysteyl-tRNA$^{Sec}$, the SECIS element of the mRNA and GTP is formed. Its formation is non-random; binding of the charged tRNA stabilizes the complex of SelB with the SECIS element.

(ii) During translation, the complex is translocated towards the ribosome, the lower helical part of the SECIS element is melted, and when the UGA arrives at the A site, SelB makes contact with the ribosome, which induces GTP hydrolysis.

(iii) The charged tRNA is released in the proximity of the A site; its release decreases the affinity of SelB for the mRNA and facilitates the dissociation of the SelB–SECIS complex.

(iv) After translation of the SECIS sequence, the RNA can refold and serve as a target for the formation of a new quaternary complex to assist the next oncoming ribosome in decoding UGA.

The model can accomodate the results of the structure/function relation of the mutant SECIS variants, like the stringent requirement of a precise distance between the UGA and the loop region. Many issues, however, remain speculative. Among them, for example, is the most relevant question whether GTP hydrolysis precedes or succeeds the release of charged tRNA.

**Specific incorporation of selenocysteine into archaeal proteins**
Selenoproteins had been identified in *Methanococcus vannielii* [60] and *Methanococcus voltae* [61] and it had been shown that the insertion of selenocysteine is directed by UGA [62]. When the first genome sequence for an archaeal microorganism, *Methanococcus jannaschii,* was published [63], a search for open reading frames containing an in-frame UGA revealed the existence of seven putative selenoproteins [64]. This was confirmed by radioactive labelling experiments using $^{75}$Se-selenite. Thus, as in bacteria and eucarya, the insertion of selenocysteine is cotranslational and directed by UGA. Two more observations were intriguing: First, there was one gene, *vhuU,* which appeared to contain two selenocysteine residues in its sequence, and second, no mRNA structure which could serve as a SECIS element could be detected in the coding part of the mRNA. Such structures, however, could be formed within the sequence of the 3′-untranslated region, and in one case in the 5′-untranslated region. They were highly conserved in organisms like *M. voltae* or *Methanopyrus kandleri*. The archaeal SECIS element was located either adjacent to the stop codon of the reading frame (Figure 4B) or

positioned further downstream. The maximal distance to the translational stop codon observed in the latter cases was 117 nucleotides [64]. The biological function of these 3′ structures as a SECIS motif could be proven meanwhile by the heterologous expression of one of the genes from *M. jannaschii* in *M. maripaludis*. Variants of the genes lacking the 3′-untranslated region or containing base changes in conserved segments of it were not expressed as selenoproteins (M Rother, Garner, A Resch, W Whitman and A Böck, unpublished results). The mechanism of selenocysteine insertion at the ribosome, therefore, is an additional feature in which archaea follow the eucaryal strategy.

Because of its genetic simplicity and easier experimental handling, the archaeal system was considered a much more facile model for unravelling the intruiging mechanism of how a sequence element located far away from a certain codon could influence its specificity of decoding. An essential step towards this goal consisted of the identification of a protein with a function homologous to that of the bacterial translation factor SelB [65]. A search of the *M. jannaschii* genome revealed Mj0495 as a possible candidate, since it exhibited the same signature motifs within the G domain as SelB (see Figure 3). The product of this open reading frame (aSelB) was purified, shown to be a GTPase with guanosine nucleotide binding properties like SelB, and demonstrated to bind selenocysteyl-tRNA$^{Sec}$ from the same organism, but preferentially to seryl-tRNA$^{Sec}$. Surprisingly, however, with the techniques employed, it did not show binding to its homologous SECIS element [65]. This inability correlated with the fact that Mj0495 contains a C-terminal extension of only 11 kDa compared to 31.5 kDa of the bacterial SelB. Clearly, in archaea at least one other macromolecular component must exist which cooperates with Mj0495 in fulfilling the function which SelB achieves in bacteria [65].

Translation factors homologous to the archaeal SelB have been identified meanwhile in mammalia [66,67] and a SECIS binding protein, SBP2, has also been isolated and shown to cooperate with the translation factor in decoding UGA with selenocysteine in the eucaryal system. [66,68,69 and Chapters 6 and 7].

**References**

1. DB Cowie, GN Cohen 1957 *Biochim Biophy Acta* 26:252
2. T Tuve, HH Williams 1961 *J Biol Chem* 236:597
3. J Heider, A Böck 1993 *Advances Microbial Physiol* 35:71
4. JL Hoffmann, KP McConell, DR Carpenter 1970 *Biochim Biophys Acta* 199:531
5. PA Young, II Kaiser 1975 *Arch Biochem Biophys* 171:483
6. S Müller, J Heider, A. Böck 1997 *Arch Microbiol* 168:421
7. WA Hendrickson, JR Horton, DM LeMaster 1990 *EMBO J* 9:1665
8. GP Mullen, RP Dunlap, JD Odon 1986 *Biochemistry* 25:5625
9. RE Huber, RS Criddle 1967 *Biochim Biophys Acta* 141:587
10. MGN Hartmannis, TC Stadtman 1982 *Proc Natl Acad Sci USA* 79:4912

11. S Müller, H Senn, B Gsell, W Vetter, C Baron, A Böck 1994 *Biochemistry* 33:3404
12. F Zinoni, A Birkmann, W Leinfelder, A Böck 1987 *Proc Natl Acad Sci USA* 84:3156
13. A Graham, HE Jenkins, NH Smith, M-A Mandrand-Berthelot, BA Haddock, DH Boxer 1980 *FEMS Microbiol Lett* 7:145
14. BA Haddock, M-A Mandrand-Berthelot 1982 *Biochem Soc Trans* 10:478
15. M-A Mandrand-Berthelot, MYK Wee, BA Haddock 1978 *FEMS Microbiol Lett* 4:37
16. YA Begg, JN Whyte, BA Haddock 1977 *FEMS Microbiol Lett* 2:47
17. M Chippaux, F Casse, C-C Pascal 1972 *J Bacteriol* 110:766
18. EL Barrett, CE Jackson, HT Fukumoto, GW Chang 1979 *Mol Gen Genet* 177:95
19. W Leinfelder, K Forchhammer, F Zinoni, G Sawers, M-A Mandrand-Berthelot, A Böck 1988 *J Bacteriol* 170:540
20. JC Cox, ES Edwards, JA DeMoss 1981 *J Bacteriol* 145:1317
21. I Chambers, J Frampton, P Goldfarb, N Affara, W McBain, PR Harrison 1986 *EMBO J* 5:1221
22. F Zinoni, A Birkmann, TC Stadtman, A Böck 1986 *Proc Natl Acad Sci USA* 83:4650
23. W Leinfelder, E Zehelein, M-A Mandrand-Berthelot, A Böck 1988 *Nature* 331:723
24. C Baron, W Westhof, A Böck, R Giegé 1993 *J Mol Biol* 231:274
25. P Tormay, R Wilting, J Heider, A Böck 1994 *J Bacteriol* 176:1268
26. C Baron, J Heider, A Böck 1990 *Nucl Acids Res* 18:6761
27. K Forchhammer, W Leinfelder, K Boesmiller, B Veprek, A Böck 1991 *J Biol Chem* 266:6318
28. K Forchhammer, A Böck 1991 *J Biol Chem* 266:6324
29. Z Veres, L Tsai, TD Scholz, M Politino, RS Balaban, TC Stadtman 1992 *Proc Natl Acad Sci USA* 89:2975
30. P Tormay, R Wilting, F Lottspeich, PK Mehta, P Christen, A Böck 1998 *Eur J Biochem* 254:655
31. H Engelhardt, K Forchhammer, S Müller, KN Goldie, A Böck 1992 *Mol Microbiol* 6:3461
32. H Dong, L Nilsson, CG Kurland 1996 *J Mol Biol* 260:649
33. C Förster, G Ott, K Forchhammer, M Sprinzl 1990 *Nucl Acids Res* 18:487
34. K Forchhammer, W Leinfelder, A Böck 1989 *Nature* 342:453
35. R Hilgenfeld, A Böck, R Wilting 1996 *Biochimie* 78:971
36. T Kawashima, C Berthet-Colominas, M Wulff, S Cusack, R Leberman 1996 *Nature* 379:511
37. C Baron, J Heider, A Böck 1993 *Proc Natl Acad Sci USA* 90:4181
38. MJ Berry, L Banu, Y Chen, SJ Mandel, JD Kieffer, JW Harney, PR Larsen 1991 *Nature* 353:273
39. MJ Berry, L Banu, JW Harney, PR Larsen 1993 *EMBO J* 12:3315
40. GE Garcia, TC Stadtman 1992 *J Bacteriol* 174:7080
41. T. Gursinsky, J Jäger, JR Andreesen, B Söhling 2000 *Arch Microbiol* 174:200
42. A Hüttenhofer, E Westhof, A Böck 1996 *RNA* 2:354
43. M Kromayer, R Wilting, P Tormay, A Böck 1996 *J Mol Biol* 262:413
44. P Tormay, G Sawers, A Böck 1996 *Molec Microbiol* 21:1253
45. J Heider, C Baron, A Böck 1992 EMBO J 11:3759
46. Z Liu, M Reches, I Groisman, H Engelberg-Kulka 1998 *Nucl Acids Res* 26:896
47. SJ Klug, A Hüttenhofer, M Kromayer, M Famulok 1997 *Proc Natl Acad Sci USA* 94:6676
48. Z Liu, M Reches, H Engelberg-Kulka 2000 *J Bacteriol* 182:6302
49. M Kromayer, B Neuhierl, A Friebel, A Böck 1999 *Mol Gen Genet* 262:800
50. MJ Berry, AL Maia, JD Kieffer, JW Harney, PR Larsen 1992 *Endocrinology* 131:1848
51. H Kollmus, L Flohé, JEG McCarthy 1996 *Nucl Acids Res* 24:1195
52. S Suppmann, BC Persson, A Böck 1999 *EMBO J* 18:2284
53. ESJ Arnér, H Sarioglu, F Lottspeich, A Holmgren, A Böck 1999 *J Mol Biol* 292:1003

54. Z Liu, M Reches, H Engelberg-Kulka 1999 *J Mol Biol* 294:1073
55. ES Poole, CM Brown, WP Tate 1995 *EMBO J* 14:151
56. S Mottagui-Tabar, A Björnsson, L Isaksson *EMBO J* 13:249
57. J Parker 1989 *Microbiol Rev* 53:273
58. A Hüttenhofer, A Böck 1998 *Biochemistry* 37:885
59. M Thanbichler, A Böck, RS Goody 2000 *J Biol Chem* 275:20458
60. S Yamazaki 1981 *J Biol Chem* 257:7926
61. E Kothe, S Halboth, I Sitzmann, A Klein 1990 in: *Microbiology and biochemistry of strict anaerobes involved in interspecies H2 transfer* (ed. Belaich et al) Plenum Press New York pp 25-36
62. S Halboth, A Klein 1992 *Molec Gen Genet* 233:217
63. CJ Bult et al 1996 *Science* 273:1058
64. R Wilting, S Schorling, BC Persson, A Böck 1997 *J Mol Biol* 266:637
65. M Rother, R Wilting, S Commans, A Böck 2000 *J Mol Biol* 299:351
66. RM Tujebajeva, PR Copeland, X-M Xu, BA Carlson, JW Harney, DM Driscoll, DL Hatfield, MJ Berry 2000 *EMBO Reports* 11:158
67. D Fagegaltier, N Hubert, K Yamada, T Mizutani, P Carbon, A Krol 2000 *EMBO J* 17:4796
68. PR Copeland, DM Driscoll 1999 *J Biol Chem* 274:25447
69. PR Copeland, JE Flechter, BA Carlson, DL Hatfield, DM Driscoll 2000 *EMBO J* 19:306

# Chapter 3. Mammalian selenocysteine tRNA

Bradley A. Carlson, F. Javier Martin-Romero, Easwari Kumaraswamy, Mohamed E. Moustafa, Huijun Zhi and Dolph L. Hatfield

*Section on the Molecular Biology of Selenium, Basic Research Laboratory, National Cancer Institute, National Institutes of Health, Bethesda, MD 20892, USA*

Byeong Jae Lee

*Laboratory of Molecular Genetics, Institute for Molecular Biology and Genetics, Seoul National University, Seoul 151-742, Korea*

**Summary:** The selenocysteine (Sec) tRNA population in mammals consists of two major isoforms that differ from each other by a single methylation group on the 2'-O-ribosyl moiety at position 34. The tRNAs are 90 nucleotides in length making them the longest eukaryotic tRNAs sequenced to date. Both tRNAs decode UGA and arise from a single copy gene. The primary transcript is generated unlike that of any known tRNA as transcription begins at the first nucleotide within the gene (i.e., the transcript lacks a 5'-leader sequence). Unlike the other 20 aminoacyl-tRNAs, Sec tRNA has an additional function of serving as the carrier molecule for synthesis of its amino acid as the isoforms are initially aminoacylated with serine. Over-expression and under-expression of the Sec tRNA population has little effect on selenoprotein biosynthesis. However, mutation of the $N^6$-isopentenyladenosine moiety at position 37 of the tRNA results in selective inhibition of Sec tRNA maturation and selenoprotein biosynthesis.

## Introduction
Selenocysteine (Sec) tRNA has been described as the key molecule [1] and the central component [2] in selenoprotein biosynthesis. Indeed, this tRNA serves as both the carrier molecule for the biosynthesis of Sec and as the donor of the synthesized amino acid to the nascent selenopeptide. Sec tRNA is first aminoacylated with serine and thus the tRNA has been designated as Sec tRNA$^{[Ser]Sec}$ [2]. In this chapter, we will discuss some of the features of this unique molecule.

## Primary and secondary structures
Sec tRNA$^{[Ser]Sec}$ has been isolated and sequenced from bovine liver [3-5], rat liver [6] and mouse and HeLa cells [7]. There are two major isoforms. Both

are 90 nucleotides in length and both contain only four modified bases. They differ by a single methyl group on the ribose at position 34 (the wobble position of the anticodon) where one form contains 5-methylcarboxymethyluridine (mcmU) and the other 5-methylcarboxymethyluridine-2'-O-methylribose (mcmUm). The shared modified bases are $N^6$-isopentenyladenosine ($i^6A$) at position 37, pseudouridine (ψ) at position 55 and 1-methyladenosine ($m^1A$) at position 58 (see Figure 1).

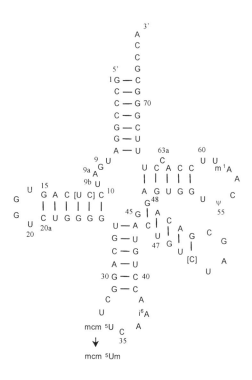

**Figure 1.** Cloverleaf model of mammalian Sec tRNA[Ser]Sec. The secondary structure of the tRNA is shown in a 7/5 based paired form (i.e., 7 base pairs in the acceptor stem and 5 base pairs in the T-stem), but there is also evidence that it may exist in a 9/4 form (see text). Bases in brackets show positions that vary in different mammals as discussed in the text.

Two different secondary structures have been proposed for Sec tRNA[Ser]Sec [1,3] and data supporting both models have been presented [8-12]. Using data generated by enzymatic and chemical probing, a 9/4 base paired structure (9 paired bases in the acceptor stem and 4 in the T-stem) has been suggested [8,9]. Certain tRNA[Ser]Sec mutants which can adopt a 9/4 secondary structure were found to be aminoacylated and to form selenocysteyl-tRNA[Ser]Sec in support of the 9/4 model [12]. Steinberg and collaborators suggested that the 7/5 base paired structure is correct based on the ability of other tRNA[Ser]Sec mutants to be aminoacylated and phosphorylated [10,11]. Sec tRNA[Ser]Sec from archaea can fold into a 9/4 [13] or an 8/5 secondary structure [14], but importantly, regardless of the structures of the eukaryotic, archaeal and eubacterial Sec tRNAs, they must disrupt and conform to those of all tRNAs used in elongation when aminoacyl-tRNAs are involved in translation [14].

**Gene copy number, transcription and maturation**
The functional gene copy number for Sec tRNA[Ser]Sec has been analyzed in a number of animals including rats [6], mice [15], humans, rabbits, cows, Chinese hamsters, frogs, zebrafish, flies and nematodes [reviewed in 16]. With the exception of zebrafish, where the gene occurs at two separate loci [17], all animals examined thus far encode a single copy of the gene in their genomes. Humans and rabbits also contain a pseudogene [16], while Chinese hamsters contain three pseudogenes [18]. The functional gene, as well as the gene product [19], in Chinese hamsters have T (U), C at positions 11 and 12 compared to C, T (U) at these tRNA[Ser]Sec positions in other mammals. Sequences of the Chinese hamster pseudogenes suggest that the functional gene may have arisen by editing at these two positions [18]. The mouse and rat functional genes have a T at position 47c compared to a C in the genes of other mammals.

Transcription of the tRNA[Ser]Sec gene has been thoroughly reviewed recently [16] and will not be further considered herein. It is important to note that the tRNA[Ser]Sec gene is the only known tRNA gene in nature that begins transcription at the first nucleotide within the coding sequence [20]. All other tRNAs are transcribed with a leader sequence that must be processed. The primary transcript of tRNA[Ser]Sec has a trailer sequence that is processed [20]. Interestingly, Sec tRNA[Ser]Sec has a triphosphate at its 5' end [20]. Most of the studies involving the transcription of tRNA[Ser]Sec since the subject was last reviewed [16] have focused on the transcription-activating factor (Staf) that specifically enhances transcription of the Sec tRNA[Ser]Sec gene [see 21,22 and references therein].

Following transcription and processing at its 3' end, the transcript undergoes maturation. Unlike most other tRNAs that may have as many as 15-17 modified bases, Sec tRNA[Ser]Sec has only four modified nucleosides on two different isoforms. The biosynthetic pathway of these modifications has been established in *Xenopus* oocytes [23,24]. The syntheses of ψ and $m^1A$ occur initially in the nucleus at positions 55 and 58, respectively. The formation of $mcm^5U$ and $mcm^5Um$ at position 34 and $i^6A$ at position 37 occurs in the cytoplasm. The $mcm^5U$ isoform is converted to the $mcm^5U$-$i^6A$ form that in turn is converted to the $mcm^5Um$ form [23,25]. Although the mcm$^5$Um isoform without i$^6$A occurs in the cytoplasm, it is not clear how this species arises as i$^6$A apparently cannot be added to this species [25].

Mutations at each modified site in Sec tRNA$^{[Ser]Sec}$ were generated and the resulting effect on modification at other positions analyzed by microinjecting these mutant tRNA$^{[Ser]Sec}$ forms into *Xenopus* oocytes [25]. This study has more clearly identified the pathway of modification biosynthesis. For example, microinjection of tRNA$^{[Ser]Sec}$ that has been mutated at position 58,

**Table 1.** Effect of mutations in Sec tRNA$^{[Ser]Sec}$ on maturation.[a]

| Mutation[b] (position) | Result of mutation[c] | Modified nucleoside formed[d] | | | | |
|---|---|---|---|---|---|---|
| | | m$^1$A | Ψ | i$^6$A | mcm$^5$U | mcm$^5$Um |
| Wild type | none | + | + | + | + | + |
| A58U | No m$^1$A | – | – | + | + | ± |
| U55G | No Ψ;TS[c] | + | – | + | + | – |
| A37G | No i$^6$A | + | + | – | + | – |
| U34G | No mcm$^5$U; No mcm$^5$Um | + | + | + | – | – |
| U16A:G17C: G18C:U19A | D-loop; TS[c] | + | + | + | + | – |

[a]Mutations were prepared in a vector encoding the tRNA$^{[Ser]Sec}$ gene, primary transcripts generated and microinjected into *Xenopus* oocytes, and the resulting maturation products analyzed for modified nucleosides [25].
[b]Designates the position within tRNA$^{[Ser]Sec}$ (see Figure 1) that mutations were made and A58U, for example, denotes that the A at position 58 was changed to a U, etc.
[c]Denotes whether the mutation(s) cause(s) the loss of a modified base and/or a change in tertiary structure (TS).
[d]Denotes whether the modified nucleoside was found in the mature, mutant tRNA$^{[Ser]Sec}$.

so that the tRNA cannot synthesize m$^1$A, revealed that this mutant tRNA does not synthesize ψ at position 55. This observation suggested that i$^6$A, or at least an A, must be at position 58 for ψ formation. Syntheses of mcm$^5$U and i$^6$A occur on the m$^1$A mutant, while conversion to the mcm$^5$Um occurs very weakly. Mutation at the ψ site (position 55), which affects tertiary structure, or mutations at sites within the D-loop, which also affect tertiary structure, show that conversion of mcm$^5$U to mcm$^5$Um is inhibited. In addition, mutation at the i$^6$A site (position 37), which apparently does not affect tertiary structure, also inhibits the methylation step. Thus, efficient methylation of mcm$^5$U requires an intact tertiary structure as well as the prior synthesis of each modified base. Formation of the modified bases at other positions, including mcm$^5$U, is not as stringently connected to precise primary and tertiary structure as is the methylation step at position 34 with the exception of ψ that requires an A or m$^1$A at position 58. The effects of the various mutations in tRNA$^{[Ser]Sec}$ on its maturation are summarized in Table 1. These studies, and those showing that the presence of the methyl group on the ribosyl moiety at position 34 dramatically alters tertiary structure [6] and that the methylation step is influenced by the intracellular selenium status [6,26,27], suggest that the mcm$^5$U and mcm$^5$Um isoforms have different roles in selenoprotein biosynthesis.

**Biosynthesis of Sec**
As noted above, tRNA$^{[Ser]Sec}$ is aminoacylated with serine and the synthesis of Sec occurs on its tRNA. Thus, serine serves as the backbone for the biosynthesis of Sec [28,29]. The identity elements that exist in tRNA$^{[Ser]Sec}$ for recognition by its synthetase for amino acid attachment correspond to those for serine and not those for Sec. The identity elements in mammalian tRNA$^{[Ser]Sec}$ for seryl-tRNA synthetase are the discriminator base (G73), which is essential for aminoacylation, the long extra arm, which plays an important role [30,31], and the acceptor, T- and D-stems, which also contribute to the recognition process [32].

After aminoacylation, seryl-tRNA$^{[Ser]Sec}$ is now ready to serve as substrate for the formation of selenocysteyl-tRNA$^{[Ser]Sec}$. Interestingly, the serylated form can serve to suppress UGA termination codons in vivo [33] and seryl-tRNA$^{[Ser]Sec}$ is therefore an authentic nonsense suppressor tRNA. The metabolic pathway for Sec synthesis on its tRNA has been completely established in bacteria [see Chapter 2] and its biosynthesis will be discussed in this chapter only briefly to provide a better understanding of what needs to be established in mammals. The bacterial enzyme, designated Sec synthase, carries out Sec synthesis in two major steps. A 2,3-elimination of water from serine occurs yielding an aminoacryl-tRNA$^{[Ser]Sec}$ intermediate which serves as the acceptor molecule for the activated form of selenium [see Chapter 2]. The selenium donor has been identified in bacteria as monoselenophosphate [see Chapter 4] which is formed from selenite and ATP by selenophosphate synthetase [see Chapters 2 and 4]. Sec synthetase then mediates the addition of selenium to the aminoacryl intermediate resulting in the generation of selenocysteyl-tRNA$^{[Ser]Sec}$ and inorganic phosphate.

In mammals, many of the details of Sec biosynthesis are lacking. A minor seryl-tRNA capable of decoding UGA [34] and forming phosphoseryl-tRNA [see 4 and refs therein] was discovered many years ago and was subsequently identified as Sec tRNA[Ser]Sec [29]. It has been suggested that phosphoseryl-tRNA may be an intermediate in selenocysteyl-tRNA[Ser]Sec biosynthesis [2]. Evidence has been published that phosphoseryl-tRNA[Ser]Sec was the true intermediate in the biosynthesis of Sec in mammals [35,36] and bacteria [37]. However, following the identification of aminoacryl-tRNA[Ser]Sec as the correct intermediate in bacteria [1,38], it was reported that phosphoseryl-tRNA[Ser]Sec was not involved in the biosynthesis of Sec and that the mammalian pathway proceeded like that in bacteria, i.e., via Sec synthase [39,40]. The intermediate in these latter studies has not been thoroughly characterized. Phosphoseryl-tRNA[Ser]Sec has been proposed more recently to be "an active storage form" that can be regenerated for Sec biosynthesis following its dephosphorylation [32]. Since seryl-tRNA[Ser]Sec can serve as an authentic suppressor of UGA stop codons [15] and phosphoseryl-tRNA has been shown to donate phosphoserine to protein [41], it would seem unlikely that the phosphorylated species would be maintained intracellularly as a storage form. It should be emphasized that the role of phosphoseryl-tRNA[Ser]Sec in Sec synthesis has not been established and the Sec biosynthetic pathway in mammals has not been thoroughly investigated.

The active form of selenium that is donated to the intermediate in the synthesis of Sec in bacteria has been shown, as noted above, to be monoselenophosphate and is mediated by selenophosphate synthetase in the presence of selenite and ATP [see Chapters 2 and 4]. Although the corresponding active selenium donor has not been identified in mammals, it seems likely that mammals utilize the same donor [42-44]. Two selenophosphate synthetase genes have been identified in mammals and they are designated *Sps1* and *Sps2* [42-44]. *Sps2* is a selenoprotein as it has a TGA codon in its open reading frame. This observation suggests that the gene product may be involved in its autoregulation [44].

## Sec tRNA[Ser]Sec population in cells and tissues
The Sec tRNA[Ser]Sec population in mammals consists of two major isoforms. One contains mcm$^5$U in the wobble position of the anticodon and the other contains mcm$^5$Um. They differ, therefore, by a single 2'-O-methylribose [16]. The relative amounts and distributions of these two isoforms vary in mammalian cells in culture [26] and in different mammalian tissues [6,27]. Their levels and distributions are influenced by selenium status. As shown in Figure 2, there is a dramatic shift in the distributions of the two isoforms in response to selenium. The figure shows the elution profiles of the two isoforms from human leukemia (HL60) cells on a reverse phase chromatographic column [26]. The earlier eluting species is mcm$^5$U and the later eluting species is mcm$^5$Um. The pattern of distribution of the two isoforms shown in panel A of Figure 2 is characteristic of a selenium

deficient status, while that shown in panel B is characteristic of a selenium sufficient status. It is not clear if selenium enhances the methylation step or reduces the turnover of the mcm$^5$Um form. However, selenium appears to stabilize the turnover of the Sec tRNA$^{[Ser]Sec}$ population in *Xenopus* oocytes and not enhance tRNA$^{[Ser]Sec}$ transcription [23].

The response of the mcm$^5$U and mcm$^5$Um isoforms to the replenishment of selenium in extremely selenium deficient rats has also been examined [27]. The levels and distributions of mcm$^5$U and mcm$^5$Um were restored to normal levels over several days, but the rate of redistribution was tissue specific. Liver and kidney restored their isoforms to a selenium sufficient pattern in about 72 hours, while that in brain and muscle occurred at a slower rate.

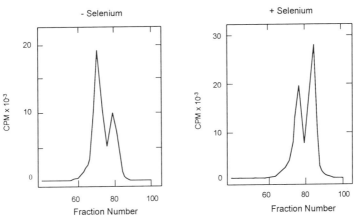

**Figure 2.** Sec tRNA$^{[Ser]Sec}$ mcm$^5$U and mcm$^5$Um isoforms from HL60 cells grown in the absence (A) or presence (B) of selenium. Total tRNA was isolated from HL60 cells, labeled with [$^3$H]-serine and fractionated over a reverse phase chromatographic column [see text and 26]. The peak that elutes first from the column is the mcm$^5$U isoform and the peak that elutes second is the mcm$^5$Um isoform.

## Over- and under-expression

The level of the Sec tRNA$^{[Ser]Sec}$ population has been over-expressed by about ten fold in mammalian cells in culture [45] and about five to six fold in transgenic mice [46]. Chinese hamster ovary cells were transfected with the mouse tRNA$^{[Ser]Sec}$ gene and varying numbers of genes were inserted into the hamster genome. The level of tRNA$^{[Ser]Sec}$ expressed from the transfected genes was directly proportional to copy number up to as many as 10 gene copies [45]. Transgenic mice were generated carrying as few as 2 and as many as 20 copies of the tRNA$^{[Ser]Sec}$ transgene, but the level of gene product did not increase proportionally to gene copy number and it varied in different tissues. For example, the level of tRNA$^{[Ser]Sec}$ increased about 3.5 times in liver, about 5 times in kidney and about 6 fold in brain and muscle in mice carrying 20 extra transgenes [46]. Interestingly, most of the tRNA$^{[Ser]Sec}$ that resulted in an increased tRNA$^{[Ser]Sec}$ population in both cells in culture and in transgenic mice was the mcm$^5$U form suggesting that methylation of this species to yield the mcm$^5$Um form was limiting. Furthermore, selenoprotein biosynthesis did not appear to be altered by the higher amounts of tRNA$^{[Ser]Sec}$ suggesting that tRNA is not limiting in this process.

Studies in which the level of the Sec tRNA$^{[Ser]Sec}$ population has been reduced by removing one copy of the wild type gene also suggest that tRNA is not limiting in selenoprotein biosynthesis. Mice that are heterozygous for a targeting vector lacking the tRNA$^{[Ser]Sec}$ gene [47] and mouse embryonic cells that are heterozygous for a similar targeting vector [15] were generated and the tRNA$^{[Ser]Sec}$ population and glutathione peroxidase levels analyzed. These mice and stem cells have approximately half the Sec tRNA$^{[Ser]Sec}$ population, but synthesize a normal amount of selenoprotein, providing further evidence that tRNA$^{[Ser]Sec}$ is not a limiting factor in selenoprotein biosynthesis. Removal of both alleles of the Sec tRNA$^{[Ser]Sec}$ gene from the mouse genome (i.e., knockout of the gene) results in embryonic lethality demonstrating that selenoprotein biosynthesis is essential to mammals [47].

**Mutant Sec tRNAs**
Making mutations in Sec tRNA$^{[Ser]Sec}$ and analyzing their biological consequences provides an alternative and important means of determining the function and structure of this unique tRNA. This approach has been fruitful in probing secondary [10-12,14] and tertiary structure [25 and S. Matsufuji, personal communication], and in determining the identity elements in tRNA$^{[Ser]Sec}$ for its aminoacylation [30-32], the identity elements in seryl-tRNA$^{[Ser]Sec}$ for its phosphorylation [see 32 and references therein] and the pathway of tRNA$^{[Ser]Sec}$ maturation [25]. In addition, mutant tRNA$^{[Ser]Sec}$ species have been used to examine other parameters of Sec tRNA$^{[Ser]Sec}$ function. For example, the anticodon of Sec tRNA$^{[Ser]Sec}$ has been changed to UAA or to UUA [48]. These altered tRNAs are capable of translating the complementary leucine codon, UUA, or the complementary termination codon, UAA, respectively, following mutation of the UGA Sec codon to UUA or UAA. This study shows that UGA is not essential for Sec insertion into the growing selenopeptide in mammalian cells provided the complementarity between anticodon and codon are preserved [see ref 49 for comparable experiments in bacteria].

Transfer RNAs that translate codons with a U in the 5'-position often have i$^6$A at position 37. The presence of i$^6$A apparently restricts wobble and prevents misreading as nonsense suppressor tRNAs lacking this modification have been shown to manifest a dramatically reduced efficiency in translating nonsense codons in bacteria and yeast [see 50-52 for review]. Sec tRNA$^{[Ser]Sec}$ lacking i$^6$A has a pronounced effect on tRNA maturation (see

Table 1 and below) and on selenoprotein biosynthesis. For example, Chinese hamster ovary cells that are co-transfected with an i$^6$A mutant tRNA$^{[Ser]Sec}$ and a selenoprotein, type 1 deiodinase, result in only a marginal inhibition of endogenous selenoprotein synthesis, but effectively inhibit (by ~80%) the deiodinase [53]. Generation of transgenic mice carrying an i$^6$A mutant tRNA$^{[Ser]Sec}$ transgene results in selective inhibition of tRNA$^{[Ser]Sec}$ maturation and selenoprotein synthesis [46]. The level of the mcm$^5$U form is reduced in these transgenic mice. Some selenoproteins such as glutathione peroxidase 1 are reduced substantially, while other selenoproteins such as thioredoxin reductase 3 are only slightly affected. The reason for the selective inhibition of selenoprotein inhibition is not clearly understood.

Preparation of other specific mutant Sec tRNAs and analyzing their effects on protein and selenoprotein biosynthesis should provide a better understanding of the cellular roles of this unique tRNA and its two major isoforms, mcm$^5$U and mcm$^5$Um.

### References

1. A Böck, K Forchhammer, J Heider, C Baron 1991 *Trends Biochem Sci* 16:463
2. DL Hatfield, IS Choi, T Ohama, J-E Jung, AM Diamond 1994 *Selenium in Biology and Human Health* RF Burk (Ed) Springer-Verlag, New York, 25
3. AM Diamond, B Dudock, DL Hatfield 1981 *Cell* 25:497
4. DL Hatfield, AM Diamond, B Dudock 1982 *Proc Natl Acad Sci USA* 79:6215
5. R Amberg, C Urban, B Reuner, P Scharff, SC Pomerantz, JA McCloskey, HJ Gross 1993 *Nucleic Acids Res* 21:5583
6. AM Diamond, IS Choi, PF Crain, T Hashizume, SC Pomerantz, R Cruz, CJ Steer, KE Hill, RF Burk, JA McCloskey, DL Hatfield 1993 *J Biol Chem* 268:14215
7. N Kato, H Hshino, F Harada 1983 *Biochem Int* 7:635
8. C Sturchler, E. Westhof, P Carbon, A Krol 1993 *Nucleic Acids Res* 21:1073
9. J Gabryszuk, A Przykorska, M Monko, E Kuligowska, C Sturchler, A Krol, G Dirheimer, JW Szarkowski, G Keith 1995 *Gene* 161:259
10. SV Steinberg, A Ioudovitch, R Cedergren 1998 *RNA* 4:241
11. A Ioudovitch, SV Steinberg 1998 *RNA* 4:365
12. T Mizutani, C Goto 2000 *FEBS Lett* 466:359
13. N Hubert, C Sturchler, E Westhof, P Carbon, A Krol 1998 *RNA* 4:1029
14. A Ioudovitch, SV Steinberg 1999 *J Mol Biol* 290:365
15. HS Chittum, WS Lane, BA Carlson, PP Roller, F-D Lung, BJ Lee, DL Hatfield 1998 *Biochemistry* 37:10867
16. DL Hatfield, VN Gladyshev, SI Park, HS Chittum, BA Carlson, ME Moustafa, JM Park, JR Huh, M Kim, BJ Lee 1999 *Comp Nat Prod Chem* JW Kelly Elsevier, New York 4:353
17. X-M Xu, X Zhou, BA Carlson, LK Kim, T-L Huh, BJ Lee, DL Hatfield 1999 *FEBS Lett* 495:16
18. X-M Xu, BA Carlson, LK Kim, BJ Lee, DL Hatfield, AM Diamond 1999 *Gene* 239:49
19. GJ Warner, CP Rusconi, IE White, JR Faust 1998 *Nucleic Acids Res* 26:5533
20. BJ Lee, P de la Pena, JA Tobain, M Zasloff, DL Hatfield 1987 *Proc Natl Acad Sci USA* 84:6384
21. M Schaub, A Krol, P Carbon 2000 *Nucleic Acids Res* 28:2114
22. K Adachi, H Saito, T Tanaka, T Oka *Endocrinology* 140:618

23. IS Choi, AM Diamond, PF Crain, JD Kolker, JA McCloskey, DL Hatfield 1994 *Biochemistry* 33:601
24. C Sturchler, A Lescure, G Keith, P Carbon, A Krol 1994 *Nucleic Acids Res* 22:1354
25. LK Kim, T Matsufuji, S Matsufuji, BA Carlson, SS Kim, DL Hatfield, BJ Lee 2000 *RNA* 6:1306
26. D Hatfield, BJ Lee, L Hampton, AM Diamond 1991 *Nucleic Acids Res* 19:939
27. HS Chittum, KE Hill, BA Carlson, BJ Lee, RF Burk, DL Hatfield 1997 *Biochim Biophys Acta* 1359:25
28. RA Sunde, JK Everson 1987 *J Biol Chem* 262:933
29. BJ Lee, PJ Worland, JN Davis, TC Stadtman, DL Hatfield 1989 *J Biol Chem* 264:9724
30. XG Wu, HJ Gross 1993 *Nucleic Acids Res* 21:5589
31. T Ohama, D Yang, DL Hatfield 1994 *Arch Biochem Biophys* 315:293
32. R Amberg, T Mizutani, X-Q Wu, HJ Gross 1996 *J Mol Biol* 263:8
33. HS Chittum, WS Lane, BA Carlson, PP Roller, FT Lung, BJ Lee, DL Hatfield 1998 *Biochemistry* 37:10866
34. DL Hatfield, FH Portugal 1970 *Proc Natl Acad Sci USA* 67:1200
35. T Mizutani, T Hitaka 1988 *FEBS Lett* 289:59
36. T Mizutani 1989 *FEBS Lett* 250:142
37. T Mizutani, N Maruyama, T Hitaka, Y Sukenaga 1989 *FEBS Lett* 247:345
38. K Forchhammer, A Böck 1991 *J Biol Chem* 266:6324
39. T Mizutani, H Kurata, K Yamada 1991 *FEBS Lett* 289:59
40. T Mizutani, H Kurata, K Yamada, T Totsuka 1992 *Biochem J* 284:827
41. T Mizutani, Y Tachibana 1986 *FEBS Lett* 207:162
42. SC Low, JW Harney, MJ Berry 1995 *J Biol Chem* 270:21659
43. IY Kim, TC Stadtman 1995 *Proc Natl Acad Sci USA* 92:7710
44. MJ Guimaraes, D Peterson, A Vicari, BG Cocks, NG Copeland, DJ Gilbert, NA Jenkins, DA Ferrick, RA Kasstelein, JF Bazan, A Zlotnik 1996 *Proc Natl Acad Sci USA* 93:15086
45. ME Moustafa, MA El-Saadani, KM Kandeel, DM Masur, BJ Lee, DL Hatfield, AM Diamond 1998 *RNA* 4:1436
46. ME Moustafa, BA Carlson, MA El-Saadani, GV Kryukov, Q-A Sun, KE Hill, JW Harney, GF Combs, L Feigenbaum, DB Masur, RF Burk, MJ Berry, AM Diamond, VN Gladyshev, BJ Lee, DL Hatfield, submitted
47. MR Bosl, K Takaku, M Oshmia, S Nishimura, MM Taketo 1997 *Proc Natl Acad Sci USA* 94:5531
48. MJ Berry, JW Harney, T Ohama, DL Hatfield 1994 *Nucleic Acids Res* 22:3753
49. J Heider, C Baron, A Böck 1992 *EMBO J* 11:3759
50. GR Björk 1998 *Modification and Editing of RNA* H Grosjean, R Benne (Eds) American Society of Microbiology Washington, D.C. 577.
51. GR Björk, T Rasmuson 1998 *Modification and Editing of RNA* H Grosjean, R Benne (Eds) American Society of Microbiology Washington, D.C. 471
52. JF Curran 1998 *Modification and Editing of RNA* H Grosjean, R Benne (Eds) American Society of Microbiology Washington, D.C. 493
53. GJ Warner, MJ Berry, ME Moustafa, BA Carlson, DL Hatfield, JR Faust 2000 *J Biol Chem* 275:28110

# Chapter 4. Selenophosphate - selenium donor for protein and tRNA

Gerard M. Lacourciere

*Laboratory of Biochemistry, National Heart, Lung, and Blood Institute, National Institutes of Health, Bethesda, MD 20892, USA*

**Summary:** Monoselenophosphate has been identified as the activated selenium donor in *Escherichia coli* that is required for both the biosynthesis of selenocysteine in selenium-dependent enzymes and for the conversion of 2-thiouridine residues to 2-selenouridine in tRNA. Selenophosphate is generated from ATP and selenide by the *selD* gene product, selenophosphate synthetase.

### Identification of selenophosphate as the activated selenium donor

In 1954, Pinsent observed gas production when anaerobically grown *Escherichia coli* cultures were supplemented with selenium [1]. Several years later, it was revealed that selenium was required for formate dehydrogenase activity in the formate-hydrogenase-lyase complex [2-4]. Cloning and sequencing of the *fdhF* gene, which encodes the 80-kilodalton formate dehydrogenase H ($FDH_H$) protein, showed it contained an in-frame TGA codon corresponding to amino acid 140 [5]. Subsequent gene fusion experiments established that expression of $FDH_H$ was dependent on selenium and the incorporation of selenium at position 140 was in the form of selenocysteine [6-8]. The finding that UGA directed the specific insertion of selenocysteine into a protein set the stage for the identification of factors needed to prevent termination at UGA and to direct the specific insertion of this unusual amino acid selenocysteine. Mutant strains of *E. coli*, which were unable to synthesize the selenocysteine containing enzymes formate dehydrogenase N and formate dehydrogenase H, were employed to identify these factors [9]. This work led to the isolation and characterization of the products of four genes, *selA*, *selB*, *selC*, and *selD*. The *selC* gene product was identified as a selenocysteyl-tRNA$^{Sec}$ that cotranslationally delivers selenocysteine to the in-frame UGA codon [10]. Initially, tRNA$^{Sec}$ is charged with serine and is converted to selenocysteyl-tRNA$^{Sec}$ by the *selA* gene product, selenocysteine synthetase [11-12]. The *selB* gene product encoded a specialized translation factor. This factor binds guanine nucleotides, selenocysteyl-tRNA$^{Sec}$ and the SECIS element, a unique stem loop structure in the mRNA located immediately downstream of UGA [13]. The binding of SelB to the stem loop stalls the ribosome and prevents termination of protein synthesis long enough for the insertion of selenocysteine to occur. *E.coli* mutants that contained a disrupted *selD* gene were unable to convert 2-thiouridine to 2-selenouridine as well as generate selenocysteyl-tRNA$^{Sec}$. This led to the assumption that the *selD* gene product, SELD, participates in selenium metabolism by delivering a reduced derivative of selenium to specific enzymes involved in selenocysteyl-tRNA$^{Sec}$ biosynthesis and in the conversion of 2-thiouridine to 2-selenouridine [14]. The in vitro biosynthesis of selenocysteyl-tRNA$^{Sec}$ in the presence of selenocysteine synthetase and SELD has been performed. The addition of an activated selenium derivative to aminoacrylyl-tRNA$^{Sec}$ required the presence of SELD, Mg·ATP, Na$_2$SeO$_3$ and dithiothreitol. Characterization of the SELD protein alone with Mg·ATP, Na$_2$SeO$_3$ and dithiothreitol revealed that the enzyme catalyzes the selenium-dependent hydrolysis of ATP. Dual labeling experiments for the SELD reaction which included [$^{75}$Se]selenite plus dithiothretiol and either [$\gamma$-$^{32}$P]ATP, [$\beta$-$^{32}$P]ATP, or [$\alpha$-$^{32}$P]ATP as substrates showed $\beta$-$^{32}$P was released as orthophosphate and the $\gamma$-$^{32}$P was released with $^{75}$Se as selenophosphate [15].

To confirm that the product of the SELD reaction is actually selenophosphate, selenophosphate was prepared chemically and compared to the enzyme product by both HPLC and NMR analysis. An enzyme catalyzed reaction performed with Na$^{77}$SeH, the NMR active isotope of selenium, and Mg·ATP as substrates was analyzed by $^{31}$P NMR. The NMR spectrum revealed the $^{31}$P resonance for selenophosphate contained two peaks which supported a direct linkage of selenium to phosphorous [16]. The ability of selenophosphate to serve as an activated selenium donor in the biosynthesis of 2-selenouridine in tRNA was tested. A partially purified enzyme from *Salmonella*, which is involved in 2-selenouridine biosynthesis, was mixed with purified SELD, Na$^{75}$SeH and Mg·ATP. The conversion of 2-thiouridine to 2-selenouridine was monitored by $^{75}$Se incorporation into tRNA. Additional reactions were performed in the presence of increasing amounts of unlabeled authentic selenophosphate. The presence of the authentic compound resulted in a dose dependent decrease in $^{75}$Se incorporation into tRNA. In a separate experiment purified $^{75}$Se labeled selenophosphate, generated by SELD, was added to the partially purified enzyme from *Salmonella* and the conversion of 2-thiouridine to [$^{75}$Se] 2-selenouridine was observed. Since this reaction occurred in the absence of both ATP and SELD, direct attack of selenophosphate on the tRNA must have occurred [17]. Taken together, the above experiments support that the activated selenium compound produced by the SELD reaction with ATP and NaSeH is monoselenophosphate. As a result the *selD* gene product is now referred to as selenophosphate synthetase (SPS).

## E. coli selenophosphate synthetase
The 37 kDa product of the *E. coli selD* gene, selenophosphate synthetase catalyzes the synthesis of monoselenophosphate, AMP, and orthophosphate in a 1:1:1 ratio from ATP and selenide (Figure 1) [18].

$$ATP + HSe^- \longrightarrow H_3SePO_3 + P_i + AMP$$

**Figure 1.** Reaction catalyzed by selenophosphate synthetase.

The homogenous enzyme from *E. coli* has been characterized in detail [18]. The enzyme has a determined specific activity of 83 nmol/min/mg ($k_{cat}$ = 3 min$^{-1}$). The determined $K_m$ values for both substrates ATP and selenide were 0.9 μM and 46 μM, respectively. AMP is a competitive inhibitor of SPS with a $K_i$ of 170 μM. In the absence of selenide, SPS catalyzes the slow hydrolysis of ATP to AMP and two orthophosphates.

## Mechanism of selenophosphate biosynthesis
The SPS selenide dependent hydrolysis of ATP results in the formation of AMP, orthophosphate and selenophosphate. Since the end product is AMP, it was proposed that the enzyme mechanism may proceed through the hydrolysis of ATP to AMP resulting in the formation of an pyrophosphoryl enzyme intermediate [19]. This intermediate would undergo a second reaction with selenide to form selenophosphate. Alternatively, the hydrolysis of ATP can occur in two steps through a phosphoryl intermediate.

To determine whether either intermediate is formed during catalysis, positional isotope exchange experiments were carried out using [γ-$^{18}$O$_4$]ATP and [β-$^{18}$O$_4$]ATP as substrates. The reactions were performed in the absence of selenide and the enzyme catalyzed migration of $^{18}$O from the β,γ-bridge position of [γ-$^{18}$O$_4$]ATP to the β-nonbridge position was monitored by $^{31}$P NMR. Reactions using [γ-$^{18}$O$_4$]ATP as substrate showed that the exchange of $^{18}$O occurred after cleavage of the β–γ-phosphoryl bond of ATP with the two $^{16}$O molecules on the β-nonbridge position. In contrast, no exchange was observed when [β-$^{18}$O$_4$]ATP was used as substrate. Taken together, these results support the formation of a phosphoryl enzyme intermediate after the initial cleavage of the γ-phosphate group of ATP. To further probe the mechanism of SPS isotope exchange, experiments were carried out with [8-$^{14}$C]ADP or [8-$^{14}$C]AMP and unlabeled ATP. In the absence of selenide, a slow enzyme catalyzed exchange was observed between ADP and ATP [20]. In the absence of selenide, and at extremely high concentrations of enzyme (160 μM), SPS hydrolyzes ATP completely to AMP. In contrast, at lower concentrations of enzyme (37 µM), SPS hydrolyzes ATP to equal amounts of ADP and AMP. These results further support the existence of a phosphorylated enzyme and ADP as intermediates in the reaction. [31]P NMR analysis of products of the selenide dependent reaction performed in $H_2^{18}O$ showed that $^{18}O$ was incorporated exclusively into the orthophosphate product indicating that the β-phosphoryl group of the bound ADP is attacked by water. In the absence of selenide, the phosphoryl enzyme intermediate is also hydrolyzed by $H_2O$. Based on these results, a catalytic model for SPS has been postulated. The first step in catalysis involves the nucleophilic attack on the γ-phosphate group of ATP which results in the formation of an enzyme-phosphoryl intermediate. The enzyme-phosphoryl intermediate undergoes nucleophilic attack by selenide resulting in the release of selenophosphate. Bound ADP is then hydrolyzed by the nucleophilic attack of a water molecule on the β-phosphoryl group of ADP to release orthophosphate and AMP as products (Figure 2).

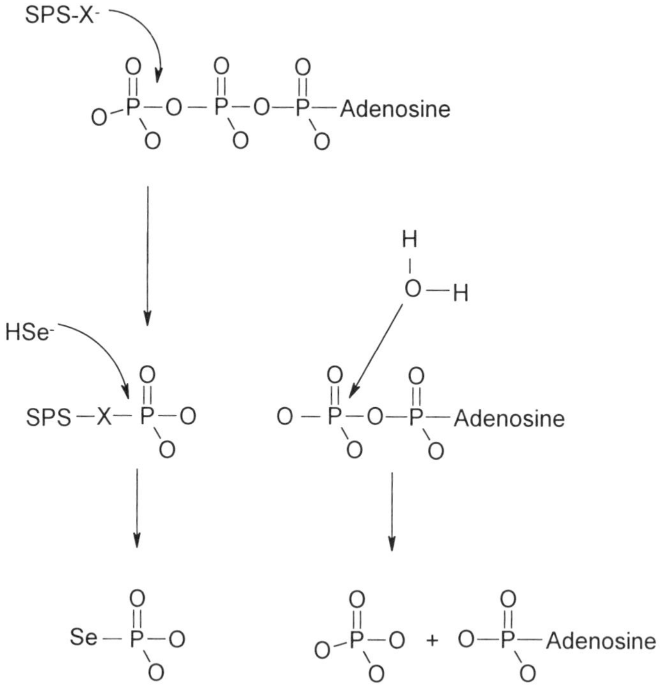

**Figure 2.** Proposed SPS mechanism.

Attempts have been made to identify protein amino acid residues which are catalytically essential in selenophosphate biosynthesis. The amino acid sequence of SPS revealed that out of the 347 amino acids seven were cysteine. Since the enzyme was sensitive to treatment with iodoacetamide [18] and 5,5'-dithiobis-(2-nitrobenzoic acid) [15] at least one cysteine residue may be essential for catalytic activity. Among likely candidates were two cysteine residues located in the N-terminal region of the protein in the sequence His-Gly-Ala-Gly-Cys$^{17}$-Gly-Cys$^{19}$-Lys-Ile-. It was originally proposed that both cysteine residues may form a disulfide bridge which could be attacked by selenide to form a selenotrisulfide on the enzyme [15]. However, determination of the number of free thiols in SPS did not support the presence of disulfide linkages [21]. It was also proposed that the N-terminal region may define an ATP binding site due to its similarity to the known conserved glycine-rich ATP binding sequence Gly-X-X-X-X-Gly-Lys (Ser/Thr) [21]. Cys$^{17}$ and Cys$^{19}$ were mutated to serine by PCR mutagenesis. The selenium dependent hydrolysis of ATP was abolished by mutation of Cys$^{17}$ suggesting it is catalytically essential. Since Cys$^{17}$ lies within a possible ATP binding region it may function as a nucleophile in the hydrolysis of ATP. In addition to the cysteine mutations, His$^{13}$ and Gly$^{18}$ were changed to Asn and Val, respectively, and Lys$^{20}$ to Arg or Gln. The Lys$^{20}$ to Gln mutation abolished activity and replacement with Arg markedly decreased activity. The retention of a trace of activity when lysine is replaced with arginine suggests a positively charged group is required for activity in this position. The His$^{13}$ to Asn replacement did not substantially alter the ATP $K_m$ and $V_{max}$ values, and the Gly$^{18}$ to Val mutation increased the ATP $K_m$ value 4-fold compared to that of the wild-type enzyme [22]. Since all mutants retain ATP binding activity, it seems unlikely that this region is a defined ATP binding sequence.

**selD homologs**
The glycine–rich sequence in the N-termial region of *E. coli* SPS has been found in *selD* homologs (Figure 3). Although the sequence is highly conserved, there is some variability at residues which align with Cys$^{17}$ in the *E.coli* protein. SPS from *Methanococcus jannaschii* [23], *Haemophilus influenzae* [24], mice [25], and humans [25] contain a selenocysteine residue which aligns with Cys$^{17}$ in the *E. coli* enzyme. A second human homolog has been identified which contains a threonine at this position [26], while a homolog from *Drosophila melanogaster* was found to contain an arginine [27]. *Caenorhabditis elegans* has a SPS homolog that contains a cysteine at the variable site as is found in *E. coli* [28]. The SPS homolog in *Dictyostelium discoideum* contains glycine at the variable position [29], while *Aquifex aeolicus* [30], which has a glycine-rich sequence with limited homology, contains a cysteine residue.

```
HuSPS2.EPQALGLSPSWRLTGFSGMKGUGCKVPQEALLKLLAGLTRP
MoSPS2.EPQTLGFSPSWRLTSFSGMKGUGCKVPQETLLKLLEGLTRP.
HuSPS1.NPESYELDKSFRLTRFTELKGTGCKVPQDVLQKLLESLQEN.
CeSPS  .DPVSNGLDEDFVLTKLTGMKGCGCKVPRNVLLQLLNTFKT.
DmSPS  .DPTAHDLDASFRLTRFADLKGRGCKVPQDVLSKLVSALQQD.
EcSPS         MSENSIRLTQYSHGAGCGCKISPKVLETILHSEQAKF.
HiSPS         MEEKIRLTQYSHGAGUGCKISPKVLGTILHSELEKF.
MjSPS       MERGNEKIKLTELVKLHGUACKLPSTELEFLVKGIVTDD.
DdSPS  .SIKDKKEELLCRLTDFTKLKGGGCKVPQAELLSLLDGIGGGI.
AaSPS          MVELLKLVRSSGCAAKVGPGDLQEILKGFNIYT.
```

**Figure 3.** Partial amino acid sequences of SPS homologs from Human, (Hu), Mouse (Mo), *Caenorhabditis elegans* (Ce), *Drosophila melanogaster* (Dm), *Escherichia coli* (Ec), *Haemophilus influenzae* (Hi), *Methanococcus jannaschii* (Mj), *Dictyostelium discoideum* (Dd), and *Aquifex aeolicus* (Aa). Residues corresponding to $Cys^{17}$ from the *E. coli* enzyme are in bold.

The variability of the amino acid corresponding to $Cys^{17}$ in the *E. coli* enzyme combined with the loss of catalytic activity in the $Cys^{17}$ to serine mutant may reflect a catalytic role for residues at this position in the glycine-rich region. Previously, it has been shown that a sulfur replacement for selenium in formate dehydrogenase H from *E. coli* resulted in more than two orders of magnitude reduction in the turnover number [31]. Organoselenols are generally more reactive compared to their sulfur analogs. The selenol of free selenocysteine has a pKa of 5.2 and is fully ionized at physiological pH compared to the thiol of free cysteine which has a pKa close to 8 and is mainly protonated [32]. The replacement of cysteine with selenocysteine in SPS may result in an enzyme with greater catalytic activity particularly if this residue functions as a nucleophile in catalysis. The gene encoding a selenocysteine containing SPS from *H. influenzae* has been cloned and over-expressed in *E. coli* [33,34]. The recombinant enzyme contained selenocysteine and was able to complement a *selD* lesion in *E. coli* [33]. Characterization of the purified recombinant enzyme revealed that the determined $K_m$ values for both substrates ATP and selenide were similar to *E. coli* SPS values, but the selenocysteine containing protein failed to exhibit enhanced catalytic activity [34]. Additionally, the over-expressed threonine-containing enzyme weakly complemented a *selD*-lesion in *E. coli*, and when transfected into mammalian cells resulted in an increased $^{75}Se$ labeling of the selenium containing deiodinases [26]. The cloned *Drosophila melanogaster selD* gene failed to complement a *selD* lesion in *E. coli* and the purified recombinant enzyme was unable to perform the selenide dependent ATP hydrolysis reaction [27]. These results, especially the inability of the selenocysteine containing enzyme to exhibit higher catalytic activity, question the catalytic relevance of the residue in the variable position.

**Delivery of selenium to SPS**

In the initial characterization of SPS, the reported $K_m$ value for the substrate selenide was 46 µM. Since selenide is extremely oxygen labile, the kinetics were repeated using more stringent anaerobic conditions. Under these conditions, the $K_m$ value for selenide was calculated as 7.3 µM, which is still above the optimal concentrations of selenite (0.1-1 µM) used in growth media. In fact, selenium concentrations of 10 µM or higher are toxic for many bacterial species. The ability to maintain concentrations of selenium below toxic levels and provide SPS with selenium are obstacles E. coli must be able to overcome. Due to the chemical similarity of selenium and sulfur, selenium can enter the cysteine biosynthetic pathway resulting in the formation of free selenocysteine and selenomethionine. These amino acids can be incorporated into proteins nonspecifically in place of cysteine and methionine [35,36]. Such substitutions occur when cultures of E. coli are grown in the presence of 0.1 µM $^{75}SeO_3^{2-}$ [36]. Potential candidates for the control of free selenium levels are the specific L-selenocysteine lyase enzymes which catalyze the pyridoxal 5'-phosphate dependent decomposition of selenocysteine to elemental selenium ($Se^0$) and alanine.

A similar role for cysteine desulfurase proteins in sulfur metabolism has been established. The NifS protein from A. vinelandii was shown to effectively mobilize sulfur from cysteine in iron-sulfur cluster formation [37,38]. Additionally, the IscS protein from E. coli was shown to mobilize sulfur from cysteine in iron-sulfur cluster formation and in the biosynthesis of biotin, thiamin, and 4-thiouridine [39,40,41,42] (Figure 4). Cultures of Clostridium sticklandii supplemented with $^3H$-, $^{14}C$-, and $^{75}Se$- labeled L-selenocysteine incorporated $^{75}Se$ derived from [$^{75}Se$]selenocysteine into selenoprotein A of the glycine reductase complex more efficiently than $^{75}SeO_3^{2-}$. In contrast, no incorporation of $^3H$ or $^{14}C$ was detected [43]. This observation supports the participation of a selenocysteine lyase which mobilizes selenium from selenocysteine. In view of the fact that the specific insertion of selenocysteine into proteins requires selenophosphate as a selenium donor, the participation of a selenocysteine lyase as a provider of the essential substrate for SPS seems likely. Selenocysteine lyase enzymes have been isolated in mice [44], pigs [45] and bacteria [46]. Three proteins, IscS, CSD, and CsdB, from E. coli have been identified which contain both cysteine desulfurase and selenocysteine lyase activities [47,48]. Each protein differed in their discrimination between cysteine and selenocysteine as substrates with CsdB exhibiting 290 times more activity with selenocysteine as a substrate.

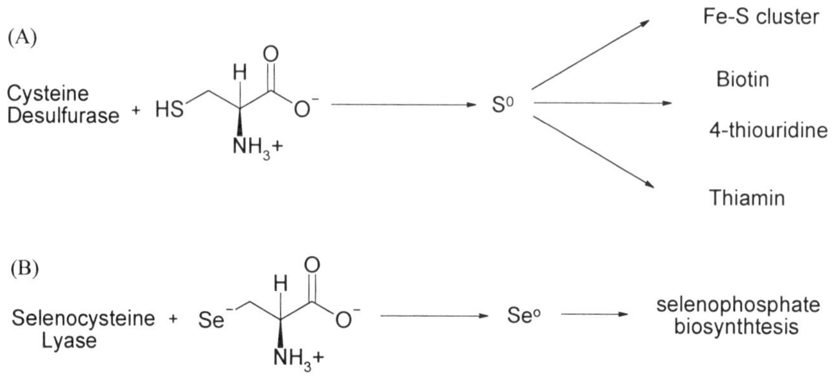

**Figure 4.** (A) Participation of cysteine desulfurase enzymes in the mobilization of sulfur for Fe-S cluster formation, biotin, thiamin, and 4-thiouridine biosynthesis. (B) Proposed participation of selenocysteine lyase enzymes in the mobilization of selenium for selenophosphate biosynthesis.

The possibility that a selenocysteine lyase may provide SPS with $Se^0$ was supported by studies which demonstrate that the NifS protein from *Azotobacter vinelandii* and L-selenocysteine could effectively replace the high level of selenide that is used in the in vitro SPS selenide dependent ATP hydrolysis assay. Despite the fact that the normal substrate for NifS is cysteine, NifS effectively mobilized $Se^0$ from selenocysteine to SPS which resulted in a higher rate of ATP hydrolysis compared with free selenide alone [49]. Similar evaluation of the three *E.coli* NifS-like proteins as selenium delivery proteins showed that SPS catalytic activity was detected indicating each protein can function as a selenium delivery protein to SPS. The measured SPS activity was found to be directly related to the amount of lyase present in each assay. As the concentration of lyase increased, SPS activity increased accordingly. SPS activity determined in the presence of lyase concentrations above 0.025 µM resulted in no additional ATP hydrolysis indicating SPS was saturated with selenium. The possibility that a complex may form between a NifS-like protein and SPS, which could result in the direct transfer of selenium, was evaluated by competition experiments performed between CsdB and SPS. The inactive mutant CsdB(R379A) [H Mihara, T Kurihara, N Esaki, unpublished results], which is unable to bind L-selenocysteine, was added to assay mixtures containing CsdB (1 µM), L-selenocysteine (10 µM), and SPS (10 µM) as a competitive inhibitor. Assuming a complex is formed between CsdB and SPS, the addition of CsdB(R379A) should displace CsdB and result in a concentration dependent decrease in SPS activity. Although no decrease in SPS activity was observed in the presence of a 5-10 molar excess of CsdB(R379A), the substitution of alanine for arginine in the mutant might also affect binding to SPS. Since CsdB exhibits both L-cysteine desulfurase and L-selenocysteine lyase activities, it is possible that binding of L-selenocysteine induces a conformational change that allows complex formation with SPS. The inability of the CsdB(R379A) mutant to bind selenocysteine, however, may prevent possible complex formation [50]. If a direct transfer of selenium from a lyase to SPS does not occur, it is possible that additional proteins may participate in selenium mobilization. The biosynthesis of thiamin and 4-thiouridine in *E. coli* requires cysteine as a sulfur source. The NifS-like protein IscS mobilizes sulfur from cysteine to a second protein ThiI which functions as a sulfurtransferase [51,52]. Moreover, the addition of ThiI to IscS, in an in vitro assay, results in an increase in cysteine desulfurase activity [40]. A selenotransferase analogous to ThiI may exist for the trafficking of selenium. The interaction of a selenotransferase with a NifS-like lyase may stimulate activity towards selenocysteine, aiding in its ability to discriminate between cysteine and selenocysteine (Figure 5).

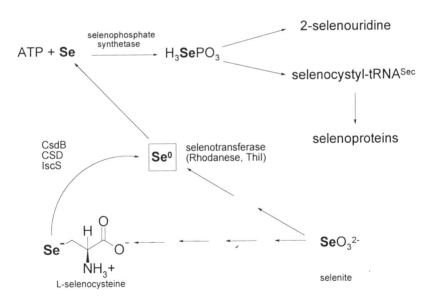

**Figure 5.** Proposed mobiliziation of selenium to SPS in *E.coli*.

## Additional functions of selenophosphate

Stability studies carried out on authentic selenophosphate revealed the rate of hydrolysis was dependent on the pH of the solution with a maximal rate of hydrolysis observed at pH ~ 7. Additionally, the rate of hydrolysis of selenophosphate was not affected by the addition of alcohols or amines. However, alcohols and amines were found to be phosphorylated in a nonselective manner [53]. The rapid hydrolysis of selenophosphate combined with its ability to phosphorylate alcohols and amines may reflect a novel function for it as an in vivo phosphorylating agent. In addition to the established roles of selenophosphate, which are essential to the biosynthesis of selenium-dependent enzymes and seleno-tRNAs, this unique metabolic component may also participate in pathways that have not been identified.

**References**

1. J Pinsent 1954 *Biochem J* 57:10
2. RL Lester, JA DeMoss 1971 *J Bacteriol* 105:1006
3. JC Cox, ES Edwards, JA DeMoss 1981 *J. Bacteriol.* 145:1317
4. A Pecher, F Zinoni, A Bock 1985 *Arch Microbiol* 141:359
5. F Zinoni, A Birkmann, TC Stadtman, A Böck 1986 *Proc Natl Acad Sci USA* 83:4650
6. F Zinoni, A Birkmann, W Leinfelder, A Böck 1987 *Proc Natl Acad Sci USA* 84:3156
7. MJ Axley, DA Grahame, TC Stadtman 1990 *J Biol Chem* 265:18213
8. TC Stadtman, JN Davis, W-M Ching, F Zinoni, A Böck 1991 *Biofactors* 3:21
9. W Leinfelder, K Forchhammer, F Zinoni, G Sawers, M-A Mandrand-Berthelot, A Böck 1988 *J Bacteriol* 170:540
10. W Leinfelder E Zehelein, M-A Mandrand-Berthelot, A Böck 1988 *Nature* 331:723
11. K Forchhammer, W Leinfelder, K Boesmiller, B Veprek, A Böck 1991 *J Biol Chem* 266:6318
12. K Forchhammer, A Böck 1991 *J Biol Chem* 266:6324
13. K Forchhammer, W Leinfelder, A Böck 1989 *Nature* 342:453
14. W Leinfelder, K Forchhammer, B Veprek, E Zehelein, A Böck 1990 *Proc Natl Acad Sci USA* 87:543
15. A Ehrenreich, K Forchhammer, P Tormay, B Veprek, A Böck 1992 *Eur J Biochem* 206:767
16. Z Veres, L Tsai, TD Scholz, M Politino, RS Balaban, TC Stadtman 1992 *Proc Natl Acad Sci USA* 89:2975
17. RS Glass, WP Singh, W Jung, Z Veres TD Scolz , TC Stadtman 1993 *Biochemistry* 32:12555
18. Z Veres, IY Kim, TD Scholz, TC Stadtman 1994 *J Biol Chem* 269:10597
19. LS Mullins, S-B Hong, GE Gibson, H Walker, TC Stadtman, FM Raushel 1997 *J Am Chem Soc* 119:6684
20. H Walker, JA Ferretti TC Stadtman *Proc Natl Acad Sci USA* 1998 95:2180
21. IY Kim, Z Veres, TC Stadtman 1992 *J Biol Chem* 267:19650
22. IY Kim, Z Veres, TC Stadtman 1993 *J Biol Chem* 268:27020
23. CJ Bult, O White, et al 1996 *Science* 373:1058
24. RD Fleischmann, MD Adams, et al 1995 *Science* 269:496
25. MJ Guimaraes, D Peterson, A Vicari, BG Cocks, NG Copeland, DJ Gilbert, NA Jenkins, DA Ferrick, RA Kastelein, JF Bazan A Zlotnik 1996 *Proc Natl Acad Sci USA* 93:15086
26. SC Low, JW Harney MJ Berry 1995 *J Biol Chem* 270:2165
27. BC Persson A Böck, H Jäckle, G Vorbrüggen 1997 *J Mol Biol* 274:174
28. The C. elegans Sequencing Consortium 1998 *Science* 282:2012
29. G Shaulsky, WF Loomis 1996 EMBL/GenBank/DDBJ databases
30. G Deckert, PV Warren et al 1998 *Nature* 392:353
31. MJ Axley, A Böck, TC Stadtman 1991 *Proc Natl Acad Sci USA* 88:8450
32. RE Huber, RS Criddle 1967 *Arch Biochem Biophys* 122:164

33. R Wilting, K Vamvakidou, A Böck 1998 *Arch Microbiol* 169:71
34. GM Lacourciere, TC Stadtman 1999 *Proc Natl Acad Sci USA* 96:44
35. GF Kramer, BN Ames 1988 *J Bacteriol* 170;736
36. S Muller, J Heider A Böck 1997 *Arch Microbiol* 168:421
37. L Zheng, DR Dean 1994 *J Biol Chem* 269:18732
38. L Zheng, RH White, VL Cash, RF Jack, DR Dean *Proc Natl Acad Sci USA* 90:2754
39. CT Lauhon, R Kambampati 2000 *J Biol Chem* 275:20096
40. R Kambampati, CT Lauhon 1999 *Biochemistry* 38:16561
41. CJ Schwartz, O Djaman, JA Imlay, PJ Kiley 2000 *Proc Natl Acad Sci USA* 97:9009
42. DH Flint 1996 *J Biol Chem* 271:16068
43. TC Stadtman, GL Dilworth, CS Chen 1979 *Proceedings of the Third International Symposium on Organic Selenium and Tellurium Compounds* D Cagniant, G Kirsch (Eds) Mety, France pp115
44. H Mihara, T Kurihara, T Watanabe, T Yoshimura, N Esaki 2000 *J Biol Chem* 257:6195
45. N Esaki, T Nakamura, H Tanaka, K Soda 1982 *J Biol Chem* 257:4386
46. P Chochat, N Esaki, K Tanizawa K Nakmura, H Tanaka, K Soda 1985 *J Biol Chem* 163:669
47. H Mihara, T Kurihara, T Yoshimura, K Soda, N Esaki 1997 *J Biol Chem* 272:22417
48. H Mihara, M Maeda, T Fujii, T Kurihara, Y Hata, N Esaki 1999 *J Biol Chem* 274:14768
49. GM Lacourciere, TC Stadtman 1998 *J Biol Chem* 273:30921
50. GM Lacourciere, H Mihara, T Kurihara, N Esaki, TC Stadtman 2000 *J Biol Chem* 275:23769
51. PM Palenchar, CJ Buck, H Cheng, T Larson, E Mueller 2000 *J Biol Chem* 275:8283
52. R Kampbampati, CT Lauhon 2000 *J Biol Chem* 275:10727
53. R Kaminski, RS Glass, TB Schroeder, J Michalski, A Skowronska 1997 *Biorg Chem* 25:247

# Chapter 5. SECIS elements

Glover W. Martin, III

*Harvard/MIT Division of Health Sciences and Technology, and the Department of Microbiology and Molecular Genetics, Harvard Medical School, Cambridge, MA 02139, USA*

Marla J. Berry

*Thyroid Division, Brigham and Women's Hospital, and the Department of Biological Chemistry and Molecular Pharmacology, Harvard Medical School, Boston, MA 02115, USA*

**Summary:** The specific, cotranslational incorporation of selenocysteine into eukaryotic proteins requires the presence of a selenocysteine insertion sequence (SECIS) element within the selenoprotein mRNA. These elements exist in two forms: a basic stem-loop, and a stem-loop containing an internal bulge. The loop and bulge contain two invariant adenosine residues, while a "core" motif containing conserved GA/AG non-Watson-Crick base pairs defines the base of the minimal element. The distance between these two conserved features is constrained to 9 to 11 base pairs, approximating one turn of an A-form double helix. The structure below the core must be open, i.e., not base paired, for optimal SECIS element activity. These criteria have allowed the development of algorithms that can predict SECIS elements, and thus selenoproteins, from nucleotide databases.

**Introduction**
Selenocysteine incorporation has in recent years come to be recognized as an important process to the overall homeostasis of organisms ranging in complexity from *Escherichia coli* to *Homo sapiens*. In eubacteria, genetic methods have availed themselves to the study of the translational recoding event which selenoprotein synthesis necessitates, and the identification of trans-acting factors involved in the process. Such genetic tractability, however, has not been as readily available for eukaryotic systems, in which, ironically, the ability to synthesize selenoproteins is thought to be absolutely essential for viability and reproduction. As a result, the investigation of selenoprotein synthesis in eukaryotes began primarily with the detailed characterization of *cis*-acting elements required for cotranslational incorporation of selenocysteine at UGA codons. This chapter will focus on the feature common to all eukaryotic selenoprotein mRNAs which allows the translational machinery to distinguish UGA termination codons from UGA

selenocysteine codons: the selenocysteine insertion sequence (SECIS) element.

**The SECIS element: functional characterization**
Two pieces of evidence pointed to the existence of distinctions between the *cis*-acting requirements for *E. coli* and eukaryotic selenocysteine incorporation. First, the cDNA for mouse cellular glutathione peroxidase (cGPx) would not give rise to a functional enzyme when expressed in *E. coli* [1]. Secondly, examination of the cGPx and type I deiodinase (D1) sequences failed to reveal homology in the regions surrounding the UGA selenocysteine codon. However, it was noted that the mRNAs of these two selenoproteins both contain rather extensive 3'-untranslated regions (3'-UTRs). For example, although the coding region for the D1 protein is contained in 771 nucleotides, the full-length cDNA is 2.1 kilobases in length, a finding that implicated a critical function for the 3'-UTR. The presence of at least some portion of the 3'-UTR was found to be critical for the recognition of UGA as a selenocysteine codon, as its removal abolished the ability of the mRNA to generate functional D1 enzyme in transiently transfected cells. At the same time, the function of mRNA in which the selenocysteine UGA codon had been replaced with UGU, encoding cysteine, was unaffected [2]. Analysis of in vitro translation products demonstrated that in the absence of its 3'-UTR, translation terminates at the UGA selenocysteine codon, proving that noncoding sequences are required for selenocysteine incorporation. When placed downstream of the D1 coding region, the 3' UTR of cGPx was able to drive synthesis of full-length, selenocysteine-containing D1 enzyme. Together, these data pointed to the presence of a common, coding sequence context-independent structure and/or sequence required for selenocysteine incorporation.

Following up on this lead, deletion mapping localized the critical region of the D1 3'-UTR to a 175 nucleotide fragment [2]. However, this region contained very little similarity at the primary sequence level to the cGPx 3' UTR. Analysis of D1 and cGPx cDNAs from multiple species revealed three short, invariant stretches of nucleotides, which yielded a consensus selenocysteine insertion sequence (SECIS) element [2]. In assessing the function of the SECIS element, it was found that it must be linked, in *cis*, to the UGA-containing open reading frame in order to function effectively [3]. Perhaps surprisingly, the length of this linkage may be extended, specifically in the case of D1, by up to 1.5 kilobases without diminution of function. Indeed, a naturally occurring SECIS element has been found to reside approximately 5 kilobases from the UGA selenocysteine codon [4]. Furthermore, a SECIS element placed in the D1 5'-UTR was unable to efficiently drive selenocysteine incorporation at a downstream UGA codon. It is unclear, however, whether the secondary structure, either alone or together with a bound protein, would have prevented the efficient initiation of translation [5-7].

In a related line of inquiry, it was observed that selenoprotein open reading frames tend not to terminate in UGA codons. Such a configuration could present a problem, as it had been shown that any given codon, when mutated to UGA could act as a selenocysteine incorporation signal, i.e., there is no codon-context component to the signal [3, 8]. One notable exception to this rule is selenoprotein W, which terminates in UGA, but also has the unusual feature of a SECIS element in extremely close proximity, 51 nucleotides, to the termination codon [9]. It is precisely this feature that precludes selenocysteine incorporation at the second UGA codon, resulting in termination of protein synthesis. It was demonstrated that as a SECIS element is moved from 111 to 60 to 51 nucleotides from a UGA codon, the selenocysteine incorporation directed by that codon is drastically reduced [10, unpublished data]. A likely explanation for this phenomenon is that steric constraints between the advancing ribosome and SECIS element recognition factors somehow preclude the formation of a productive interaction among factors required for selenocysteine insertion at the UGA codon.

In *E. coli*, the SECIS element and UGA selenocysteine codon may be seen as a complete unit, since they are constrained to lie adjacent to each other in order to function. That is, these two features are recognized together, simultaneously, as a signal for selenocysteine incorporation. In contrast, eukaryotic SECIS elements are found outside the coding region and with very clear advantages. In particular, one SECIS element may function to direct selenocysteine incorporation at multiple UGA codons. Therefore, while likely a sub-optimal configuration, the formal possibility of a eukaryotic SECIS element functioning inside a coding region was tested and proven to be extremely inefficient [10].

## The SECIS element: structure and conserved features

Taken together, these observations established the physical and functional limitations of eukaryotic SECIS element function. However, they do little to assist in the visualization of SECIS element structure. Alignment of the D1 and cGPx 3'-UTRs revealed the existence of three short regions of nucleotide invariance, consisting of AUGA, AAA, and UGAU [2]. As more selenoprotein sequences have become available, these consensus sequences have been modified to RUGA, AA, and GA, respectively [11]. The initial characterization of the 175 nucleotide D1 SECIS element included a computer FOLD analysis of potential secondary structures which may form in this region [2]. These coarse predictions placed the conserved AA residues in an apical loop and the RUGA and GA in an internal bulge. When this rather large region was pared down to its essential parts, the D1 SECIS element consisted of a 45 nucleotide stem-loop structure, bounded by the RUGA and GA motifs, inclusive, and containing the two conserved A residues in the 5' edge of the apical loop (Figure 1). A key to the acceptance of this structure as the minimal eukaryotic SECIS element was its presence in the cDNAs of selenoproteins from a wide variety of eukaryotic organisms [12].

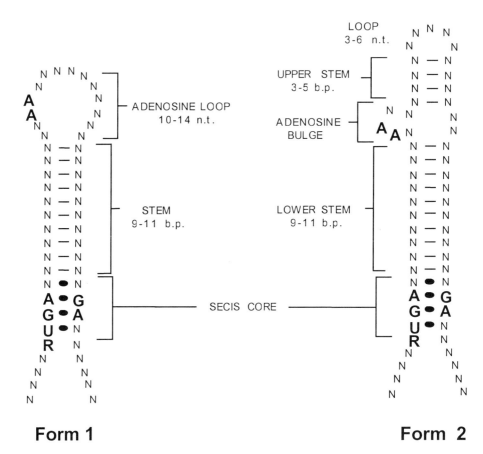

**Figure 1.** Anatomy of the two classes of eukaryotic SECIS elements. The critical features are labeled. See text for discussion.

Compensatory mutagenesis of the putative stem nucleotides, while not absolutely confirming the register of base pairing in the stem, have demonstrated that the overall stem loop structure is correct [3, 10], as have enzymatic and chemical probing experiments [13, 14]. Walczak et al. used physical methods to demonstrate that the GA dinucleotides in the 5' and 3' invariant regions pair with each other through non-Watson-Crick interactions to form a novel motif which appears to be central to the function of the SECIS element (see below and Figure 1) [15].

However, it was discovered that this characterization was not quite complete. Studies carried out by Shen et al. demonstrated that sequences below the stem of the minimal cGPx SECIS element, or flanking the conserved AUGA and UGAU, were critical to SECIS function [16]. Further investigation revealed that when compared to each other structurally, the sequences flanking the SECIS elements were nonconserved, but all were predicted to be unpaired. For example, the D1 SECIS element stem terminated in a structural motif consisting of the conserved AUGA and UGAU nucleotides, which was followed by an open, non-paired structure. Walczak et al. demonstrated by enzymatic and chemical structure probing that this prediction was correct [13], revealing the accessible character of both this region as well as the apical loop. Further in vivo experiments showed conclusively that replacing the native sequence in this region with unrelated nucleotides had no effect on SECIS element function, subject to a single constraint - the open structure must be maintained. Replacement of native sequence with nucleotides having a high potential to form base-pairs, caused a loss of SECIS element function [12]. The loss of function was mirrored by the strength of the introduced base-pairing interactions.

While the minimal D1 SECIS element was effectively defined as the sequences between, and including, the two extreme sets of invariant nucleotides, the term "minimal" may be overly broad in this context. It was proven that sequences beyond these are not required for SECIS element function; however, it was not clear that all the sequences included in this region are essential for function. Grunder-Culemann et al. addressed this issue by assessing the tolerance of the D1 SECIS element for changes in stem and apical loop sizes [17]. The data showed that the window of acceptable stem lengths was quite narrow, with SECIS element activity diminishing by more than 95% by subtracting two or adding three base pairs. Therefore, the optimal D1 SECIS stem length lies at nine predicted Watson-Crick base pairs which corresponds to a spacing of 12 nucleotides between the AUGA and AAA motifs. Whether the fact that this distance approximates a single turn of an A-form double helix is significant is an intriguing prospect that requires further investigation. Activity of the D1 SECIS element was reduced by approximately 50% and 90% by deletions of four and five nucleotides from the apical loop, respectively [17]. These data support the hypothesis that the size of the D1 SECIS element has likely been optimized over time for stability as well as specificity and avidity of protein binding. However, further studies have yet to confirm that these observations are extendable to the SECIS elements of other selenoprotein mRNAs.

It was serendipitous that the first SECIS elements to be used as models for mutagenesis and structural studies were those of D1 and cGPx. The predicted forms that these elements take on are similar, consisting of basic stem-loop structures. The third mammalian selenoprotein to be identified and cloned was selenoprotein P (SelP), a plasma protein of unknown function in which the majority of circulating selenium is bound [18]. SelP was of particular interest because its mRNA was also the first to be found to contain multiple UGA selenocysteine codons, as many as 17, depending upon the species of origin. Furthermore, the mRNA harbors two distinct SECIS elements, which bear no relation to each other. Comparison of the two SelP SECIS elements showed that the 5' element is slightly longer than its 3' counterpart, having an extended apical loop [19]. As more selenoprotein cDNA sequences became available, Low and Berry noticed that SECIS element sequences could be separated into two classes, based on the presence or absence of the extended apical loop [20]. Manual modification of software-generated secondary structure predictions led to the proposal that the extended apical loop could form into an internal loop, containing the conserved A residues and apical stem and loop (Figure 1).

Interestingly, the form 2 hypothesis explained a certain anomalous result in which lengthening the D1 SECIS element by three base-pairs caused a recovery of wild type activity in the midst of a trend of decreasing activity with increasing stem length (unpublished observations). When reinvestigated, it was discovered that the particular stem extension that had been introduced shifted the register of base pairing and allowed the formation of a short apical stem loop, effectively converting the D1 SECIS element form 1 to form 2. That such interconversion is possible, without loss of function, was confirmed by the addition of nucleotides predicted to extend the D1 SECIS element with an apical stem loop [17]. Similarly, the simple removal of the apical stem-loop of the 5'-SelP SECIS element caused a 50% loss of activity, although there is no reason to believe that full activity could not be recovered with other structural modifications. Finally, disruption of the apical stem of the 5'-SelP SECIS element abolished its biological activity and compensatory mutagenesis allowed a partial recovery of function.

Recently, these in vivo data have been complemented by further in vitro enzymatic structure probing experiments [11]. The data clearly show the difference in sensitivities of nucleotides in the apical regions of form 1 and form 2 SECIS elements to reagents that have double- and single-stranded cleavage specificities. In addition, the elements of human SelD and SelX appear to readily convert between the SECIS element forms in a dynamic equilibrium. This information does not contribute significantly to our understanding of the roles of the two different classes of SECIS elements. When tested for activity, consistent differences between elements of the two forms have not been found [12], therefore one class is not inherently more effective than the other. The fact that SECIS-binding protein 2 (SBP2; see below) binds to elements of either class with similar affinity [21] tends to support this conclusion; however, there may be other, as yet unidentified SECIS-binding proteins which may exhibit such behavior. Data presented by Fagegaltier et al. also suggest that the conserved A residues are more accessible in form 2 elements [22]. Since the SBP2-SECIS element interaction has been shown to be independent of these nucleotides [23], it seems possible that another SECIS-binding protein is responsible for distinguishing the two forms of SECIS element.

**Applications and Future directions**
Taken together, SECIS element research has yielded a very detailed portrait of the RNA structure required for selenocysteine incorporation in eukaryotes. Researchers have already used this information to design software-based database search algorithms in an attempt to identify potential selenoprotein cDNAs in the GenBank. Kryukov et al. employed an algorithm in which the primary sequence requirements were first identified, followed by secondary structural analysis and free energy calculation [24]. In a second stage of the search, functional parameters of the putative SECIS elements were assessed, such as position relative to the open reading frame and UGA selenocysteine codon. Using this strategy, all previously known selenoproteins were identified, strongly validating the algorithm. Importantly, this approach yielded two novel selenoproteins, designated SelT and SelR, subsequent analyses of which showed that the proteins do indeed contain selenium. Selenoproteins SelX, SelN and SelZ were identified by Lescure et al. [25] using a similar strategy after the SECIS elements had been validated by their ability to drive selenocysteine incorporation into cGPx. Given the near ubiquity of SelR and SelX or their cysteine-containing analogs in organisms ranging from *E. coli* to fungi, plants and man, it is likely that these proteins serve an important role in processes critical to the sustenance of life. Furthermore, virally encoded selenoproteins have now been shown to confer an advantage to the virus bearing the gene for cGPx [26].

Another use of detailed information about the SECIS element is now in its infancy. Copeland et al. [21, 23] have identified and cloned SECIS-binding protein-2 (SBP2) which interacts with SECIS elements. While there had been reports previously of proteins having this activity, none were shown convincingly to bind to SECIS elements with high specificity [27-29]. Even more important, it was shown that SBP2 binding mediates the activity of the SECIS element. Both in vivo and in vitro, SBP2 binding enhances selenocysteine incorporation in a SECIS element-dependent manner. It is not yet clear whether additional proteins bind the SECIS element at the same time; however, this is a possibility, as mutations to the SECIS "core" affect SBP2 binding, while disrupting the AA motif in the loop does not. With identification of SBP2 and other such proteins, the use of x-ray crystallographic and chemical probing techniques can assign amino acid-nucleotide contacts in the SBP1-SECIS interaction. Such work will begin to address some of the unanswered questions regarding the functional effects of certain mutations in the SECIS element.

This leads to a final point. Although SECIS element structure and function have been examined in great detail, there are some very clear and pressing issues that have yet to be resolved. First, while there is very compelling information pointing to the existence of the tetrapurine core motif, conclusive evidence derived from physical methods has not yet been obtained. It is expected that as technology advances, and NMR becomes even more refined, the positions of the atoms in a full SECIS element as it exists in solution may be assigned with more certainty. Secondly, the cloning of SECIS-binding protein-2 (SBP2) will allow the opening of many areas of inquiry regarding SECIS element function. Specifically, the identification of SBP2-SECIS points of contact may consist of the information that brings coherence to much of the previously obtained mutagenesis data. In addition, the availability of this information should give researchers the ability to employ crystallographic techniques to further refine our current model of the structure of the SECIS element in action.

**References**

1. C Rocher, C Faucheu, F Hervé, C Bénicourt, JL Lalanne 1991 *Gene* 98:193
2. MJ Berry, L Banu, Y Chen, SJ Mandel, JD Kieffer, JW Harney, PR Larsen 1991 *Nature* 353:273
3. MJ Berry, L Banu, JW Harney, PR Larsen 1993 *EMBO J* 12: 3315
4. C Buettner, JW Harney, PR Larsen 1998 *J Biol Chem* 273:33374
5. M Kozak 1986 *Proc Natl Acad Sci USA* 83:2850
6. EA Leibold, HN Munro 1988 *Proc Natl Acad Sci USA* 85:2171
7. TA Rouault, MW Hentze, SW Caughman, JB Harford, RD Klausner 1988 *Science* 241:1207
8. Q Shen, FF Chu, PE Newburger 1993 *J Biol Chem* 268:11463
9. SC Vendeland, MA Beilstein, J-Y Yeh, W Ream, PD Whanger 1995 *Proc Natl Acad Sci USA* 92:8749
10. GW Martin 3rd, JW Harney, MJ Berry 1996 *RNA* 2:171
11. D Fagegaltier, A Lescure, R Walczak, P Carbon, A Krol 2000 *Nucleic Acids Res* 28:2679
12. GW Martin 3rd, JW Harney, MJ Berry 1998 *RNA* 4:65
13. R Walczak, E Westhof, P Carbon, A Krol 1996 *RNA* 2:367
14. R Walczak, N Hubert, P Carbon, A Krol 1997 *Biomed Environ Sci* 10:177
15. R Walczak, P Carbon, A Krol 1998 *RNA* 4:74
16. Q Shen, JL Leonard, PE Newburger 1995 *RNA* 1:519
17. E Grundner-Culemann, GW Martin 3rd, JW Harney, MJ Berry 1999 *RNA* 5:625
18. KE Hill, RS Lloyd, JG Yang, R Read, RF Burk 1991 *J Biol Chem* 266:10050
19. KE Hill, RS Lloyd, RF Burk 1993 *Proc Natl Acad Sci USA* 90:537
20. SC Low, MJ Berry 1996 *Trends Biochem Sci* 21:203
21. PR Copeland, DM Driscoll 1999 *J Biol Chem* 274:25447
22. D Fagegaltier, N Hubert, P Carbon, A Krol 2000 *Biochimie* 82:117
23. PR Copeland, JE Fletcher, BA Carlson, DL Hatfield, DM Driscoll 2000 *EMBO J* 19:306

24. GV Kryukov, VM Kryukov, VN Gladyshev 1999 *J Biol Chem* 274:33888
25. A Lescure, D Gautheret, P Carbon, A Krol 1999 *J Biol Chem* 274:38147
26. W Zhang, CS Ramanathan, RG Nadimpalli, AA Bhat, AG Cox, EW Taylor 1999 *Biol Trace Elem Res* 70:97
27. Q Shen, R Wu, JL Leonard, PE Newburger 1998 *J Biol Chem* 273:5443
28. Q Shen, PA McQuilkin, PE Newburger 1995 *J Biol Chem* 270:30448
29. N Hubert, R Walczak, P Carbon, A Krol 1996 *Nucleic Acids Res* 24:464

# Chapter 6. SECIS binding proteins

Paul R. Copeland and Donna M. Driscoll

*Department of Cell Biology, Lerner Research Institute, Cleveland Clinic Foundation, Cleveland, OH 44195, USA*

**Summary:** The 3'-untranslated regions (UTRs) of selenoprotein mRNAs contain a stable stem-loop termed the selenocysteine insertion sequence (SECIS) element. Among the *trans*-acting factors that are thought to be required for selenocysteine (Sec) insertion are RNA binding proteins that interact with the SECIS element. In addition to several proteins which bind non-specifically, this element is specifically bound by SECIS binding protein 2 (SBP2), and this interaction is required for Sec insertion. SBP2 is a novel 94 kDa protein that contains an RNA binding domain found in ribosomal proteins and eukaryotic translation termination release factor 1 (eRF-1). SBP2 specifically interacts with the conserved AUGA element in the SECIS element core, but binding is not affected by mutations to the conserved AAA motif. Mutation analysis of SBP2 mutants indicates that it has distinct functional and RNA binding domains. While the mechanism of SBP2 action remains to be determined, we have recently observed that it is stably associated with ribosomes in vivo and in vitro, and that this association requires both the RNA binding and functional domains. These recent data suggest that SBP2 may be involved in defining a subset of ribosomes capable of translating selenoprotein mRNAs.

**Introduction**
In both prokaryotes and eukaryotes, the co-translational incorporation of Sec into selenoproteins requires the recoding of a UGA stop codon as one specific for Sec. The pathway for Sec insertion in the *Escherichia coli* formate dehyrogenase isozymes has been elegantly elucidated using biochemical and genetic approaches. Sec insertion in prokaryotes requires a SECIS element in the coding region of the selenoprotein mRNA and SelB, a novel elongation factor that binds to both the SECIS element and Sec-tRNA$^{Sec}$. In eukaryotes, Sec incorporation requires the assembly of what we will term herein the Selenocysteine Insertion Complex (SIC). That the SIC is in fact a stable complex remains to be seen, so for the purposes of this chapter it should be simply defined as the *cis* and *trans* components required

for the recoding of an in-frame UGA as Sec. To date the SIC appears to consist of the Sec codon (UGA), the selenocysteine insertion sequence (SECIS) in the 3'-UTR of the selenoprotein mRNA, the Sec-tRNA$^{Sec}$, the Sec-specific elongation factor eEFSec, and the 3'-UTR binding protein SECIS binding protein-2 (SBP2). It is not known, however, if this complex is sufficient. The focus of this chapter will be on the history and current theory concerning the role of SECIS binding proteins (SBPs) in the insertion of Sec.

## SBPs

Since the discovery that Sec insertion required sequences in the 3'-UTR, the SECIS element has become an obvious target in a search for RNA binding proteins that would function to link the UGA codon to the distant *cis*-acting element. The early definition of specific conserved elements within the UTR as functionally important was significant as it provided a basis for the search for binding proteins based on a detailed map of potential interacting points [1-4]. These elements include an AUGA in the middle of the 5' portion of the stem, an AAA motif in the terminal loop, and a GA opposite the AUGA sequence. Each of these has been shown to be required for Sec insertion, but only the AUGA/GA elements have been demonstrated to be involved in protein binding.

The first report of a SECIS binding protein came from Shen et al. [5] who observed a complex by electrophoretic mobility assay (EMSA) after incubation of the glutathione peroxidase (GPx) 3'-UTR with COS-1 cell extracts. The RNA-protein complex was sensitive to deletion of the last six nucleotides of the basal stem of the 3'-UTR, a region of the SECIS element that is not conserved and is not required for SECIS function. UV cross-linking analysis suggested that the molecular weight of this binding protein is 71 kDa. The same group later identified this binding activity as dbpB, a member of the Y-box family of DNA and RNA binding proteins [6]. It has been corroborated that dbpB can bind the SECIS element [7], but its binding is not affected by debilitating SECIS mutations, and its role in Sec insertion, if any, remains elusive.

The second SBP activity was described by Hubert et al. [8] as one or a pair of complexes resolved by EMSA in HeLa S100 extracts. The type I iodothyronine deiodinase SECIS yielded two complexes while the GPx displayed only one. The specificity of the RNA-protein complexes was determined by comparison to 5S rRNA and antisense SECIS RNAs, but it is not known whether the protein requires any of the conserved SECIS elements for binding. UV cross-linking analysis in these extracts indicated that the binding activity has an apparent molecular weight of 65 kDa, and this protein was the first to be named "SBP". We thus consider this protein to be "SBP1", although its identity and function in Sec insertion remain a mystery. A more recent study verified that this binding protein is distinct from dbpB [9].

The third and most recently described SBP (SBP2) was identified by our group as a 120 kDa protein from rat testicular extracts that bound the phospholipid hydroperoxide glutathione peroxidase (PHGPx) 3'-UTR [10]. This binding activity was eliminated by mutations in the conserved AUGA element which is now considered to be the SECIS "core" consisting of non-Watson/Crick base pairs in a GA quartet with bases on the 3' side of the stem [11-13]. We have recently undertaken a RNA footprinting analysis with SBP2 and the PHGPx and GPx SECIS elements. Preliminary results indicate that, as expected, SBP2 contacts the SECIS element on both sides of the SECIS core, but does not contact the conserved AAA motif [JE Fletcher, PR Copeland, DM Driscoll, A Krol, unpublished results]. Our search for factors that bind this region have so far been unsuccessful and its role in Sec insertion is unknown.

**Known properties of SBP2**
We purified SBP2 by RNA affinity chromatography from rat testicular S100 extracts where the selenoprotein PHGPx is known to be overexpressed [7]. Purified SBP2 was subjected to tryptic digestion and peptide sequencing by mass spectrometry, the results of which eventually allowed us to identify the corresponding cDNA [14]. By means of both Western and UV cross-linking analysis, SBP2 is only easily detectable in testis extracts. However, using UV cross-linking followed by immunoprecipitation, we found that the protein is expressed at low levels in other tissues including rat liver and aortic smooth muscle cells [PR Copeland, C Gerber, DM Driscoll, unpublished results]. It is likely, therefore, that our utilization of testis as a source of SBPs was fortunate as searches for this factor in other tissue types or cell lines may have been fruitless due to its low abundance and the lack of a sensitive assay. Northern analysis of mRNAs from multiple rat tissues indicates the presence of three transcripts encoding SBP2. Two of these, 3.5 and 4.4 kb fragments, are present in all of the tissues tested. An abundant 2.5 kb transcript was detected only in testis and possibly at a very low level in liver [14]. Thus there appears to be a correlation between SBP2 expression and the presence of the small transcript. Theoretically, SBP2 will be required in most tissues if it is a common factor in Sec incorporation, so it is possible that the larger transcripts are responsible for regulated expression while the smaller transcript is required for "overexpression."

The SBP2 cDNA encodes a novel protein with a predicted molecular weight of only 94 kDa [14]. The discrepancy between observed (120 kDa) and predicted molecular weights is unlikely to be the result of a specific modification as it persists in all expression systems tested including *E. coli*. Based on cDNA sequence analysis, SBP2 does not share homology with elongation factors. Thus in mammals, the function of SelB is carried out by at least two distinct polypeptides, SBP2 which binds to the SECIS element, and eEFSec which binds to the Sec-tRNA$^{Sec}$.

| SECIS binding | | |
|---|---|---|
| SBP2 | rat | AKTKRR**L**V**L**G**LREV**L**KH**L**K**LR**K**L**K**C**II**ISPNC |
| hSLP | human | AKARRR**L**V**M**G**LREV**T**KH**M**K**L**N**K**I**KC**VII**SPNC |
| dSLP | *Drosophila* | ARAHPR**L**V**L**G**V**REA**LA**R**L**R**INK**V**K**L**L**F**LATDC |
| Release Factors | | |
| eRF1 | yeast | SQDTGK**F**CYG**I**DD**TL**KA**L**D**L**GAVEK**LIV**FENL |
| eRF1 | *Arabidopsis* | SQDTGK**Y**VFG**V**ED**TL**KA**L**E**M**GAVET**LIV**WENL |
| eRF1 | *Xenopus* | SQDTGK**Y**CFG**V**ED**TL**KA**L**E**M**GAVE**ILIV**YENL |
| eRF1 | human | SQDTGK**Y**CFG**V**ED**TL**KA**L**E**M**GAVE**ILIV**YENL |
| Ribosomal Proteins | | |
| RPL30 | human | VMKSGK**Y**V**L**G**Y**KQ**TL**K**M**I**R**QGKAK**L**V**IL**ANNC |
| RPL30 | yeast | VIKSGK**Y**T**L**G**Y**KS**TV**KS**L**RQGKS**KLIII**AANT |
| RPL30 | *T. cruzi* | AQDTGK**I**V**M**GARKS**I**QYAKMGGAK**LIIV**ARNA |
| RPL30 | *M. vannielii* | AVDTGN**V**V**L**GTKQA**I**KN**I**KHGEGK**LVII**AGNC |
| RPL30 | *S.acidocald* | LLRSGK**V**I**L**GTRK**TL**K**LL**K**T**GK**V**KG**VVV**SSTL |
| RPS6 | *H. marismortui* | ARDTGA**V**KKGTNE**T**TKS**I**ERGSAE**L**V**F**VAEDV |
| RPL4A | yeast | SPKPYA**V**K**Y**G**L**NHV**VAL**IENKKAK**L**V**LI**ANDV |
| RPL7A | rice | AKKPIV**V**K**Y**G**L**NGV**TY**L**I**EQSKAQ**L**V**VI**AHDV |
| RPL7A | human | TKRPPV**L**RAGVN**T**V**TT**LVENKKAQ**L**V**VI**AHDV |
| RPS12 | *T. brucei* | ARETNG**L**I**C**G**L**SEV**TR**A**L**DRRTAH**L**C**VL**ADDC |
| RPS12 | rat | ALIHDG**L**A**R**G**I**REA**AK**A**L**DKRQAH**L**C**VL**ASNC |
| YIF4 | *B. subtilis* | ANRARK**V**V**S**G**E**D**LV**IKE**I**RNARAK**L**V**LL**TEDA |
| Miscellaneous/Unknown Function | | |
| RimK | *E. coli* | TSDLID**M**VGGAP**L**V**VK**LVEGTQG**I**G**VVL**AETR |
| DOM34 | yeast | NKDDDK**A**W**Y**G**E**KE**VV**KAAEYGA**I**S**YLLL**TDKV |
| DOM34R | yeast | SKDDNK**A**W**Y**G**A**EE**TE**RAAK**L**DA**I**E**TLLI**TDSB |
| NHP2 | yeast | ASKAKN**V**K**R**G**V**KE**VV**KA**L**RKGEKG**LVVI**AGDI |

**Figure 1**. Alignment of L30 RNA binding domain. Conserved residues are in boldface.

Sequence analysis revealed that SBP2 shares a motif with the translation termination release factor eRF-1 as well as several ribosomal proteins. This 32 amino acid motif was originally hypothesized to play a role in RNA binding [15], and has recently been demonstrated to be required for ribosomal protein L30 RNA binding activity [16]. L30 is involved in its own post-transcriptional regulation by virtue of its ability to bind L30 pre-mRNA (regulating its splicing) and L30 mRNA (regulating its translation) [17,18]. By a comparison of the sequence and structure of the binding sites found in its mRNA and the known structure of the 28S rRNA, it has recently been determined that L30 binds to a related sequence in 28S rRNA [19]. Although initial structural studies of the L30 mRNA target did not predict the presence of a non-Watson/Crick paired GA quartet, this motif is found in the structure of the 28S rRNA binding site [20]. An alignment of the sequences contained in this motif for several eukaryotic and prokaryotic proteins is shown in Figure 1. Our own analysis of mutants in this region of SBP2 indicates that it is required for SECIS element binding as a mutation of the conserved Gly residue (G669) eliminates both SBP2 function and SECIS binding activity [PR Copeland, VA Stepanik, DM Driscoll, unpublished results]. The fact that SBP2 shares this domain with factors that interact directly with the ribosome is intriguing and may suggest a common type of target within the rRNA. This relationship also bears on our current thinking about the mechanism of SBP2 action, which is discussed below.

Although there are no known homologs of SBP2, a protein of unknown function identified as human hypothetical protein KIAA0256 possesses 46% identity to SBP2 at the amino acid level and nearly 100% similarity in the RNA binding region. Here we will refer to this protein as human SBP2-like protein (hSLP). A small open reading frame in Drosophila also contains a region similar to SBP2 and hSLP, but no SBP2 relatives have been found in other lower eukaryotes or prokaryotes. Despite the high degree of conservation in the RNA binding domain, hSLP only weakly binds the PHGPx 3'-UTR, and it is not able to support Sec insertion [PR Copeland, VA Stepanik, DM Driscoll, unpublished results] suggesting that sequences outside this domain may be required for the specificity of binding. For this reason, and because human and mouse ESTs that are highly homologous to SBP2 at the nucleic acid level are present in the database, we believe that hSLP is not the human version of SBP2, which remains to be identified. The role of SLPs in Sec insertion is currently under investigation, but it is certainly conceivable that they perform functions outside that realm. Thus, we conclude that SBP2 may belong to a family of proteins that may serve a variety of processes involving RNA-protein interactions.

## SBP2 Function
Since the primary sequence of SBP2 has not provided concrete information about its possible mechanism of action, we have begun to map the domain structure of this protein by truncation and site-directed mutagenesis. Based on early work with partial SBP2 cDNA clones, it was clear at the outset that

the N-terminal half (amino acids 1-398) was not required for SBP2 function [14]. By constructing a series of deletion mutants in the C-terminal half of SBP2, we have recently mapped the domain structure of SBP2 and found that the RNA binding domain and "functional" domain are distinct and do not overlap [PR Copeland, VA Stepanik, DM Driscoll, unpublished results] as shown in Figure 2.

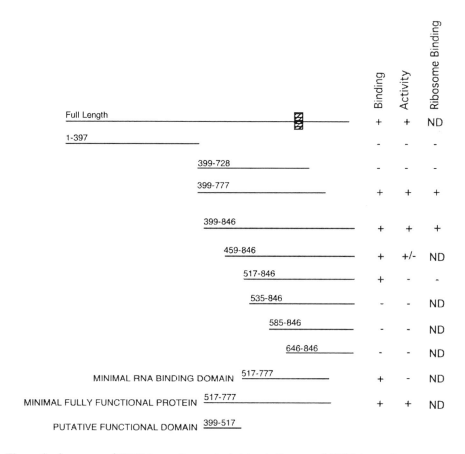

**Figure 2.** Summary of SBP2 truncation mutant data. A diagram of SBP2 truncation mutants is shown. RNA binding, functional and ribosome binding activity are summarized to the right of each mutant. The putative functional domain as determined from the mutant data is indicated at the bottom. The hatched box on the full length diagram represents the conserved L30 RNA binding domain. Constructs that were not tested in the ribosome binding assay are marked as "ND."

We define SBP2 function as its ability to stimulate PHGPx synthesis in vitro. In our standard assay, the addition of excess SBP2 to the reaction results in a ~20-fold stimulation. We have concluded from this data that the minimally functional portion of SBP2 contains amino acids 399-777. Interestingly, the region of the protein that is responsible for the size discrepancy overlaps with part of the RNA binding domain. Our results do confirm that the RNA binding domain includes the conserved motif mentioned below, but the functional domain bears no resemblance to any known motifs.

A key factor in our ability to analyze SBP2 function arose from our use of a modified in vitro translation system in rabbit reticulocyte lysate that is capable of efficient selenoprotein synthesis. Success in the in vitro assay depended entirely on the concentration of input mRNA. At the standard level (4 µg/ml), a small amount of full-length product was detectable, but it was apparently the product of non-specific stop codon suppression as it was not codon or SECIS element dependent [PR Copeland, JE Fletcher, DM Driscoll, unpublished results]. At a level of mRNA 10 times lower than normal, codon and SECIS element dependent translation was observed. These results suggest that there are multiple *trans*-acting factors which are required for Sec insertion that may be "swamped out" at higher mRNA concentrations. SBP2 appears to be limiting in reticulocyte lysate since the addition of SBP2 to the reaction enhanced PHGPx synthesis over 20-fold [14]. In addition, the amount of PHGPx synthesis was further increased by raising the incubation temperature from the standard 30°C to 37°C. This is consistent with the fact that the binding activity of SBP2 is maximal at the higher temperature [7]. Using this system, we were able to inactivate Sec incorporation by depleting SBP2 from rabbit reticulocyte lysate with anti-SBP2 antibodies. PHGPx synthesis was restored by adding recombinant SBP2 back to the reaction, thus demonstrating its requirement for Sec insertion.

**SBP2 and the ribosome**
As discussed above, SBP2 contains a novel RNA-binding domain that is found in several ribosomal proteins. Through the use of glycerol gradient sedimentation, we have recently found that SBP2 is stably associated with 80S ribosomes [PR Copeland, VA Stepanik, DM Driscoll, unpublished results]. An analysis of SBP2 mutants indicates that both the RNA binding domain and the functional domain are required for this association. Interestingly, the point mutation at G669 that eliminates SECIS binding activity has no effect on ribosome association. This is a surprising result considering that the basis of the SBP2 ribosome interaction appears to be at the level of 28S rRNA. Several 28S rRNA structures are similar enough to the SECIS element to theoretically support SBP2 binding, and this hypothesis is currently being tested. However, if the ribosome interaction is

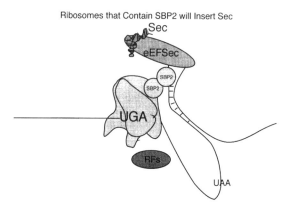

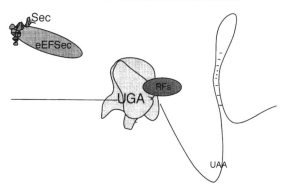

**Figure 3.** A model for Sec insertion. The top panel illustrates a situation where SBP2 is bound to the ribosome and has thus made that ribosome competent for Sec insertion. The bottom panel shows a ribosome lacking SBP2 that will terminate translation at the Sec codon.

based on RNA, then one would expect the G669 mutant to also prevent ribosome binding. This paradox could be resolved if the SECIS and 28S rRNA binding sites on SBP2 overlap but are not identical. In this model, SBP2 would exist in the SIC as at least a dimer where one RNA binding domain would interact with the ribosome and the other with a SECIS element (see Figure 3). This hypothesis is supported by our earlier data that found endogenous, partially purified SBP2 to exist as a large complex as determined by gel filtration [7]. In addition, purified recombinant SBP2 also sediments as an apparent complex in glycerol gradients [PR Copeland, VA Stepanik, DM Driscoll, unpublished results].

## The missing link

Now that the Sec-tRNA$^{Sec}$ specific elongation factor, eEFSec has been identified [21,22], one may be tempted to conclude that our understanding of the SIC is complete. eEFSec has been shown to interact with SBP2 in transfected cells, possibly in an RNA-dependent manner [22]. A simple model for Sec insertion might include the SBP2/SECIS/ribosome complex providing a platform for the specific association of eEFSec near the ribosomal A site. But this model does not explain the necessity of the AAA motif that is found in the terminal loop or an internal loop within all known SECIS elements. It is unlikely that it is stabilizing the entire structure of the SECIS element as it is not known to be involved in intramolecular contacts within the 3'-UTR [11,23]. In addition, SBP2 binding is not affected by a deletion of the motif, indicating that the SBP2/SECIS interaction is not sufficient [10]. The importance of the loop is further underscored by recent mutational analysis that indicates that the size of the loop as well as the spacing between the SECIS core and the loop are critical for Sec insertion [24]. One would expect, therefore, that another binding factor may specifically interact with the terminal loop, but no such factor has been reported and our attempts to identify a binding activity with specificity for the AAA motif have thus far been unsuccessful. It is of course possible that another factor is not required and that this motif is involved in direct contacts with the ribosome and/or the Sec-tRNA$^{Sec}$.

## A model for SBP2 function

One of the key issues regarding Sec insertion is that of the role of translation termination in regulating the process. There is little doubt that the specific selection of UGA as a Sec codon involves a competition with the termination machinery, but direct evidence for this is lacking. We believe that the major question is whether the SIC is actively or passively involved in preventing termination. In the prokaryotic system, a SelB mediated pause at the UGA codon appears to be exclusive of termination because noncognate Sec-tRNA$^{Sec}$ severely interferes with UGA suppression by an authentic suppressor tRNA suggesting that the SelB/SECIS/Sec-tRNA$^{Sec}$ complex is blocking access to the ribosomal A site [25]. We have recently completed a polysome analysis which suggests that the same sort of pausing event is occurring on eukaryotic selenoprotein mRNAs [26]. We have not been able to detect a direct interaction between SBP2 and the eukaryotic release factors, eRF1 and eRF3 in vitro [PR Copeland, DM Driscoll, unpublished results]. To get at the question of active versus passive inhibition of termination, the first order of business is to determine if SBP2 is necessary for the block of termination.

The competition with termination also requires a consideration of the efficiency of Sec insertion. In prokaryotes, the process has been reported to be very inefficient, i.e., the termination reaction is only partially blocked by the Sec insertion complex [25]. By this definition, fully efficient Sec insertion at a UGA codon would yield the same amount of protein as Cys incorporation at a UGU (Cys) codon. The question of efficiency and processivity (incorporation of Sec into multiple in-frame UGA codons) in eukaryotes has been investigated, and it was concluded that the process is inefficient [27,28]. One caveat of these studies is that they used transiently transfected cells that greatly overexpress selenoprotein mRNAs. Furthermore, this conclusion seems to be in conflict with nature as it is clear that at least some tissues (e.g., testis) are able to synthesize large quantities of selenoproteins [29]. In no case has efficiency been measured in terms of endogenous protein and mRNA levels, and we know from our own work that excess mRNA can shut down Sec insertion in vitro and in transfected cells [PR Copeland, J Fletcher, DM Driscoll, unpublished results]. In terms of processivity, the fact that zebrafish SelP, which contains 17 in frame UGA codons [22], is made at all seems to suggest that the reaction is at least somewhat processive. Many studies have also shown that the context of the UGA codon can affect the efficiency of both termination and Sec insertion [30-33], suggesting that the regulation of efficiency may be specific for each selenoprotein mRNA. Together these studies present fairly clear evidence that the question of in vivo efficiency remains unanswered.

As of now, our model of Sec insertion is necessarily simple. Figure 3 illustrates our current thinking. The major difference between this model and those generated in the past is the idea that SBP2 is not stably associated with the SECIS element until after translation initiation. In this manner, ribosomes loaded with SBP2 will pause at the UGA and the presumed SBP2/SECIS/eEFSec/Sec-tRNA$^{Sec}$ complex will actively deliver Sec to the growing peptide chain. At this time there is no evidence to suggest that SBP2 is actively involved in blocking release factor access by direct binding, so the lack of termination may be explained by a steric block mediated by the relative large complex required for Sec insertion.

**SBP2 as a "master regulator"**
In any multi-factorial system, it is the limiting component that is the most effective target for regulation. While the relative abundance of each of the SIC components is not known, it is known that SBP2 is limiting in reticulocyte lysates [14], while the Sec-tRNA$^{Sec}$ and eEFSec are not [PR Copeland, DM Driscoll, unpublished results]. As mentioned above, the addition of SBP2 to an in vitro translation assay programmed with PHGPx mRNA results in a ~20-fold increase in synthesis. If this level of regulation occurs in vivo, then it is clear that simply changing the amount of SBP2 can have a dramatic effect on intracellular selenoprotein concentration. In this light, therefore, the study of SBP2 expression and the regulation of its expression become critical in the understanding of differential selenoprotein expression. While one level of SBP2 regulation is bound to occur at the transcriptional level, the fact that there appear to be three SBP2 mRNA isoforms suggests that SBP2 is also regulated at the translational level. Indeed, preliminary work in our lab has shown that the smallest SBP2 mRNA does not contain the >600 nucleotide 3'-UTR found in the larger two transcripts [PR Copeland, DM Driscoll, unpublished results]. Considering the growing body of evidence that translation is regulated by various 3'-UTRs [34], these data suggest that translational regulation via the SBP2 3'-UTR may be responsible for reducing the amount of protein made in cells that lack the smaller transcript.

Beyond the regulation of SBP2 expression, there are many possible levels of regulation of SBP2 activity. We have already shown that SBP2 binding activity is competed approximately 10 times more efficiently with the PHGPx versus the GPx 3'-UTR [10]. Further evidence of binding differences are suggested by differences in the ethylnitrosourea interference patterns between PHGPx and GPx [JE Fletcher, PR Copeland, DM Driscoll, A Krol, unpublished results]. We fully expect, therefore, to find distinct differences in binding affinities for various 3'-UTRs, and that these affinities will correlate with the efficiency of selenoprotein synthesis. It is tempting to speculate that SBP2 may be involved in establishing or maintaining the hierarchy of selenoprotein expression that has been observed under selenium deficient conditions [35]. In this case, the synthesis of certain selenoproteins decreases in a very specific order during selenium depletion with PHGPx and GI-GPx being the most resistant to down regulation. This shut-off mechanism appears to involve differential mRNA stability as PHGPx and GI-GPx mRNA remain stable during selenium deficiency while GPx mRNA is quite unstable under these conditions. It is possible, therefore, that SBP2 binding affinity is part of what determines the efficiency of synthesis and possibly even the longevity of the mRNA. Since GPx has been demonstrated to be a target of nonsense mediated decay during selenium depletion [36,37], it will be of interest to determine the interplay between the known components of the nonsense mediated decay system and SBP2.

## Conclusions

In contrast to the system in prokaryotes where one protein (SelB) binds both the SECIS element and delivers the Sec-tRNA$^{Sec}$, at least two proteins perform this function in eukaryotes (SBP2 and eEFSec). We have evidence that SBP2 is a multi-domain protein that may be involved in preventing translation termination at the Sec codon while also interacting with eEFSec as the latter presents Sec-tRNA$^{Sec}$ to the ribosomal A site. Based on our model that SBP2 is stably associated with ribosomes, we expect that the amount of SBP2 in the cell will be a primary determinant of that cell's capacity to synthesize selenoproteins. Limitations of other factors would add further levels of control to what has clearly been established as a highly regulated system.

### References

1. MJ Berry, L Banu, YY Chen, SJ Mandel, JD Kieffer, JW Harney, PR Larsen 1991 *Nature* 353:273
2. MJ Berry, L Banu, JW Harney, PR Larsen 1993 *EMBO J* 12:3315
3. Q Shen, FF Chu, PE Newburger 1993 *J Biol Chem* 268:11463
4. Q Shen, JL Leonard, PE Newburger 1995 *RNA* 1:519
5. Q Shen, PA McQuilkin, PE Newburger 1995 *J Biol Chem* 270:30448
6. Q Shen, R Wu, JL Leonard, PE Newburger 1998 *J Biol Chem* 273
7. PR Copeland, DM Driscoll 1999 *J Biol Chem* 274:25447
8. N Hubert, R Walczak, P Carbon, A Krol 1996 *Nucleic Acids Res* 24:464
9. D Fagegaltier, N Hubert, P Carbon, A Krol 2000 *Biochimie* 82:117
10. A Lesoon, A Mehta, R Singh, GM Chisolm, and DM Driscoll 1997 *Mol Cell Biol* 17:1977
11. R Walczak, E Westhof, P Carbon, A Krol 1996 *RNA* 2:367
12. R Walczak, N Hubert, P Carbon, A Krol 1997 *Biomed Environ Sci* 10:177
13. R Walczak, P Carbon, A Krol 1998 *RNA* 4:74
14. PR Copeland, JE Fletcher, BA Carlson, DL Hatfield, DM Driscoll 2000 *EMBO J* 19:306
15. EV Koonin, P Bork, C Sander 1994 *Nucleic Acids Res* 22:2166
16. H Mao, SA White, JR Williamson 1999 *Nat Struct Biol* 6:1139
17. FJ Eng, JR Warner 1991 *Cell* 65:797
18. B Li, J Vilardell, JR Warner 1996 *Proc Natl Acad Sci USA* 93:1596
19. J Vilardell, SJ Yu, JR Warner 2000 *Mol Cell* 5:761
20. RR Gutell, MN Schnare, MW Gray 1990 *Nucleic Acids Res* 18 Suppl:2319

21. D Fagegaltier, N Hubert, K Yamada, T Mizutani, P Carbon, A Krol 2000 *EMBO J* 19:4796
22. RM Tujebajeva, PR Copeland, X-M Xu, BA Carlson, JW Harney, DM Driscoll, DL Hatfield, MJ Berry 2000 *EMBO Reports* 1:1
23. D Fagegaltier, A Lescure, R Walczak, P Carbon, A Krol 2000 *Nucleic Acids Res* 28:2679
24. E Grundner-Culemann, GW Martin, 3rd, JW Harney, MJ Berry 1999 *RNA* 5:625
25. S Suppmann, BC Persson, A Bock 1999 *EMBO J* 18:2284
26. JE Fletcher, PR Copeland, DM Driscoll 2000 *RNA* 6:1573
27. MJ Berry, AL Maia, JD Kieffer, JW Harney, PR Larsen 1992 *Endocrinology* 131:1848
28. MT Nasim, S Jaenecke, A Belduz, H Kollmus, L Flohe, JE McCarthy 2000 *J Biol Chem* 275:14846
29. A Roveri, M Maiorino, C Nisii, F Ursini 1994 *Biochim Biophys Acta* 1208:211
30. KK McCaughan, CM Brown, ME Dalphin, MJ Berry, WP Tate 1995 *Proc Natl Acad Sci USA* 92:5431
31. WP Tate, ES Poole, ME Dalphin, LL Major, DJ Crawford, SA Mannering 1996 *Biochimie* 78:945
32. MY Pavlov, DV Freistroffer, V Dincbas, J MacDougall, RH Buckingham, M Ehrenberg 1998 *J Mol Biol* 284:579
33. W Wen, SL Weiss, RA Sunde 1998 *J Biol Chem* 273:28533
34. AB Sachs, G Varani 2000 *Nat Struct Biol* 7:356
35. R Brigelius-Flohe 1999 *Free Radic Biol Med* 27:951
36. PM Moriarty, CC Reddy, LE Maquat 1998 *Mol Cell Biol* 18:2932
37. X Sun, PM Moriarty, LE Maquat 2000 *EMBO J* 19:4734

# Chapter 7. Towards a mechanism for selenocysteine incorporation in eukaryotes

John B. Mansell and Marla J. Berry

*Thyroid Division, Brigham and Women's Hospital, and the Department of Biological Chemistry and Molecular Pharmacology, Harvard Medical School, Boston, MA 02115, USA*

**Summary:** The recent identification and cloning of two key proteins involved in the cotranslational incorporation of selenocysteine in eukaryotes has allowed the investigation of the mechanism by which the internal UGA codons of selenoprotein mRNAs are recoded from a canonical reading as termination, to encode selenocysteine. These proteins, selenocysteine insertion sequence (SECIS) binding protein 2 (SBP2), and eukaryotic selenocysteyl-tRNA$^{[Ser]Sec}$-specific elongation factor (eEFsec), represent the first trans-acting protein factors clearly demonstrated to be involved in selenocysteine incorporation in eukaryotes. This chapter will focus on the cloning of eEFsec, in addition to presenting a discussion of prevalent current models describing the mechanism of selenocysteine incorporation in eukaryotes.

## Introduction

Selenocysteine incorporation occurs cotranslationally at UGA codons in prokaryotes, eukaryotes, and archaea. The recoding of UGA codons to specify selenocysteine, rather than termination, is signaled by the presence of specific secondary structures in selenoprotein mRNAs, termed selenocysteine insertion sequence or SECIS [1] elements. Although we have a clear understanding of what comprises a SECIS element, the mechanism by which the translational recoding it signals is achieved in eukaryotes is poorly understood, and is based predominantly on comparison with, and inference from, that in prokaryotes.

Selenocysteine incorporation in prokaryotes has been characterized in considerable detail [2]. In *Escherichia coli*, tRNA$^{[Ser]Sec}$, encoded by the selC gene, has an anticodon complementary to UGA, and is initially charged with serine by seryl-tRNA synthetase. This seryl-tRNA$^{[Ser]Sec}$ precursor is then converted to selenocysteyl-tRNA$^{[Ser]Sec}$ by the selA gene product, selenocysteine synthase. This reaction requires the activated selenium donor species monoselenophosphate, formed through the action of the selD gene product, selenophosphate synthetase.

Several unique structural features of tRNA$^{[Ser]Sec}$ render it poorly recognized by the standard elongation factor (EF), EF-Tu [3], in turn precluding it from functioning as a nonspecific UGA suppressor. Instead, selenocysteyl-tRNA$^{[Ser]Sec}$ is bound by the selenocysteine-specific elongation factor, SELB. SELB consists of an amino-terminal domain with high homology to EF-Tu, and a carboxy-terminal extension [1,4]. SELB exhibits specificity not only for tRNA$^{[Ser]Sec}$, but also for the presence of selenocysteine on the tRNA: it does not appreciably bind the seryl-tRNA precursor [4]. This discrimination on the part of the elongation factor prevents misincorporation of serine at UGA selenocysteine codons. The carboxy-terminal extension of SELB is a SECIS RNA-binding domain [1,5]. In prokaryotes, the SECIS elements are located in the coding region, immediately downstream of the UGA codons they serve. Binding to the SECIS element tethers the protein to the mRNA, thereby recruiting the selenocysteyl-tRNA$^{[Ser]Sec}$-SELB complex to the ribosome as it approaches the UGA codon.

Unlike prokaryotes, in eukaryotes and archaea SECIS elements are located in the 3'-untranslated region (UTR), or in one case in archaea in the 5'-UTR [6], at considerable distances from the UGA codons they serve. In eukaryotes, SECIS elements have been shown to recode the entire message, functioning for any upstream in-frame UGA [7-9], provided a minimal spacing requirement is met [10].

Given the differences in the architecture of prokaryotic and eukaryotic selenoprotein mRNAs, it has been assumed that the mechanism of selenocysteine incorporation within these two kingdoms also differs. After extensive efforts over the last decade, some of the details of the mechanism in eukaryotes are finally being revealed, primarily through the identification of the protein factors catalyzing cotranslational selenocysteine insertion. In striking contrast to the relative ease of identifying eukaryotic homologs of the selC and selD genes, the eukaryotic functional homolog of SELB has proved much more difficult to identify. A long-standing question in the field of selenoprotein biology has been what is the nature of the protein or proteins that catalyze selenocysteine incorporation directed from the 3'-UTR? Specifically, do there exist in eukaryotes and archaea bifunctional proteins equivalent to prokaryotic SELB, or multiple proteins carrying out these functions?

**Identification of the essential trans-acting factors in eukaryotes**
With the completion of the genome sequence of the archaeon, *Methanococcus jannaschii*, three elongation factor genes were identified, two bearing homology to the standard elongation factors, eEF1 and eEF2, and the third unassigned. Selenoproteins had previously been identified in this organism and a tRNA$^{[Ser]Sec}$ gene was also present in the genome

sequence. It was therefore plausible that the unassigned EF might function as a selenocysteyl-tRNA-specific elongation factor [11].

Intriguingly, unlike the prokaryotic SELB, the archaeon sequence contained no C-terminal extension. Conserved secondary structures proposed to function as SECIS elements were identified in the 3'- and 5'-UTRs of the archaea selenoprotein genes, suggesting that archaea and eukaryotes might employ similar recoding mechanisms. The unassigned EF was subsequently shown to exhibit specificity for selenocysteyl-tRNA$^{[Ser]Sec}$ [12], but did not bind the archaeal SECIS elements. A separate SECIS-binding activity was identified in archaea cell lysates, but has not yet been purified. In parallel, studies of the factors involved in eukaryotic selenocysteine insertion were finally making progress. Eukaryotic SECIS elements had previously been characterized in detail and shown to exhibit specific sequence and structural features, including a conserved G-A, A-G tandem purine pair at the base of a stem-loop, with the stem consisting of 10 to 12 base-pairs [10,13-16]. This tandem purine pair is required for selenocysteine insertion. In 1999 and early 2000, Copeland and Driscoll reported purification, cloning and characterization of a mammalian protein, SECIS-binding protein 2 (SBP2) which exhibits specificity for the purine pair in eukaryotic SECIS elements [17,18]. Immunodepletion of SBP2 from reticulocyte lysates with specific antibodies abolished selenocysteine incorporation in vitro, while addition of recombinant SBP2 restored selenocysteine insertion, clearly demonstrating SBP2 protein is required for selenocysteine incorporation in vitro. Importantly, SBP2 contains neither elongation factor homology nor activity.

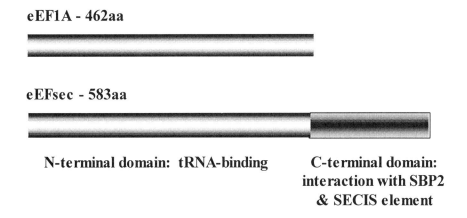

**Figure 1.** Schematic showing the carboxy-terminal extension of the murine eEFsec, relative to murine eEF1A. Proposed functional domains of eEFsec are indicated.

With the archaeal SELB sequence and the information that SBP2 exhibited SECIS specificity, but not EF function, the search for eukaryotic SELB homologs without the prerequisite for a SECIS-binding domain ensued. Sequential database searches with the putative archaeal SELB sequence led to identification of selenocysteyl-tRNA-specific EF candidates in *Caenorhabditis elegans*, *Drosophila melanogaster*, and finally the murine and human EST databases [19,20]. The sequence of the murine EST predicts a protein of 583 amino acids, significantly larger than the 462 for eEF1A (Figure 1). Alignment of the four eEFsec sequences along with the murine and human eEF1A sequences reveals a high degree of conservation in the amino-terminal EF domain, but also regions unique to the eEFsec family. Insertions are present in the eEF1A sequences that are not found in the eEFsec family, as reported previously for the bacterial SELB family [1]. In bacterial EF-Tu, these regions are required for interaction with the guanine nucleotide exchange factor, EF-Ts. As bacterial SELB has a ten-fold lower affinity for GDP than GTP [21], exchange of GDP for GTP occurs spontaneously without the requirement for a separate nucleotide exchange factor [22]. The absence of similar regions in the eEFsec family suggests guanine nucleotide exchange also occurs in the absence of any extrinsic factor in eukaryotes. Furthermore, the carboxy-terminal extensions of the eEFsec family share regions of significant similarity, some are restricted to the eukarya, but others also include the archaeal protein.

The tRNA-binding specificity of the murine factor was determined in a nitrocellulose filter binding assay using purified recombinant protein and purified radiolabeled $^{3}$H-seryl-tRNA or $^{75}$Se-selenocysteyl-tRNA. Selenocysteyl-tRNA$^{[Ser]Sec}$ exists as two isoforms which differ by the presence or absence of a 2'-O-ribose methylation at U34, the anticodon wobble base [23]. Binding by the factor was highly specific for the two isoforms of Sec-tRNA$^{[Ser]Sec}$, whereas binding to either the serylated precursor, seryl-tRNA$^{[Ser]Sec}$, or to the standard tRNA, seryl-tRNA$^{Ser}$, occurred at background levels. This capacity to discriminate between the correctly charged selenocysteyl-tRNA$^{[Ser]Sec}$ and the serylated precursor is of functional significance, for, as with the prokaryotic SELB, it prevents the misincorporation of serine at UGA codons specifying selenocysteine. Binding to methionyl-tRNA$^{Met}$ or to the only other non-standard tRNA, initiator methionyl-tRNA$^{MetI}$, also represented background levels.

The recombinant factor was shown to have a $K_d$ for GTP of 0.11 mM, while competition with unlabeled GDP indicated a $K_d$ approximately 3 fold higher, again consistent with previous findings for SELB. In conjunction with the comparative sequence analysis, this suggests guanine nucleotide exchange proceeds without the requirement for a GTP exchange factor. The fulfillment of the two criteria for a selenocysteyl-tRNA-specific EF, namely

GTP binding, and specificity of the factor for both the appropriate tRNA and the correct amino acid, led to its designation as eEFsec.

**Interactions between eEFsec, SBP2, and the SECIS element**
That the two distinct functions of SELB, elongation factor and SECIS-binding, are performed in eukaryotes by two separate proteins suggested that the two eukaryotic proteins would functionally interact.

We hypothesize that selenocysteine incorporation directed from the 3'-UTR involves binding of SBP2 to the SECIS element, followed by the recruitment of eEFsec-tRNA$^{[Ser]Sec}$ by the SBP2-SECIS complex. Assembly of the complex at the SECIS element, followed by delivery of eEFsec-tRNA$^{[Ser]Sec}$ to a UGA codon occupying the ribosomal A site, would allow translation of UGA codons at any site in the open reading frame, without strict context requirements. This model predicts the association of SBP2 and eEFsec, either through direct protein-protein interactions, protein-RNA interactions or through other interacting partners. To address these possibilities, coimmunoprecipitation studies involving eEFsec and SBP2 were performed, using proteins expressed by cotransfection in a human cell line, HEK-293. eEFsec was subcloned into a mammalian expression vector with a FLAG$^{TM}$ epitope tag introduced at the amino terminus, and cotransfected with plasmids expressing a mammalian selenoprotein and SBP2, the latter to increase selenoprotein synthesis over the endogenous level. Immunoprecipitation of transfected cell homogenates with antisera against SBP2 resulted in the coprecipitation of eEFsec as detected by western blotting with FLAG antibody. The amount of eEFsec in the coprecipitate increased significantly with SBP2 cotransfection, but eEFsec was also detected in the absence of cotransfected SBP2, presumably due to association with endogenous SBP2, which is present at significant levels in this cell line. When immunoprecipitation reactions were incubated in the presence of RNase A, the amount of eEFsec recovered decreased by 75%, indicating that complex formation was enhanced by the SECIS element or another RNA ligand. However, RNase treatment did not completely abolish coprecipitation, suggesting that there is an RNA-independent interaction which is augmented by the presence of RNA, or that association is dependent on a partially nuclease resistant RNA.

To investigate the role of SECIS RNA in complex formation, electrophoretic mobility shift assays (EMSA) with in vitro transcribed $^{32}$P-labeled SECIS elements and purified bacterially expressed recombinant proteins were employed. Incubation of a wild type SECIS element with either SBP2 or eEFsec retarded the mobility of the SECIS element. The mobility patterns of the two complexes were clearly distinguishable, indicating the two proteins bind the SECIS element independently. Incubation with both proteins resulted in an enhancement of the amount of shifted RNA. Binding specificity was next tested using $^{32}$P-labeled SECIS mutants consisting of either a point mutation in the conserved essential sequence, AUGA (mutated to AUCA), or inversion of the first two nucleotides (to UAGA). Binding of the mutant SECIS elements by SBP2 was not detected in the gel shift assay. In contrast, eEFsec exhibited binding to both mutants. Strikingly, when both proteins were added together, binding to the mutant SECIS elements was not seen. Thus, the presence of SBP2 prevents eEFsec binding to the SECIS mutants, conferring specificity for the wild type sequence. This result suggests that interaction with SBP2 alters the conformation of the eEFsec RNA binding domain.

It was not clear from the EMSA patterns whether the mobility shift observed when both proteins were present was due to the formation of a new complex involving the RNA and both proteins, or resulted from the superimposition of the two individual RNA-protein complexes. To investigate this, the wild type SECIS element was incubated with SBP2 in the absence or presence of eEFsec. After UV crosslinking, the RNA-protein complexes were resolved by nondenaturing polyacrylamide gel electrophoresis and analyzed for SBP2 by Western blotting. A fraction of the SBP2-SECIS complex was shifted upward upon the addition of eEFsec, indicating that a new complex containing both proteins had formed on the RNA. Formation of complexes in the absence of the RNA was not detectable.

To further assess binding specificity, competition studies utilizing a nitrocellulose filter binding assay were performed with unlabeled SECIS RNAs. Addition of a 50-fold excess of unlabeled wild type SECIS element to the labeled wild type element prior to incubation with eEFsec resulted in 70% inhibition of binding of the labeled RNA. Similarly, competition for binding by SBP2 or by the combination of the two proteins was inhibited by 60% and 67%, respectively. A 50-fold excess of the wild type element inhibited binding of eEFsec to the labeled AUCA mutant by ~70%, and to the UAGA mutant by >90%. A 50-fold excess of unlabeled AUCA mutant competed binding of eEFsec to the labeled AUCA mutant by 70% and to the UAGA mutant by 85%. The same molar excess of unlabeled AUCA SECIS mutant decreased binding of labeled wild type SECIS by eEFsec by only ~20%, indicating a lower affinity of eEFsec for the mutant than for the wild type element. Yeast tRNA decreased binding to the wild type and AUCA mutant elements by less than 10%, but inhibited binding to the UAGA mutant by ~70%, indicating a weak or relatively non-specific interaction with the latter.

Finally, the functions of the two proteins were assessed in vivo by cotransfection in HEK-293 cells with a cDNA encoding selenoprotein P from the zebrafish, *Danio rerio*. This cDNA contains 17 selenocysteine codons and two SECIS elements, making the expressed protein a sensitive indicator of effects on selenoprotein synthesis. Previous studies of selenoprotein P purified from rat plasma identified, in addition to the full-length protein, heterogeneous products resulting from premature termination at the UGA codons [24]. A similar phenomenon has been observed with transiently expressed rat and zebrafish selenoprotein P [25, and unpublished]. Selenocysteine incorporation was assessed following transfection by addition of $^{75}$Se to the media, and analysis of the resultant labeled selenoprotein P. In the absence of added factors, a diffuse band representing full-length and UGA-terminated species was seen. Transfection of eEFsec alone resulted in a shift upward, indicating more full-length protein was synthesized. Transfection of SBP2 resulted in a dramatic decrease in the ratio of full-length protein to premature termination product. Cotransfection of both factors reversed the SBP2 effect, shifting the balance again towards full-length protein. One possible explanation for this result is that the overexpression of SBP2 may perturb the stoichiometry of the factors required for selenocysteine incorporation, resulting in more efficient insertion at some UGA codons, but perhaps less so at others. The coexpression of eEFsec apparently restores this balance.

## Efficiency of selenocysteine incorporation in eukaryotes

Two related questions in the field of eukaryotic selenoprotein synthesis bear directly on our understanding of the mechanism of selenocysteine insertion; namely, how efficient is selenoprotein synthesis in vivo, and to what extent do selenocysteine incorporation and termination compete at any given UGA codon? Selenocysteine incorporation has been reported to be inefficient in all systems studied. Termination occurs in *Escherichia coli* selenoproteins [26, JB Mansell, WP Tate, unpublished results], in rabbit reticulocyte in vitro translation reactions [18,27-29], in transiently transfected mammalian cells [19,25,28,30] and in baculovirus-insect cell expression systems [31; C Buettner, unpublished results]. In mammalian cells, overexpression of selenoprotein mRNAs by transfection of increasing amounts of selenoprotein-encoding plasmid increases the ratio of termination product to full-length protein [28; E Grundner-Culemann, MJ Berry, unpublished results]. Cotransfection of components of the selenocysteine incorporation pathway, either tRNA$^{[Ser]Sec}$ [28], selenophosphate synthetase [32], or SBP2 [18; JW Harney, MJ Berry, unpublished results], partially reverses this effect, increasing selenocysteine incorporation. Selenium supplementation also increases incorporation [28,33]. However, the levels of full-length selenoprotein do not approach those of the corresponding cysteine-mutant protein under any of these conditions, implying that selenocysteine incorporation may be inherently inefficient. Attempts at overexpression might thus exacerbate any inherent inefficiency in this process. Termination at selenocysteine codons has also been observed in intact animals.

Purification of selenoprotein P (SelP) from rat plasma revealed multiple isoforms of the protein, shown by carboxypeptidase sequencing to comprise full-length and prematurely UGA-terminated species [24]. The amounts of truncated products increased upon dietary selenium limitation, but premature termination was even observed in animals maintained on a selenium sufficient diet.

Termination of translation in eukaryotes is catalyzed by the release factors (RFs) eRF1 and eRF3, both of which were identified and cloned in eukaryotes only within the last five years [34-36]. eRF1 confers recognition of all three termination codons [37], while eRF3 functions as an eRF1 and ribosome-dependent GTPase [38]. If selenocysteine incorporation were in direct competition with termination, we would predict that the levels or activities of the release factors in cells would affect selenocysteine incorporation efficiency, such that increased termination efficiency would result in decreased synthesis of full-length selenoproteins. However, little is known about levels of the release factors in different tissues or cell lines, or about circumstances that might affect release factor levels or activity.

The number of ribosomes on any mRNA is a function of the size of the open reading frame (ORF), and the rate of translation initiation. The spacing of ribosomes on mRNAs has been shown to average approximately one every 100 nucleotides [39]. Following the assumption that UGA selenocysteine codons direct incorporation rather than specifying termination, the predicted ORF sizes and numbers of ribosomes for the type 1 deiodinase (D1) and glutathione peroxidase (GPx) mRNAs are ~750 nucleotides (7 - 8 ribosomes) and ~600 nucleotides (6 ribosomes), respectively. Polysome profile analysis showed that the majority of D1 mRNA was found in polysomes consisting of ~1- 3 ribosomes, and GPx mRNA was predominantly associated with one ribosome. In both cases, the number of ribosomes is far fewer than predicted by the sizes of their ORFs, corresponding instead to that predicted for the shorter, UGA-terminated ORF [40]. In contrast, a D1 mRNA in which the UGA codon was replaced with a UGU cysteine codon was predominantly associated with 8-10 ribosomes. Similar results have been observed in studies investigating polysome loading on phospholipid hydroperoxide glutathione peroxidase mRNA [41]. Thus, the presence of a selenocysteine codon decreases the polysome loading on D1 mRNA, relative to that seen with the substitution of a cysteine codon, conferring a 'translational penalty' on the selenoprotein mRNA, suggesting that decoding of the UGA codon by the eukaryotic selenocysteine insertion machinery is an inherently inefficient process. Whether this is the result of direct competition with the alternate decoding event at the UGA codon, termination, or reflects a significant difference in the rates of eEF1A and eEFsec-mediated decoding, has yet to be characterized in detail.

## Towards a model for selenocysteine incorporation in eukaryotes

What are the implications of the observation that in eukaryotes the two known functions of prokaryotic SELB reside not in a single protein, but are separated amongst two distinct proteins? Does this separation confer some mechanistic advantage? Is it necessary to accommodate the distal placement of the SECIS in the 3'-UTR, itself required to relieve the coding sequence adjacent to the UGA codon from the constraint of forming a secondary structural element capable of binding the elongation factor? Or does it point to an alternative, less stochastic mechanism in which ribosomes are persistently reprogrammed from a canonical reading of the UGA codon?

Recent analyses show that the affinity of SELB for the bacterial SECIS is increased when selenocysteyl-tRNA$^{[Ser]Sec}$ is bound [22]. This suggests that following selenocysteine peptide bond formation, the lower affinity of SELB for the SECIS allows it to dissociate from the mRNA, in turn allowing ribosomal translocation to continue. In eukaryotes, given the placement of the SECIS element outside the coding region, translation of the entire coding region does not require the dissociation of SBP2 from the SECIS element. Thereby, the expression of the two functions of SELB in two separate proteins provides a mechanism for the rapid exchange of charged for uncharged tRNA-elongation factor complex. Here, SBP2 remains associated with the SECIS, and this complex recruits de novo eEFsec-selenocysteyl-tRNA$^{[Ser]Sec}$-GTP, rather than the eEFsec moiety remaining complexed with the SBP2-SECIS and undergoing both GDP for GTP exchange, and deacyl for acyl-tRNA$^{[Ser]Sec}$ exchange, in situ. Such a mechanism would allow a single SECIS element to efficiently serve multiple, distant UGA codons. The intervening mRNA would be looped out, accommodating the variable spacing between UGA and SECIS element - from a few hundred base pairs to nearly 5 kb - found in eukaryotic selenoprotein mRNAs [7]. The minimal spacing requirement of approximately 60 nucleotides may reflect the spacing needed to loop back and properly orient a SECIS RNA-protein complex at the ribosome (Figure 2).

An alternative, though not mutually exclusive function of the SECIS element has been suggested to explain the ability of a single element to serve multiple UGAs. In this model, the SECIS element delivers a signal to or confers a modification upon upstream ribosomes, presumably through a SECIS-bound factor or factors, such that the ribosomes acquire the ability to read UGA as selenocysteine throughout the transit of that message [42]. Copeland and Driscoll have shown that SBP2 is able to bind the ribosome and propose a model in which SBP2 remains associated with the ribosome, and it may be this persistent association that reprograms the ribosome to recode UGA as selenocysteine (see Chapter 6). Intriguingly, SBP2 contains a domain homologous to the yeast SUP1 omnipotent suppressor of translation termination, suggesting it may serve an anti-termination function. This function may act in concert with the eEFsec-recruiting activity to confer both selenocysteine incorporation and suppression of termination to ribosomes translating the UGA codons within eukaryotic selenoprotein mRNAs. There is some evidence for the perturbation of normal termination during the translation of selenoprotein mRNAs. If we invoke a simple competitive model where a UGA codon in the ribosomal A site may be decoded either by a eEFsec-selenocysteyl-tRNA$^{[Ser]Sec}$ complex, or by eRF1, increasing the cellular level of eRF1 would be predicted to favor increased termination. Indeed, earlier experiments investigating the effect of the sequence context surrounding the UGA codon on the efficiency of selenocysteine incorporation showed that if the UGA codon were placed in a sequence context favorable for termination, selenocysteine incorporation was decreased, consistent with such a model [43]. However, preliminary experiments analyzing the effect on selenoprotein synthesis of coexpressing eRF1 and eRF3 have yielded the somewhat counterintuitive result of increasing selenocysteine incorporation [E Grundner-Culemann, MJ Berry, unpublished results].

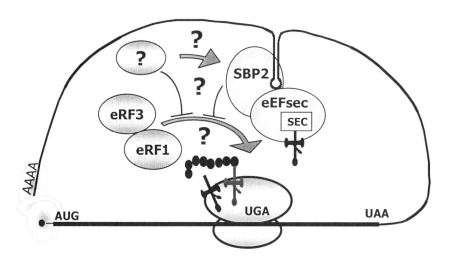

**Figure 2.** A model for eukaryotic selenocysteine incorporation showing the interaction between SBP2, eEFsec and the SECIS element. Any additional protein factors that may be required for the formation of this complex, or for its productive interaction with the ribosome, are yet to be characterized.

That the legitimate termination codon of selenoprotein W is a UGA, yet supports 'normal' termination, has been ascribed to the spacing of its SECIS element within the critical lower limit. This may indicate that any putative anti-termination activity is associated with, or requires the presence of, the SECIS element-protein complex. Such observations may suggest that ribosomes translating eukaryotic selenoprotein mRNAs may escape the rules governing canonical translation.

It is of significant importance to our understanding of the mechanism of selenocysteine incorporation in eukaryotes to determine how the presence, and absence, of selenocysteyl-tRNA$^{[Ser]Sec}$, eEFsec, SBP2, and the SECIS, affect the kinetics of their respective interactions. Similarly, the identification of additional factors involved in complex formation at the SECIS, or during the interaction of eEFsec-selenocysteyl-tRNA$^{[Ser]Sec}$ complex with the ribosome, are of critical importance to the elucidation of this important translational recoding event.

**References**

1.  M Kromayer, R Wilting, P Tormay, A Böck 1996 *J Mol Biol* 262:413
2.  A Böck 2000 *Biofactors* 11:77
3.  C Baron, A Böck 1991 *J Biol Chem* 266:20375
4.  K Forchhammer, W Leinfelder, A Böck 1989 *Nature* 342:453
5.  A Huttenhofer A Böck 1998 *Biochemistry* 37:885
6.  R Wilting, S Schorling, BC Persson, A Böck 1997 *J Mol Biol* 266:637
7.  MJ Berry, L Banu, JW Harney, PR Larsen 1993 *Embo J* 12:3315
8.  KE Hill, RS Lloyd, RF Burk 1993 *Proc Natl Acad Sci USA* 90:537
9.  Q Shen, FF Chu, PE Newburger 1993 *J Biol Chem* 268:11463
10. GW Martin, 3rd, JW Harney, MJ Berry 1996 *RNA* 2:171
11. PP Dennis 1997 *Cell* 89:1007
12. M Rother, R Wilting, S Commans, A Böck 2000 *J Mol Biol* 299:351
13. R Walczak, E Westhof, P Carbon, A Krol 1996 *RNA* 2:367
14. R Walczak, P Carbon, A Krol 1998 *RNA* 4:74
15. GW Martin, 3rd, JW Harney, MJ Berry 1998 *RNA* 4:65
16. E Grundner-Culemann, GW Martin, 3rd, JW Harney, MJ Berry 1999 *RNA* 5:625
17. PR Copeland, DM Driscoll 1999 *J Biol Chem* 274:25447
18. PR Copeland, JE Fletcher, BA Carlson, DL Hatfield, DM Driscoll 2000 *EMBO J* 19:306
19. RM Tujebajeva, PR Copeland, XM Xu, BA Carlson, JW Harney, DM Driscoll, DL Hatfield, MJ Berry 2000 *EMBO R* 2:158
20. D Fagegaltier, N Hubert, K Yamada, T Mizutani, P Carbon, A Krol 2000 *EMBO J* 19:4796
21. R Hilgenfeld, A Böck, R Wilting 1996 *Biochimie* 78:971
22. M Thanbichler, A Böck, RS Goody 2000 *J Biol Chem* 275:20458
23. AM Diamond, IS Choi, PF Crain, T Hashizume, SC Pomerantz, R Cruz, CJ Steer, KE Hill, RF Burk, JA McCloskey, DL Hatfield 1993 *J Biol Chem* 268:14215
24. S Himeno, HS Chittum, RF Burk 1996 *J Biol Chem* 271:15769
25. RM Tujebajeva, JW Harney, J Berry 2000 *J Biol Chem* 275:6288
26. S Suppmann, BC Persson, A Böck 1999 *EMBO J* 18:2284
27. MJ Berry, L Banu, YY Chen, SJ Mandel, JD Kieffer, JW Harney, PR Larsen 1991 *Nature* 353:273
28. MJ Berry, JW Harney, T Ohama, DL Hatfield 1994 *Nucleic Acids Res* 22:3753
29. JE Jung, V Karoor, MG Sandbaken, BJ Lee, T Ohama, RF Gesteland, JF Atkins, GT Mullenbach, KE Hill, AJ Wahba, DL Hatfield 1994 *J Biol Chem* 269:29739
30. MT Nasim, S Jaenecke, A Belduz, H Kollmus, L Flohe, JE McCarthy 2000 *J Biol Chem* 275:14846
31. IY Kim, MJ Guimaraes, A Zlotnik, JF Bazan, TC Stadtman 1997 *Proc Natl Acad Sci USA* 94:418

32. SC Low, JW Harney, MJ Berry 1995 *J Biol Chem* 270:21659
33. R Brigelius-Flohe, B Friedrichs, S Maurer, R Streicher 1997 *Biomed Environ Sci* 10:163
34. L Frolova, X Le Goff, HH Rasmussen, S Cheperegin, G Drugeon, M Kress, I Arman, AL Haenni, JE Celis, M Philippe, J Justesen, L Kisselev 1994 *Nature* 372:701
35. G Grentzmann, D Brechemier-Baey, V Heurgue, L Mora, RH Buckingham 1994 *Proc Natl Acad Sci U S A* 91:5848
36. O Mikuni, K Ito, J Moffat, K Matsumura, K McCaughan, T Nobukuni, W Tate, Y Nakamura 1994 *Proc Natl Acad Sci USA* 91:5798
37. DS Konecki, KC Aune, WP Tate, CT Caskey 1977 *J Biol Chem* 252:4514
38. L Frolova, X Le Goff, G Zhouravleva, E Davydova, M Philippe, L Kisselev 1996 *RNA* 2:334
39. JE Bergmann, HF Lodish 1979 *J Biol Chem* 254:11927
40. GW Martin, 3rd, MJ Berry 2001 *Genes to Cells* in press
41. JE Fletcher, PR Copeland, DM Driscoll 2000 *RNA* 6:1573
42. JF Atkins, A Böck, S Matsufuji, RF Gesteland 1999 *The RNA World* RF Gesteland, TR Cech, JF Atkins (Eds) Cold Spring Harbor Laboratory Press, Cold Spring Harbor, New York p637
43. KK McCaughan, CM Brown, ME Dalphin, MJ Berry, WP Tate 1995 *Proc Natl Acad Sci USA* 92:5431

# Chapter 8. Regulation of selenoprotein expression

Roger A. Sunde

*F21C Nutrition Cluster Leader, Professor of Nutritional Sciences and of Biochemistry, University of Missouri, Columbia, MO 65211, USA*

**Summary:** Studies on selenoprotein expression are revealing three distinct patterns of selenium regulation. In rats, selenium deficiency can result in >90% decreases in both glutathione peroxidase-1 (GPx1) activity and mRNA, whereas glutathione peroxidase-4 activity and mRNA decrease 60% and <10%, respectively. Our recent work with thioredoxin reductase reveals activity decreases of ~90% but with mRNA decreases <30%. This selenium regulation of gene expression could occur potentially at any of six points. Regulation of selenoprotein translation by selenium, mediated by Sec-tRNA availability, regulates the level of all selenoproteins. For optimum translation of a selenoprotein, its mRNA appears to require an optimum UGA location, an optimum UGA context and SECIS element(s) in the 3'-untranslated region (3'-UTR) which optimize Sec incorporation. The unique and specific regulation of GPx1 expression by selenium is mediated by regulation of GPx1 mRNA stability, and involves nonsense-mediated mRNA decay. This regulation, too, requires a functional SECIS element and a UGA codon, and the UGA must be followed by an intron. Our recent results indicate that GPx1 mRNA in selenium-deficient rat liver is moderately abundant, and that GPx1 mRNA in rat liver increases more than 20-fold with selenium supplementation. We hypothesize that this selenium regulation of GPx1 expression is a major component of GPx1 function in higher animals, and that in this role, GPx1 serves as a biological selenium buffer that maintains modest selenium stores for future selenoprotein synthesis.

**Introduction**
The discovery that glutathione peroxidase is a selenium-dependent enzyme [1] was followed almost immediately by the demonstration that the expression of glutathione peroxidase (GPx) activity in rats and other animals was highly regulated by selenium status [2]. Initially, this regulation was viewed as similar to changes in metalloproteins or metal-dependent enzyme activities observed with other mineral deficiencies. In subsequent years, however, GPx expression proved to be extremely useful for determining

selenium status and for the establishment of selenium requirements [3-5]. What has emerged in the past 13 years is that selenium regulation of glutathione peroxidase-1 (GPx1, EC 1.11.1.9) expression is novel and unique, relative to the regulation by metals of most identified markers of mineral status. This chapter will (i) review our work characterizing selenium regulation of GPx1 expression, (ii) review the regulation of other selenium-dependent proteins and enzymes, (iii) discuss the biochemical mechanism for this regulation, and lastly (iv) discuss the potential role of this regulation in the biological function of GPx1. The impact of this regulation on transcript abundance will also be discussed.

In hindsight, it should have been obvious that selenium regulation of GPx1 was different from most markers of mineral nutrient status. In selenium deficiency, liver GPx1 activity falls exponentially, dramatically, and it decreases to zero 21 days after weanling rats are switched to a selenium-deficient diet [2]. Similar, but less dramatic falls in GPx1 activity are observed in other tissues such as erythrocytes. Secondly, graded dietary supplementation of selenium raises GPx1 activity sigmoidally such that it reaches a plateau at 0.1 µg/g diet; additional dietary selenium does not increase liver GPx1 activity above this plateau. Lastly, the minimum dietary selenium requirement necessary to reach the GPx1 activity plateau is remarkably constant across a wide range of species, suggesting that a common molecular mechanism is responsible for this tight regulation [4]. This regulation makes GPx1 the most sensitive parameter for changes in selenium status over the deficient to adequate range [4], and has made GPx1 arguably the parameter of choice for assessment of selenium status and selenium requirements.

**Control points of regulation**
The genome of an organism provides a fixed blueprint for development and metabolism. Control of gene expression by nutrient status can play an important role in modulating the impact of the genome. Perhaps even more important, feedback loops sensing mineral status and modulating expression are likely to provide key homeostatic regulation of intracellular and organism-wide levels of mineral nutrients such as selenium [6]. Nutrients in general, and selenium specifically, can impact gene expression potentially at six points: transcription, nuclear processing, nuclear export, translation, mRNA stability, and protein turnover [6]. All selenoproteins are regulated by selenium at the level of translation. GPx1 expression is regulated significantly by selenium status at two of these levels -- translation and mRNA stability. In addition, transcriptional regulation, independent of selenium, is mediated by factors such as age, gender and tissue type [6]. A key footnote to this discussion is that regulation here refers (i) to changes in

profound biological endpoints that arise specifically due to differences in state of an organism, such as nutrient status or gender or age, and (ii) to changes that are mediated by specific molecular mechanisms; small but statistically significant differences are excluded in this discussion.

**Glutathione peroxidase-1 regulation**
We set out in 1985 to examine the underlying mechanism responsible for the uniformity of dietary selenium requirements. Our first approach was to examine selenium regulation of GPx1 protein as well as activity levels. Using anti-GPx1 antibodies, we found that weanling rats fed a selenium-deficient diet has a rapid exponential decline in GPx1 protein as well as GPx1 activity, with half-lives in liver of 5.2 and 2.8 days, respectively. This clearly indicates that more than just loss of the selenium cofactor is responsible for this decrease in activity [7]. For selenium repletion, GPx1 protein as well as activity requires larger doses and longer time periods than for maintenance [8]. We now know that this occurs because other selenoproteins have first priority for selenium in selenium-deficient rat liver (see below).

The cloning of GPx1 [9] next gave us opportunity to assess the impact of selenium status on GPx1 mRNA levels. We found that selenium deficiency has a dramatic effect on GPx1 mRNA levels in liver [10] which fall to approximately one-tenth of those found in selenium-adequate animals. In progressive selenium deficiency, we saw a coordinated dramatic exponential drop in GPx1 mRNA ($t_{1/2}$ = 3.2 d) as well as GPx1 activity ($t_{1/2}$ = 3.3 d) and GPx1 protein ($t_{1/2}$ =5.0 d) [11]. These experiments thus begin to substantiate that an underlying molecular mechanism is likely to be responsible for selenium regulation of GPx1 mRNA levels [12].

For these animal studies, we use as our model the young, rapidly growing weanling rat fed selenium-deficient torula yeast diets (0.008 µg selenium/g) or crystalline amino acid-based diets (0.002 µg selenium/g), and supplemented with graded levels of selenium as $Na_2SeO_3$ for 28 days. With male rats, both liver GPx1 activity and mRNA levels responded sigmoidally to increasing dietary selenium concentration [13,14], with GPx1 activity and mRNA reaching plateaus at 0.1 and 0.05 µg selenium/g diet, respectively (Figure 1). To compare these regulatory response curves, we use graphical analysis to determine a break point between the steepest part of the response curve and the plateau for each parameter. We have defined **"plateau break point"** [15] as the intersection of the line tangent to the steepest slope of the response curve and the line through the plateau region. This plateau break point can then be used as a quantitative means to determine the minimum dietary selenium required for maximal response.

In female rats, which have more than two-times the level of GPx1 mRNA as well as activity, the selenium regulation curves are very similar to male rats [15]. Liver GPx1 activity in selenium-deficient female rats is 2% of the level found in females fed selenium-adequate diets, and the plateau breakpoint is 0.1 μg selenium/g diet. Selenium-deficient liver GPx mRNA levels are 11 to 17% of selenium adequate levels, and the plateau breakpoint is reached at 0.05 μg selenium/g diet. Plotting the change in GPx1 activity in liver as a function of liver GPx1 mRNA levels shows that GPx1 mRNA levels respond to increasing selenium status before GPx1 activity; GPx1 mRNA reaches half maximum at 0.027 μg selenium/g diet whereas half maximal GPx1 activity requires 0.075 μg selenium/g diet [15]. At concentrations greater than 0.1 μg selenium/g diet, this relationship breaks down, showing that above 0.1 μg selenium/g diet, selenium status no longer regulates either GPx1 mRNA or GPx1 activity.

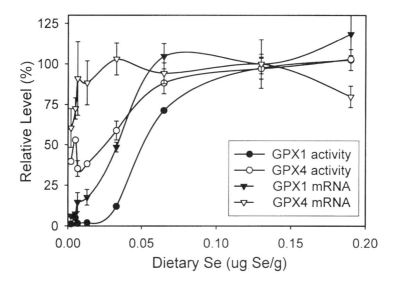

**Figure 1.** Relative level of liver GPx1 and GPx4 activity and mRNA in rats fed a selenium-deficient diet (0.002 μg selenium/g) and supplemented with graded levels of selenium. Levels are expressed as a percent relative to rats fed 0.13 μg selenium/g [14].

Erythrocyte GPx1 activity in selenium-deficient rats is 40% of selenium-adequate levels. While erythrocyte GPx1 activity continues to increase with increasing selenium, a breakpoint can be identified for this activity as well at 0.1 μg selenium/g diet. This increasing level of erythrocyte GPx1 activity in rats as they age is independent of selenium status and thus likely due to age-related transcriptional up-regulation.

Interestingly, plasma glutathione peroxidase (GPx3) activity in selenium-deficient rats is 7-8% of the levels found in selenium-adequate animals. Plasma GPx3 reaches a plateau break-point at 0.07 µg selenium/g diet [15], and thus shows a distinct response curve for this kidney-derived plasma selenoenzyme.

When cells are cultured in selenium-deficient media (2 nM selenium) and supplemented with graded levels of selenium, similar GPx1 activity and mRNA response curves are observed [16]. Under these conditions, however, GPx1 activity only falls to about 10-20% of the plateau level, and GPx1 mRNA levels only fall to about 50% of plateau levels [17]. The discrepancy between response curves in cultured cells and in intact animals is not understood; one hypothesis is that higher intracellular oxygen tension in cultured cells reduces the concentration of precursor forms of selenium ($H_3SePO_3$ or Sec-tRNA$^{-Sec}$) used for selenium incorporation.

## Glutathione peroxidase-4 regulation

The discovery of a monomeric second selenium-dependent glutathione peroxidase, phospholipid hydroperoxide glutathione peroxidase (GPx4; EC 1.11.1.12) [18], with distinct enzyme properties and tissue distribution, offered a new avenue to investigate selenium regulation. After a partial clone was identified [19], we isolated a full-length cDNA clone from pig and rat [20], and used these clones to evaluate the effect of dietary selenium status on GPx1 and GPx4 activity and mRNA levels in male and female rats. These two GPxs have 40% nucleotide and amino acid sequence identity. In selenium-deficient male rats fed an ultra-low selenium-deficient diet (0.002 µg selenium/g) [14], liver GPx1 activities are 1% of selenium adequate animals (Figure 1, Table 1). There is little increase in GPx1 activity below 0.033 µg selenium/g diet, and then there is a sharp increase. The plateau break point occurs at 0.1 µg selenium/g diet. Similar response curves also appear in heart, kidney, and lung [14]. In contrast, GPx4 activity only decreases to 41% of selenium adequate GPx4 activities in liver, and the plateau breakpoint occurs at 0.065 µg selenium/g diet. Again, similar response curves are observed in other tissues. These experiments clearly show the differential regulation of GPx1 and GPx4 by selenium in the same tissue.

When mRNA levels are quantitated, GPx1 mRNA is reduced to 7% of selenium-adequate levels and also responds sigmoidally to increasing dietary selenium concentration such that GPx1 mRNA levels in liver reach the break point at 0.065 µg selenium/g diet. In contrast, liver GPx4 mRNA is not significantly affected by dietary selenium, with the plateau break point for GPx4 mRNA occurring before 0.013 µg selenium/g diet (Figure 1). Similar response curves are observed for other tissues as well.

**Iodothyronine deiodinase regulation**
Similar experiments have been conducted to evaluate the effect of dietary selenium on iodothyronine deiodinase-1 (D1) activity and mRNA in liver. Arthur and colleagues [21] saw similar effects on GPx1 and GPx4 activities and mRNA levels in male hooded Wister rats fed diets containing 3 ng selenium/g diet. Liver D1 activity in selenium-deficient animals is 5% of that observed in selenium-adequate animals, and D1 mRNA levels are 50% of that observed in selenium-adequate animals (Table 1). As dietary selenium increases, the increases in D1 activity are parallel to the increases in GPx1 mRNA levels. The D1 mRNA response curve is hyperbolic rather than sigmoidal with a plateau breakpoint at about half of that observed for GPx4 activity, GPx1 mRNA, and D1 activity. These measures are complicated by TSH-mediated modulation of D1 transcription as part of the feedback system to maintain $T_3$ levels; for instance, thyroid D1 mRNA increases 40% in selenium deficiency [21].

Table 1. Patterns of down-regulation of selenoproteins in selenium-deficient rats.[a]

| Selenoprotein | Protein level (% decrease) | RNA level (% decrease) | Reference |
|---|---|---|---|
| *Pattern 1*: | | | |
| GPx1 | 99 | 90 | [14] |
| *Pattern 2*: | | | |
| GPx4 | 59 | <10 | [14] |
| *Pattern 3*: | | | |
| D1 | 95 | 50 | [21] |
| SelP | 90 | 33/<10 | [23]/[24] |
| TR1 | 90 | <30 | [25] |

[a]Percentage decrease of selenoprotein protein levels and mRNA levels in selenium-deficient rat liver, expressed as the percent decrease relative to selenium-supplemented controls. Protein was assessed by enzyme activity or by immunoprecipitation, and mRNA by northern blotting or ribonuclease protection analysis.

**Selenoprotein P regulation**
Selenium deficiency in rats causes a decrease in plasma selenoprotein P (SelP) concentration to less than 10% of adequate animals [22], and selenium supplementation of deficient rats results in an increase in SelP concentration ahead of the increases in plasma GPx3 activity and liver GPx1 activity, (when expressed on a percentage basis relative to rats supplemented with 0.5 µg selenium/g diet). At 4.5 weeks after feeding a selenium-deficient diet to rats, Burk and Hill [23] found reduced levels of GPx1, D1, and SelP mRNA levels in liver (Table 1). The impact of selenium deficiency

on SelP mRNA levels is that SelP mRNA levels fall to 67% of levels found in selenium-adequate animals whereas liver GPx1 mRNA levels are 19% of adequate levels. Expressed another way, the relative decreases in mRNA are 80% for D1 and 40% for SelP relative to the decrease for GPx1. These studies indicate that in selenium deficiency, SelP mRNA is less affected than D1 mRNA, which is less affected than GPx1 mRNA. Importantly, when we use ribonuclease protection analysis (RPA) to simultaneously quantitate changes in selenoprotein mRNA levels in our rat model [24], liver SelP mRNA levels are not decreased at all relative to selenium-supplemented controls, whereas GPx1 mRNA falls to 10-15%. This further indicates that SelP regulation is not significantly mediated by alterations in mRNA levels.

**Thioredoxin reductase regulation**
To better understand selenium regulation of selenoprotein expression, we have recently cloned rat thioredoxin reductase-1 (TR1) cDNA and used RPA to characterize the regulation of TR1 relative to other selenoenzymes in rat liver. Liver TR activity falls to 10% of selenium-adequate levels as compared to GPx1 (1-2%) and GPx4 (45%) [25]. In contrast, the pattern of mRNA regulation of TR is more similar to GPx4 (selenium-deficient levels are 80% of selenium-adequate levels) than to GPx1 (10-15% of selenium-adequate levels) (Table 1). These studies further emphasize that GPx1 regulation is unique relative to other selenoproteins.

In summary, selenium regulation of selenoproteins in our well-regulated rat liver model reveals three patterns of selenium regulation (Table 1). GPx1 is unique as both activity and mRNA levels are decreased ~90%; GPx4 represents a second pattern with modest (60%) decreases in activity and little change in mRNA. TR1 and SelP represent a third distinct pattern with dramatic decreases in activity/protein, but with little change in mRNA. Selenium regulation of D1 appears to be more similar to this third pattern, but often with reported modest changes in mRNA levels. This hierarchy of selenium regulation of selenoprotein mRNA level is (most to least regulation): GPx1 >> D1 > TR1 > SelP, GPx4. Experiments focused on the underlying mechanism suggest that selenium regulation of translation is important for the GPx4 and the TR1/SelP patterns, whereas mRNA stability as well as translational control is important in selenium regulation of GPx1.

**Selenium regulation of translation**
The hierarchy of selenium regulation of selenoprotein translation is complicated by underlying alterations in absolute level of mRNA transcripts (see below). Nonetheless, it appears that at the same tissue, GPx1 translation is most regulated and GPx4 least regulated, with the hierarchy (most to least selenium regulation): GPx1 > D1, SelP, TR1 > GPx4. The most logical

hypothesis to explain this hierarchy is that GPx4 mRNA is best optimized to compete for selenium incorporation when selenium is in limiting supply. What follows is a discussion of evidence in support of this hypothesis.

Laudable progress has been made recently to identify the components and mechanism involved in selenocysteine (Sec) incorporation into eukaryotic selenoproteins; these developments are reviewed in Chapters 5-7 in this book. The central features are that Sec insertion occurs co-translationally, the Sec is synthesized while esterified to its cognate tRNA, tRNA$^{-Sec}$, using inorganic selenophosphate and serine, and the position of Sec in the peptide backbone of selenoproteins is encoded by in-frame UGA codons in the mRNA. The last unique and necessary component for eukaryotic selenoprotein translation is the selenocysteine insertion sequence (SECIS) element in the 3'-UTR [26].

With sufficient selenium and the other necessary components, Sec-tRNA$^{-Sec}$ will compete sufficiently well with termination release factors for binding at UGA such that full length selenoproteins are synthesized. The identification of all necessary and sufficient components of the full Sec translation complex is probably not complete, but includes the mRNA with UGA codon and 3'-UTR SECIS [Chapter 5], Sec-tRNA$^{-Sec}$ [27 and Chapter 3], SECIS binding protein-2 (SBP2) [28 and Chapter 6], Sec-tRNA$^{-Sec}$-binding protein (EF$_{Sec}$) [Chapter 7] and GTP. In transfected cells, selenium supplementation of deficient media (~0.2 nM selenium) to 50 nM selenium increases Sec translation as measured by relative deiodinase activity 66-fold [29], showing the importance of selenium status on translation efficiency. In contrast, increasing Sec-tRNA$^{-Sec}$ levels has only modest impact (0-40% increase) irrespective of selenium status. Selenium concentration clearly appears to be the preeminent factor modulating Sec translation. Without selenium and therefore insufficient Sec-tRNA$^{-Sec}$, UGA will be interpreted as a stop codon, thus limiting selenoprotein translation. Availability of Sec-tRNA$^{-Sec}$, therefore, rather than free Sec or inorganic selenium compounds, appears to control translation of selenoproteins.

One assumption in this discussion is that limited availability of selenium uniformly will reduce the concentration of Sec-tRNA$^{-Sec}$ available for synthesis of all selenoproteins. As yet, there is no experimental evidence suggesting that subcellular differences in selenium concentration or differences in subcellular localization of translation can account for differential availability of the Sec translation factors.

Thus, we must conclude that differences in selenium regulation of selenoproteins arise because of cis-acting differences in their mRNAs. As discussed below, differences in SECIS element, UGA context or UGA/SECIS separation are likely to be why GPx4 mRNA out-competes other selenoprotein mRNAs for translation when selenium is limiting.

Telling and important experiments on differential selenoprotein mRNA translation were conducted by Berry and colleagues [30] as they identified and characterized the SECIS element. Using D1 chimeric constructs with 3'-UTRs from different selenoprotein mRNAs, they showed that different SECIS elements can alter the translational efficiency of D1 in cultured cells. Relative to the D1 SECIS, the relative efficiency with a 3'-UTR containing both SelP SECIS loops is 3.7, the efficiency of the first SelP stem-loop (SelP-loop 1) is 2.9 [30], and the efficiency of the GPx1 stem-loop 0.42 [31]. Subsequent studies by this group have found that the SECIS stem-loops can be divided into two forms [32]. The classical form 1 stem-loop is found in D1, GPx1 and SelP-loop2. The new form 2, with an additional stem positioned on the loop, is more common amongst selenoproteins, and includes stem-loops in GPx4, SelP-loop1 and TR1. Thus, the underlying suggestion is that differences in selenoprotein mRNA SECIS elements alter relative affinities for the Sec translation complex and thus confer differential translatability, especially when selenium is limiting. Additional studies will be necessary to determine if stem-loop form/affinity is the necessary and sufficient feature of GPx4 mRNA that facilitates retention of activity in selenium deficiency.

A second possible cause of differential Sec translatability might be distance between the UGA and the SECIS element. Berry and colleagues [33] have found that distances of <111 nt curtail Sec translation. This only has apparent biological impact, however, for selenoprotein W, which employs UGA as its authentic stop codon, but concomitantly has such a short spacing tether (~55 nt) that Sec incorporation does not occur at its stop codon.

A third cause of differential translation has been postulated to be UGA codon context. To investigate the impact of UGA codon context, systematic mutations of the base immediately following the UGA (referred to as the 4$^{th}$ base) have been made [34]. This 4$^{th}$ base is hypothesized to play a role in release factor recognition. These in vitro transfection experiments revealed that purines following the UGA codon increase termination frequency. Recent elegant studies with lacZ-luciferase fusion constructs [35] also found small effects (0.4 to 2.1%) on Sec incorporation due to mutations of the 4$^{th}$ base. Similar small effects on Sec translation are also observed when codons preceding the UGA are mutated.

We have also conducted experiments to systematically study the role of UGA position and context on GPx1 expression. Wen et al. [36] constructed a genomic GPx1 expression vector that would result in over-expression of $^{75}$Se-labeled recombinant protein in COS-7 cells at higher levels than for endogenously encoded selenoproteins. When the GPx1 UGA is mutated to cysteine (UGC) and ten separate codons are mutated to UGA, Wen found that UGA codons located in the middle of the open reading frame most efficiently direct Sec incorporation. Differences in $4^{th}$ base have relatively little effect, in contrast. UGA codons located close to the 5'-start or 3'-termination codons are much less efficient, but insertion of a green fluorescent protein coding region as a spacer raises efficiency of Sec incorporation at these positions to that of the wild-type UGA position. Thus, the UGA must be >21 nt from the AUG-start and >204 nt from the SECIS element for optimum Sec incorporation. In these studies, one particular UGA codon position, normally restricted in Sec insertional efficiency because it is located 81 nt from the SECIS element, had its Sec insertional efficiency increased 10-fold when this distance is increased by 711 nt.

Collectively, these studies suggest that the SECIS element, UGA/SECIS spacing, local UGA context and/or RNA secondary structure may be part of the mechanism that modulates Sec insertion during translation. All of these studies [30-36] used recombinant cells in culture and all but Wen and colleagues [36] had selenoprotein synthesis rates far lower than endogenous rates. As cautioned by Berry [26], however, extrapolation from these cell culture experiments to selenoprotein synthesis in intact organisms may be dangerous.

## Selenium regulation of GPx1 mRNA stability

The singular pattern of selenium regulation of GPx1 is mediated by the dramatic effect of selenium status on GPx1 mRNA levels. To investigate the *cis*-acting nucleic acid sequence requirements for selenium regulation of GPx1 mRNA levels, Chinese hamster ovary (CHO) cells were transferred with GPx1 cDNA constructs in which 3'-UTR regions of GPx1 are deleted or mutated [16]. In this cultured-cell model, GPx1 wild-type transfects result in at most 5-fold increases in GPx1 activity, and 2-3 fold increase in GPx1 mRNA relative to endogenous GPx1 levels. Transfection with reporter genes, in contrast, results in dramatic over-expression of mRNA and protein, clearly demonstrating that the necessary factors for selenium incorporation are limiting or kinetically dampened in cultured cells such that translation of Sec-containing polypeptides is reduced by a factor of 100 or more. In this model, the breakpoint for regulation of GPx1 activity occurs at 50-100 nM selenium. In low selenium media, transfected GPx1 mRNA levels are significantly decreased to 66% of the levels observed in selenium-supplemented cells, and the plateau break point for the regulation of GPx1 mRNA levels occurs at approximately 4 nM selenium. Similar to the translation experiments of Berry, deletion of portions of the 3'-UTR also completely eliminates selenium regulation of GPx1 mRNA, point mutations that alter the structure of the eSECIS element result in loss of selenium regulation of GPx1 mRNA, but mutations that restore the stem base-pairing

*Regulation of selenoprotein expression* 91

of the eSECIS restore selenium regulation [16]. These studies indicate that a functional SECIS element in the 3'-UTR is required for selenium regulation of GPx1 mRNA stability.

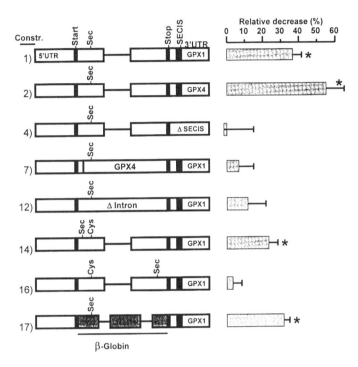

**Figure 2.** Impact of chimeric GPx1 constructs on selenium regulation of GPx1 mRNA in stably transfected CHO cells. The indicated constructs are shown with GPx4 substitutions (2,7), deletions (4,12), mutations (14,16) and replacement of the GPx coding region with β-globin containing a Sec codon. Also shown is the decrease in GPx1 chimeric mRNA levels in cells cultured in selenium-deficient media relative to the same pool of transfectants in selenium-supplemented media [16].

In a second series of experiments, chimeric GPx1 DNA constructs in which specific regions of the GPx1 gene are mutated, deleted or replaced by comparable regions from unregulated GPx4 were used [17] (Figure 2). Isolated stable CHO transfectants that were pooled two weeks after transfection, divided into selenium-deficient (2 nM selenium) or selenium-adequate (200 nM selenium) medium and grown for an additional 4 days. Transfected mRNA levels were specifically determined by RNase protection assay. Analysis of chimeric GPx1/GPx4 constructs (Figure 2) shows that the GPx4 3'-UTR can completely replace the GPx1 3'-UTR in selenium regulation of GPx1 mRNA! This indicates that the GPx1 3'-UTR and SECIS alone cannot fully explain the unique selenium regulation of GPx1 mRNA.

In contrast, replacement of the GPx1 coding regions with corresponding GPx4 coding regions diminishes or eliminates selenium regulation of the transfected GPx1 mRNA (Figure 2) [17]. Further analysis of the GPx1 coding region demonstrated that the GPx1 Sec codon (UGA) and the GPx1 intron sequences are required for full selenium regulation of transfected GPx1 mRNA levels. Mutations which moved the GPx1 Sec codon to three different positions within the GPx1 coding region suggest that the mechanism for selenium regulation of GPx1 mRNA requires a Sec codon within exon 1. Lastly, we found that addition of the GPx1 3'-UTR to β-globin mRNA can convey significant selenium regulation to β-globin mRNA levels when a UGA codon is placed within exon 1 (Figure 2). These studies clearly show that selenium regulation of GPx1 mRNA requires a functional SECIS in the 3'-UTR and a Sec codon followed by an intron [17].

Nonsense codons are known to destabilize many different mRNA species, particularly when located upstream from an intron (reviewed by [37,38]). Maquat and colleagues [39], who have been studying the mechanism of nonsense-mediated decay in β-globin, used GPx1 to help characterize where nonsense-mediated decay of mRNA occurs. They demonstrated that GPx1 mRNA degradation in rat hepatocytes occurs in the cytoplasm and that this degradation is likely to occur via nonsense-mediated decay of mRNA. In studies with recombinant constructs in cells [39], media selenium concentration raises wild type GPx1 mRNA levels but does not affect mRNA level when the UGA codon is replaced with the nonsense codon UAA or the cysteine codon UGC.

Collectively, both approaches [17,39] establish that GPx1 mRNA is degraded by a nonsense-mediated mRNA decay mechanism when Sec is not available for translation. In selenium deficiency, insufficient Sec-tRNA$^{-Sec}$ effectively causes interpretation of UGA as a nonsense codon rather than the Sec codon. In summary, Sec-tRNA$^{-Sec}$ concentrations control gene expression of GPx1 not only by limiting translation but also by modulating GPx1 mRNA stability; selenium regulation of selenoprotein expression at the mRNA stability level thus utilizes a selenium-specific switch. This selenium-specific mechanism thus rejoins general biology and employs nonsense-mediated mRNA decay of mRNA to accomplish this regulation.

**Selenoprotein transcript abundance**
The studies on regulation of translation either have been conducted in recombinant cells using co-transfected markers to correct for differences in transcription, or have been conducted in intact cells or animals using assays such as western blotting or enzyme activity to assess relative translation without any correction for differences in transcript abundance. We have now

conducted an initial study, in our well-regulated rat model using intact selenium-deficient and selenium-adequate rats, to evaluate the relative contribution of mRNA abundance versus translational efficiency to overall regulation of GPx1 expression [SL Weiss, Sachdev and RA Sunde, unpublished results]. GPx1, GPx4 and glyceraldehyde-3-phosphate dehydrogenase (GAPDH) transcripts per cell in rat liver were quantitated using RNase protection assay. Surprisingly, we found that GPx1 transcripts in selenium deficiency are moderately abundant and similar in abundance to GAPDH and other selenoprotein mRNAs; selenium supplementation increases GPx1 mRNA so that it is 30-fold higher than GAPDH mRNA. Secondly, in the same animals, we quantitated translation by assessing $^{75}$Se incorporation into gel-purified GPx1, GPx4 and TR. We found that translational efficiency of GPx1 mRNA is still half of that of GPx4, even when corrected for differences in transcript abundance. More importantly, translational efficiency of GPx1 mRNA decreases to 4-6% in selenium deficiency. The net effect is that selenium regulation of GPx1 mRNA stability appears to switch GPx1 mRNA from nonsense-mediated degradation to translation. This regulatory switch -- a selenium thermostat -- can explain why GPx1 expression is the optimum parameter for assessment of selenium status.

If GPx1 mRNA is the only mammalian selenoprotein with a selenium-specific switch controlling mRNA stability, what is the cause for the modest decreases in other selenoprotein mRNAs in selenium deficiency? The apparent selenium-specific down-regulation of mRNA levels for SelP, D1 and even GPx4 may be explained by ribosomal pausing that must occur when Sec-tRNA$^{-Sec}$ concentrations are limiting for protein synthesis. The competition between termination factors and a limited supply of Sec-tRNA$^{-Sec}$ would predispose these selenoproteins for early termination. And this idle mRNA could then be subjected to more rapid degradation via normal decapping and exonuclease hydrolysis. This suggests that there are two components involved in selenium regulation of mRNA levels: 1) non-specific effects relative to whether or not sufficient selenium (Sec-tRNA$^{-Sec}$) is available for translation; and 2) GPx1 mRNA-specific selenium regulation. This promises to be an exciting area of future research.

**Other factors controlling gene expression**
There is no evidence that selenium status has any effect on transcription initiation rates for any selenium-dependent gene [22,40,41]. Similarly, there is no evidence for altered rate of nuclear processing or for export of GPx1 mRNA from the nucleus [39,42]. There are clear developmental changes in selenoprotein expression independent of selenium [24], as well as marked differences in tissue distribution, and gender differences is some species [6]

presumably due to tissue-specific transcription factors. Perhaps of interest to those that speculate that deficiency of one "antioxidant nutrient" will result in compensatory increases in expression of other antioxidant factors, we have found that vitamin E deficiency does not alter any aspect of GPx1 gene expression in our rat model [6]. These examples do illustrate that it is critical to know that a marker of nutrient status is not modulated by other conditions before assuming that apparent altered gene expression (arrays) indicates insufficient dietary nutrient status.

The final control point for gene expression is protein degradation. There is no solid evidence, to date, for selenium deficiency increasing the degradation of GPx1 or any other selenoprotein.

## Biological selenium buffer

So what is the survival advantage of having a peroxidase whose concentration is profoundly regulated by the status of its cofactor? The discovery that selenium was a component of GPx1 logically explained the antioxidant functions of selenium [43]. The discovery of multiple GPx enzymes and genes, however, indicates that this perception was clearly an over-simplification. The demonstration that the GPx1 knockout mouse is viable, and grows and reproduces the same as congenic wildtype controls [44], further indicates that GPx1 is not the crucial antioxidant enzyme as originally proposed. The hierarchy of protection of these enzymes against decreases due to dietary selenium deficiency shows that GPx4, SelP and D1 all are protected from selenium deficiency relative to GPx1. Secondly, when mRNA changes are carefully compared relative to changes in GPx1 mRNA, it is clear that GPx1 mRNA levels fall by an order of magnitude or more, whereas mRNA levels for GPx4, TR1, D1 and SelP typically do not change or decrease by only a factor of 2. The unique and specific regulation of GPx1 mRNA suggests that this may be an important aspect of the physiological role of GPx1.

We propose that the important and major role of GPx1 in the liver, and perhaps other tissues is part of the homeostatic mechanism that keeps free concentration of selenium low, that diverts selenium to more important biological functions of selenium in times of deficiency, and that can reversibly bind excess selenium over the deficient to adequate range. In other words, the major role of GPx is to serve as a **"biological selenium buffer"** [45]. The impact of this buffering capacity is an expansion of the dietary selenium range or media concentration range between selenium deficiency and selenium toxicity. This function is more appropriately called a biological selenium buffer, rather than a "selenium store" or "selenium sink" because it indicates the dynamic homeostatic nature and because it indicates GPx1's active role in modulating selenium flux between incorporation into other selenoenzymes and incorporation into GPx1 [4].

We have conducted some initial studies using the knockout mouse, wild type controls, and GPx1 heterozygotes using our typical weanling rodent paradigm [Ferguson-Kohout and Sunde, unpublished]. We found that the only effect of knocking-out the GPx1 gene is the expected dramatic reduction in liver selenium, and the consequent development of depressed growth due severe selenium deficiency when weanling knockout mice are fed a selenium-deficient diet (normally only seen in second-generation selenium-deficient rats [46]). Gene duplication in evolution may have preserved the important selenium-dependent peroxidase activities in the form of GPx4 and the other GPxs. Selective selenium regulation of GPx1 expression would result in storage of selenium in times of plenty, and result in release of this selenium as GPx1 turns-over when selenium is limiting. Stored selenium from GPx1 would thus be reincorporated preferentially into selenoproteins whose mRNAs lack the cis-acting elements necessary for selenium regulation of mRNA stability.

**References**

1. JT Rotruck, AL Pope, HE Ganther, AB Swanson, DG Hafeman, WG Hoekstra 1973 *Science* 179:588
2. DG Hafeman, RA Sunde, WG Hoekstra 1974 *J Nutr* 104:580
3. National Research Council 1983 *Selenium in Nutrition* National Academy Press Washington, DC
4. RA Sunde 1997 *Handbook of Nutritionally Essential Mineral Elements* BL O'Dell, RA Sunde (Eds) Marcel Dekker New York p 493
5. RA Sunde 2000 *Biochemical and Physiological Aspects of Human Nutrition* MH Stipanuk (Ed) W.B. Sanders New York p 782
6. RA Sunde, JK Evenson 2000 *Trace Elements in Man and Animals 10* AM Roussel, RA Anderson, AE Favier (Eds) Plenum Publishers New York p 21
7. SAB Knight, RA Sunde 1987 *J Nutr* 117:732
8. SAB Knight, RA Sunde 1988 *J Nutr* 118:853
9. I Chambers, J Frampton, PS Goldfarb, N Affara, W McBain, PR Harrison 1986 *EMBO J* 5:1221
10. MS Saedi, CG Smith, J Frampton, I Chambers, PR Harrison, RA Sunde 1988 *Biochem Biophys Res Commun* 153:855
11. RA Sunde, MS Saedi, SAB Knight, CG Smith, JK Evenson 1989 *Selenium in Biology and Medicine* A Wendel (Ed) Springer-Verlag Heidelberg, Germany p 8
12. RA Sunde 1990 *Annu Rev Nutr* 10:451
13. SL Weiss, JK Evenson, KM Thompson, RA Sunde 1997 *J Nutr Biochem* 8:85
14. XG Lei, JK Evenson, KM Thompson, RA Sunde 1995 *J Nutr* 125:1438
15. SL Weiss, JK Evenson, KM Thompson, RA Sunde 1996 *J Nutr* 126:2260
16. SL Weiss, RA Sunde 1997 *J Nutr* 127:1304
17. SL Weiss, RA Sunde 1998 *RNA* 4:816
18. F Ursini, M Maiorino, C Gregolin 1985 *Biochim Biophys Acta* 839:62
19. R Schuckelt, R Brigelius-Flohé, M Maiorino, A Roveri, J Reumkens, W Strassburger, L Flohé 1991 *Free Rad Res Commun* 14:343

20. RA Sunde, JA Dyer, TV Moran, JK Evenson, M Sugimoto 1993 *Biochem Biophys Res Commun* 193:905
21. G Bermano, F Nicol, JA Dyer, RA Sunde, GJ Beckett, JR Arthur, JE Hesketh 1995 *Biochem J* 311:425
22. RF Burk, KE Hill 1994 *J Nutr* 124:1891
23. KE Hill, PR Lyons, RF Burk 1992 *Biochem Biophys Res Commun* 185:260
24. RA Sunde, KM Thompson, JK Evenson, SL Weiss 1998 *Proc Nutr Soc* 57:155A
25. KB Hadley, KM Thompson, MT Rademaker, JK Evenson, RA Sunde 1999 *FASEB J* 13:A248
26. SC Low, MJ Berry 1996 *Trends Biochem Sci* 21:203
27. DL Hatfield, IS Choi, T Ohama, J-E Jung, AM Diamond 1994 *Selenium in Biology and Human Health* RF Burk (Ed) Springer-Verlag New York, Inc. New York p 25
28. PR Copeland, JE Fletcher, BA Carlson, DL Hatfield, DM Driscoll 2000 *EMBO J* 19:306
29. MJ Berry, JW Harney, T Ohama, DL Hatfield 1994 *Nucleic Acids Res* 22:3753
30. MJ Berry, L Banu, JW Harney, PR Larsen 1993 *EMBO J* 12:3315
31. MJ Berry, L Banu, Y Chen, SJ Mandel, JD Kieffer, JW Harney, PR Larsen 1991 *Nature (London)* 353:273
32. E Grundner-Culemann, GW Martin, III, JW Harney, MJ Berry 1999 *RNA* 5:625
33. GW Martin, III, JW Harney, MJ Berry 1996 *RNA* 2:171
34. KK McCaughan, CM Brown, ME Dalphin, MJ Berry, WP Tate 1995 *Proc Natl Acad Sci USA* 92:5431
35. MT Nasim, S Jaenecke, A Belduz, H Kollmus, L Flohé, JEG McCarthy 2000 *J Biol Chem* 275:14846
36. W Wen, SL Weiss, RA Sunde 1998 *J Biol Chem* 273:28533
37. A Jacobson, SW Peltz 1996 *Annu Rev Biochem* 65:693
38. E Nagy, LE Maquat 1998 *Trends Biochem Sci* 23:198
39. PM Moriarty, CC Reddy, LE Maquat 1998 *Mol Cell Biol* 18:2932
40. MJ Christensen, KW Burgener 1992 *J Nutr* 122:1620
41. H Toyoda, S Himeno, N Imura 1990 *Biochim Biophys Acta* 1049:213
42. M Sugimoto, RA Sunde 1992 *FASEB J* 6:A1366 (abs.)
43. WG Hoekstra 1975 *Fed Proc* 34:2083
44. A Spector, Y Yang, YS Ho, JL Magnenat, RR Wang, W Ma, WC Li 1996 *Exp Eye Res* 62:521
45. RA Sunde 1994 *Selenium in Biology and Human Health* RF Burk (Ed) Springer-Verlag New York, NY p 45
46. KM Thompson, H Haibach, RA Sunde 1995 *J Nutr* 125:864

# Part II

Selenium-containing proteins

# Chapter 9. Identity, evolution and function of selenoproteins and selenoprotein genes

Vadim N. Gladyshev

*Department of Biochemistry, University of Nebraska, Lincoln, NE 68588, USA*

**Summary:** UGA has a dual role in the genetic code serving as a signal for termination and a codon for selenocysteine (Sec). Sec appears to have been added to the already existing genetic code and its use has accumulated during evolution of eukaryotes culminating in vertebrates. Sec-containing proteins have diverse functions and lack a common amino acid motif or consensus sequence. Twenty two known eukaryotic selenoproteins may be subdivided into distinct selenoprotein groups on the basis of the location and functional properties of Sec. A set of criteria, designated Mammalian Selenoprotein Gene Signature (MSGS), allow recognition of selenoproteins through identification of SECIS elements and homology analyses of Sec-flanking areas. Identification of new selenoprotein sequences may lead to an understanding of many biological and health-related properties of selenium.

**Introduction**
The discovery of selenium as a component of glutathione peroxidase in 1973 [1] marked the first example of a natural selenium-containing protein. In subsequent years, the list of selenoproteins has been steadily growing and these proteins were identified in representatives of bacteria, archaea and eukaryotes. In each of these domains of life, the major biological form of selenium in proteins was shown to be selenocysteine (Sec). This amino acid was first recognized as an internal component of glycine reductase selenoprotein A [2] and later was shown to be encoded by TGA in this and other selenoproteins [3-5]. Sec is now known as the 21$^{st}$ amino acid in protein as it has its own codon and Sec-specific biosynthetic and insertion machinery [6]. A stem-loop structure, designated Sec insertion sequence (SECIS) element [7], is present in 3'-untranslated regions (UTRs) of eukaryotic selenoprotein genes (Figure 1), or in the coding regions immediately downstream of TGA codons in bacterial selenoprotein genes, to specify in-frame TGA as Sec.

## Prokaryotic Selenoproteins

### Bacterial selenoproteins

The current list of bacterial Sec-containing proteins (Table 1) includes formate dehydrogenases, hydrogenases, selenophosphate synthetases (SPS), selenoproteins A and B of glycine reductase, sarcosine reductase and betaine reductase complexes, proline reductase and peroxiredoxin. The majority of these proteins appear to be involved in catabolic or detoxification processes and Sec is coordinated to metals in several of these enzymes. One of the selenoproteins, Sec-containing SPS, participates in Sec biosynthesis and therefore may be considered as an autoregulatory enzyme.

*Formate dehydrogenase*
Sec-containing formate dehydrogenases are molybdopterin-dependent enzymes, in which Sec is coordinated, through its selenium atom, to a molybdenum atom [8]. Other ligands of molybdenum are sulfur atoms of two molybdopterin molecules and the sixth ligand is hydroxyl group or water that is replaced with a substrate in the course of catalytic reaction [9]. The role of the selenium ligand in formate dehydrogenases is to fine tune the molybdenum center and orchestrate the transfer of electrons and protons from formate to downstream acceptors. Replacement of Sec with Cys resulted in ~1000 fold decreased activity of the enzyme [10], and other Sec mutations completely inactivated this enzyme.

*Hydrogenase*
In selenium-containing hydrogenases, Sec is coordinated to a nickel atom in the enzyme active center. Like in formate dehydrogenases, Sec is involved in the transfer of electrons and protons and in tuning the metal site [11].

*Selenophosphate synthetase*
Selenophosphate synthetase catalyzes ATP-dependent synthesis of selenophosphate, which in turn serves a function of a selenium donor compound for biosynthesis of Sec [12]. Several bacterial selenophosphate synthetases are selenoproteins containing Sec in their N-terminal sequences [13].

*Selenoproteins A and B*
Selenoproteins A and B are components of glycine reductase, sarcosine reductase and betaine reductase complexes in *Eubacterium acidaminophilum* [14,15]. The three complexes contain the same selenoprotein A polypeptide and distinct substrate-specific selenoprotein B polypeptides (each complex has its own selenoprotein B).

*Other bacterial selenoproteins*
An unexpected recent addition to a list of bacterial selenoproteins is peroxiredoxin [15]. This antioxidant protein is a thiol-dependent peroxidase, a function that appears to be more consistent with eukaryotic selenoproteins. It will be interesting to learn if the use of Sec in a bacterial peroxiredoxin is a typical phenomenon or an exception.

Certain bacteria also contain selenoproteins in which selenium is present in the form of a dissociable cofactor that can be inserted into proteins posttranslationally. Examples of such proteins include nicotinic acid hydroxylase [16], xanthine dehydrogenase [17] and carbon monoxide dehydrogenase [18]. Selenium is coordinated to molybdenum in the active centers and is essential for catalytic activity of these enzymes.

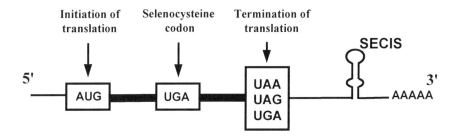

**Figure 1.** Structure of eukaryotic selenoprotein genes. Selenoprotein mRNAs contain the Sec-encoding UGA codon in the open reading frame and the SECIS element in the 3'-untranslated region. The SECIS element dictates recognition of in-frame UGA as a Sec codon rather than as a stop signal.

## Archaeal selenoproteins

Known Sec-containing proteins in archaea (Table 1) resemble bacterial selenoproteins and include formate dehydrogenase, hydrogenase, heterodisulfide reductase, formylmethanofuran dehydrogenase and selenophosphate synthetase. To date, only some of these selenoproteins were biochemically characterized.

In spite of a similarity between known bacterial and archaeal selenoproteins, the mechanism for Sec incorporation appears to be different in these organisms. In fact, Sec insertion in archaea resembles that of eukaryotes in that Sec is incorporated in response to SECIS elements located in 3'-UTRs of archaeal selenoprotein genes [19].

**Table 1.** Sec-containing proteins in prokaryotes.

**Bacteria**
Formate dehydrogenases
Hydrogenases
Selenoprotein A[a]
Selenoprotein B[a]
Proline reductase
Selenophosphate synthetase
Peroxiredoxin

**Archaea**
Formate dehydrogenases
Hydrogenases
Heterodisulfide reductase
Formylmethanofuran dehydrogenase
Selenophosphate synthetase

[a]Selenoprotein A (encoded by a single gene) and complex-specific selenoproteins B (encoded by different genes) are components of glycine, sarcosine and betaine reductase complexes.

## Eukaryotic selenoproteins

The largest number of selenoproteins is found in vertebrate genomes and the list of known eukaryotic selenoproteins is rapidly growing. In contrast to prokaryotes, eukaryotic selenoproteins of known function (e.g., thioredoxin reductases, glutathione peroxidases and deiodinases) participate in redox pathways linked to anabolic and regulatory metabolism. No sequence homology is seen when prokaryotic selenoproteins are compared to eukaryotic selenoproteins except for selenophosphate synthetase, which participates in Sec biosynthesis. Thus, the direct relation between prokaryotic and eukaryotic selenoproteins is limited and suggests independent origins for selenoproteomes (all selenoproteins in an organism) in these organisms.

Twenty two known eukaryotic selenoproteins (Table 2) are organized in 12 families on the basis of sequence homology (homologous selenoproteins have similar names, i.e., four glutathione peroxidases, three thioredoxin reductases, etc.). In more general terms, selenoproteins may be subdivided into groups based on the location of Sec in selenoprotein polypeptides. Two selenoprotein groups emerge from this analysis. The first (GPx) group is the most abundant and includes proteins in which Sec is located in the N-terminal portion of a relatively short (80-250 amino acid residues) functional domain. Secondary structure prediction reveals the presence of both $\alpha$-helices and $\beta$-sheets in proteins of this group. Many, if not all of these proteins, are characterized by the thioredoxin fold. In the proteins of the GPx group, Sec is either oxidized during catalysis to selenenic acid or forms predicted selenosulfide bonds. Several proteins of this group contain the CxxU or UxxC motifs that resemble active centers of thioredoxins and glutaredoxins. This selenoprotein group includes GPx isozymes, SelW, SelP, SelT, zebrafish SelW2, SelT2 and SelPb, and *Drosophila* BthD.

**Table 2.** Selenocysteine-containing proteins in eukaryotes.

| Selenoproteins[a] | Sec-flanking sequences[b] |
|---|---|
| **Group I (GPx group)** | |
| 1. Cytosolic glutathione peroxidase | ...VLLIENVASLUGTTVRDYTQM |
| 2. Gastrointestinal glutathione peroxidase | ...AVLIENVASLUGTTTRDFTQL... |
| 3. Plasma glutathione peroxidase | ...YVLFVNVASYUGLTGQYIELN... |
| 4. Phospholipid hydroperoxide glutathione peroxidase | ...VCIVTNVASQUGKTEVNYTQL... |
| 5. Selenoprotein P | ...VTVVALLQASUYLCIIEASKL... |
| 6. Selenoprotein Pb[a'] | ...VVVVALLKASUHFCLTQAARL... |
| 7. Selenoprotein W | ...LAVRVVYCGAUGYKSKYLQLK... |
| 8. Selenoprotein W2[a'] | ...VQIKVEYCGGUGYEPRYQELK... |
| 9. Selenoprotein T | ...PLLKFQICVSUGYRRVFEEYM... |
| 10. Selenoprotein T2[a'] | ...PVLKFQYCISUGYSKVFQEYS... |
| 11. BthD[a''] | ...PVLYVEHCRSURVFRRRAEEL... |
| **Group II (TR group)** | |
| 12. Cytosolic thioredoxin reductase (TR1) | ...SGASILQAGCUG |
| 13. Thioredoxin reductase expressed in testis (TR2) | ...SGLDITQKGCUG |
| 14. Mitochondrial thioredoxin reductase (TR3) | ...SGLDPTVTGCUG |
| 15. G-rich[a''] | ...MSCNMPAGGGUG |
| **Group III (other selenoproteins)** | |
| 16. Thyroid hormone deiodinase 1 | ...PLVLNFGSCTUPSFMFKFDQF... |
| 17. Thyroid hormone deiodinase 2 | ...PLVVNFGSATUPPFTSQLPAF... |
| 18. Thyroid hormone deiodinase 3 | ...PLVLNFGSCTUPPFMARMSAF... |
| 19. Selenoprotein R (Selenoprotein X) | ...DGPKPGQSRFUIFSSSLKFVP... |
| 20. Selenoprotein N | ...LWGALDDQSCUGSGRTLRETV... |
| 21. 15 kDa selenoprotein | ...YAGAILEVCGUKLGRFPQVQA... |
| 22. Selenophosphate synthetase 2 | ...RLTGFSGMKGUGCKVPQEALL... |

[a] All known eukaryotic selenoproteins are listed: [a'] Zebrafish selenoproteins - several zebrafish selenoproteins (GPx1, GPx4, SelT, SelW2) are represented by two genes that evolved by genome duplication and only one of these duplicated genes is shown in the table; and [a''] *Drosophila* selenoproteins. All other sequences represent human proteins.
[b] Sec-flanking sequences (10 amino acids upstream and 10 amino acids downstream of Sec) are shown for all known eukaryotic proteins. Selenocysteines (U) are highlighted and Cys and Gly residues in the vicinity of Sec are shown in bold.

The second group of eukaryotic selenoproteins is characterized by the presence of Sec in C-terminal sequences. In these proteins, the location of

Sec in conformationally flexible C-terminal sequences ensures accessibility of this amino acid residue. The current list of such proteins includes three mammalian thioredoxin reductases (TRs) and *Drosophila* G-rich protein. Other eukaryotic selenoproteins are currently placed in the third group that consists of three deiodinase isozymes, SelR, SelN, SPS2 and the 15 kDa selenoprotein. Four of these proteins contain a motif in which Sec is separated from Cys by a single amino acid residue. However, further studies are needed to subdivide these proteins into additional functional groups.

Comparison of Sec-flanking sequences for eukaryotic selenoproteins indicates the abundance of Gly, Cys and basic amino acids. In particular, the majority of selenoproteins contain Gly adjacent to Sec, but the basis for the preference of this Sec-flanking amino acid residue is not fully understood. Next, the known eukaryotic selenoproteins are briefly discussed.

**Glutathione peroxidases**

The four mammalian GPx isozymes catalyze glutathione dependent degradation of various hydroperoxides [20]. GPx1, a cytosolic enzyme, is the most abundant selenoprotein in mammals. It is expressed in every cell type and is thought to be one of the major antioxidant proteins in mammals. GPx2 is the closest homolog of GPx1, but its expression is limited to the gastrointestinal tract. GPx3 is a secreted glycoprotein and is the second most abundant selenoprotein in plasma (after SelP). The natural electron donor for GPx3 is not known since levels of reduced glutathione in plasma appear to be insufficient to contribute to hydroperoxide reduction by this enzyme. GPx1, GPx2 and GPx3 are homotetramers composed of ~22 kDa polypeptides. In contrast, GPx4 is ~20 kDa monomer that has been localized to both cytosol and mitochondria. GPx4 is also called phospholipid hydroperoxide glutathione peroxidase (PHGPx) as it can specifically reduce phospholipid hydroperoxides. In the testes, GPx4 is transformed from a soluble enzyme into the inactive structural component of a mitochondrial shell upon sperm maturation [21]. A fifth glutathione peroxidase, GPx5, has also been identified, but expression of this enzyme is restricted to the epididimys, and this protein contains Cys in place of Sec. The glutathione peroxidases are further discussed in Chapter 14.

**Thioredoxin reductases**

Mammalian thioredoxin reductases belong to a class of NADPH-dependent pyridine nucleotide disulfide oxidoreductases [22]. These proteins are homodiners of ~55 kDa subunits and each subunit also contains a single FAD molecule. In TRs, Sec is located in the C-terminal redox motif, Gly-Cys-Sec-Gly, which is essential for enzyme activity. It was proposed that this motif serves a function of a protein-fused glutathione analog [23]: during catalysis, the Sec redox center is reduced by the N-terminal dithiol active center of the enzyme, followed by conformational changes that expose the reduced C-terminal motif and allow its interaction with the active center of a substrate (e.g., thioredoxin).

TR1, the best-studied enzyme in the TR family, is a cytosolic enzyme and is also the most abundant TR isozyme in mammalian cells [24]. TR2, also called TGR, was characterized as a fusion of the C-terminal TR domain and the N-terminal glutaredoxin domain [23]. This fusion allows TR2 to acquire glutathione reduction function while maintaining specificity for thioredoxin reduction. This enzyme and its mRNA were only detected in testes and TR2 expression is puberty-dependent in this organ. TR3 is known as a mitochondrial enzyme [25], although alternative splicing forms of this protein have been detected that could potentially be targeted to other cellular compartments [26]. In the TR family, TR1 and TR2 are highly homologous, whereas TR3 is a more distantly related enzyme. The thioredoxin reductases are further discussed in Chapter 15.

### Thyroid hormone deiodinases
Deiodinases catalyze activation or inactivation (or both) of thyroid hormones (T3 and T4) [27]. Three isozymes are known, DI1, DI2 and DI3, which exhibit tissue-specific expression and appear to be dimers of 29-31 kDa subunits. Sec is located in the middle of deiodinase sequences and is essential for enzyme activity. The thyroid hormone deiodinases are further discussed in Chapter 16.

### Selenophosphate synthetases
Two selenophosphate synthetases are known, of which only SPS2 is a selenoenzyme [12,28]. In this protein, like in bacterial SPS, Sec is located in the N-terminal motif conserved in all known SPSs. SPS1 contains threonine in place of Sec (or arginine in the *Drosophila* sequence). The mechanism of action of SPS in bacteria is discussed in Chapter 4.

### Selenoprotein P
SelP is the only characterized mammalian selenoprotein that contains more than one Sec residue [29]. The SelP sequence is composed of several sequence blocks: the N-terminal signal peptide that is responsible for protein secretion, domain containing the UxxC motif in its N-terminal portion and the His-rich motif in its C-terminal portion, and a C-terminal Sec-rich sequence. Human, mouse and rat SelP contain 10, bovine 12, and zebrafish SelPa 17 Sec residues. Interestingly, a second zebrafish SelP was also found that contains only a single Sec. This protein lacks a C-terminal Sec-rich sequence suggesting that the function of the N-terminal domain may be separated from that of the C-terminal domain in SelP proteins containing multiple Sec residues [30]. SelP is a glycoprotein and is also the major

plasma selenoprotein accounting for 60% of plasma selenium. SelP is further discussed in Chapter 11.

**Selenoprotein W**
SelW is a small selenoprotein containing a conserved N-terminal CxxU redox motif. Its function is not known [31]. SelW is further discussed in Chapter 12.

**Selenoprotein T**
SelT, like SelW, is a protein of unknown function having a conserved N-terminal CxxU motif [32]. SelT, as well as SelR [32], SelN [33], G-rich and BthD [34], were discovered *in silico* using SECISearch and other computer programs.

**Selenoprotein R**
SelR (also called SelX) contains Sec in the C-terminal portion [32]. This protein has no homology to known proteins and its function is not known. However, phylogenetic profiles and domain fusion analyses implicate SelR in methionine sulfoxide reduction (unpublished data).

**15 kDa selenoprotein**
The 15 kDa selenoprotein has an N-terminal signal peptide that directs this protein to the endoplasmic reticulum. In the endoplasmic reticulum, it tightly binds UDP-glucose glycoprotein glucosyltransferase, a protein whose function is quality control of protein folding [35]. The specific role for the 15 kDa selenoprotein in this process is not known. The selenoprotein does not have sequence homology to known proteins. Two polymorphic sites were identified in 3'-UTR of the 15 kDa selenoprotein gene. One polymorphism is located in the SECIS element and was shown to influence, in a selenium-dependent manner, efficiency of Sec incorporation into the protein [36]. The protein was implicated in cancer prevention because of its differential expression in normal and tumor samples, the location of the 15 kDa selenoprotein gene in a region of chromosome 1 that is commonly deleted or mutated in human cancers and the presence of natural polymorphisms that regulate selenoprotein expression. The 15 kDa selenoprotein is further discussed in Chapter 13.

**Selenoprotein N**
Only a partial cDNA sequence for this protein is known [33]. SelN exhibits no homology to known proteins. Sec is present in the C-terminal half of the protein and Cys is adjacent to Sec.

## G-rich
This 12 kDa selenoprotein was found by a computational screen of the entire *Drosophila* genome [34]. Its function is not known. G-rich contains a C-terminal penultimate Sec flanked by Gly residues.

## BthD
The 28 kDa *Drosophila* BthD contains Sec in a CxxU motif in its N-terminal sequence [34]. Homology analyses indicate that the protein is composed on two domains. The N-terminal Sec-containing domain has homologs in other organisms, including *Drosophila*, whereas no homologs of the C-terminal domain could be detected by database searches.

## Evolution of Sec insertion
Although Sec-containing proteins have been identified in the three major domains of life (bacteria, archaea and eukaryotes), certain representatives of these organisms lack selenoprotein genes. Recent availability of complete sequences of many genomes allows assessment of selenoprotein genes encoded in these genomes. Interestingly, about half of the completely sequenced genomes in each of the three domains of life appears to lack selenoprotein genes. In these genomes, neither genes that are conserved among organisms that contain the Sec insertion system (e.g., SPS, Sec tRNA, etc.) nor known selenoproteins can be found.

Evolutionary implications of this observation are interesting and currently not fully understood. Several proposals were advanced to explain evolution of Sec in the genetic code. One proposal on how Sec originated, accumulated and was lost during evolution predicted that Sec was encoded by UGA in primitive anaerobic organisms and the use of Sec had been extensive [37,38]. It was further predicted that the appearance of oxygen in the atmosphere in the later stages of evolution selected against the use of Sec due to lability of this amino acid in the presence of oxygen. However, certain anaerobic and well-protected organisms conserved Sec-containing proteins and the Sec insertion system. The aerobic lability and the development of SECIS elements to dictate Sec insertion left TGA codons unprotected, resulting in the adaptation to use UGA as a termination signal. It was predicted that the initial gain of the Sec function may have been linked to the evolution of the TGN codons that inserted Cys in the primordial world. It was suggested that during development of the genetic code, the presence of a purine in the third position of the codon could indiscriminately code for Sec or Cys, while a pyrimidine in that position could conserve the Cys insertion function. It was also suggested that TGG was trapped by the newly evolved tryptophan and TGA evolved into a Sec codon.

An alternative hypothesis suggests that Sec evolved in later stages of evolution, after the other 20 amino acids with their initially specified codons evolved [39]. This proposal is consistent with the observation that many eukaryotic selenoproteins, such as glutathione peroxidases and thioredoxin reductases, are adapted to function in antioxidant systems. In addition, in contrast to the time when the first proposal was made, the number of known eukaryotic selenoproteins now exceeds that of prokaryotes.

It is now clear that Sec is strikingly different from the other 20 amino acids in its basic biosynthetic characteristics and its incorporation into protein. These differences have been described elsewhere in this book and can be summarized as follows: 1) the codon for Sec, UGA, serves a dual function in the genetic code which is the insertion of Sec into protein and the termination of protein synthesis (and the only other codon in the genetic code that serves a dual function is AUG which codes for the initiation of protein synthesis and the insertion of Met at internal positions of proteins); 2) a Sec insertion sequence (SECIS) element (Figure 2), which is a stem-loop structure located downstream of the Sec UGA codon, is the only known mRNA structure that specifies insertion of an amino acid and as such may be considered as a unique extension of the genetic code; 3) Sec is biosynthesized on its tRNA from serine which is attached to the tRNA by seryl-tRNA synthetase, an enzyme that most likely predated the use of Sec in protein as it must have evolved early for serine fidelity; 4) Sec tRNAs contain relatively few modified bases compared to other tRNAs; and 5) Sec has its own specific elongation factor in bacteria, archaea [see Chapter 2] and eukaryotes [see Chapter 7], while the other twenty amino acids share a common elongation factor. These properties argue that Sec biosynthesis and its insertion into selenopolypeptides are unique compared to the other amino acids and that Sec was likely added to the already existing 20-amino acid-genetic code.

Irrespective of when the incorporation of Sec into protein originated, this amino acid serves as an example of numerous unique modifications that emerged for its specific use within the universal genetic code. It will be interesting to see if other examples of such novel variations in the genetic code will be found.

**Comparative analysis of eukaryotic selenoproteomes**
The largest number of selenoproteins in any organism, in which these proteins have been characterized, was found in zebrafish. Although all identified zebrafish selenoproteins are homologous to mammalian selenoproteins, zebrafish accumulated multiple homologs of many mammalian selenoproteins through gene and genome duplications. For example, the zebrafish genome encodes at least three homologs of mammalian SelW and SelT, two homologs of SelP, GPx1 and GPx4 [30], and has two copies of the Sec tRNA gene [40]. In addition, zebrafish contains a selenoprotein (SelPa) with 17 Sec residues, the largest number of selenocysteines in any known protein.

The general tendency for the evolution of new selenoprotein genes from lower eukaryotes to vertebrates appears to be the increase in the number of selenoproteins with vertebrate selenoproteins currently accounting for all known eukaryotic selenoproteins. Several selenoproteins (e.g., deiodinases, selenoprotein P, selenoprotein N) are known that are exclusively found in vertebrates. However, perhaps certain eukaryotic organisms evolved unique selenoproteins that are either absent or replaced by non-selenoproteins in vertebrates. Further studies on eukaryotic genomic sequences should reveal whether such proteins exist. If found, the location of Sec in these proteins may indicate an important active or regulatory site in a protein sequence. This information may help in functional characterization of these proteins as well as homologous proteins, in which Sec is replaced with other amino acids.

An interesting feature for the majority of known selenoproteins is the occurrence of homologs that contain Cys in place of Sec. This tendency is especially pronounced for eukaryotes where lower eukaryotes often contain Cys homologs of vertebrate selenoproteins. A likely explanation for this phenomenon is that there is a great deal of similarity in chemical properties between Sec and Cys. These amino acids, as well as serine, differ by only a single atom. However, while Ser is completely protonated, and the majority of cysteines are also protonated at physiological pH, Sec, having a pKa of ~5.5, is ionized. This property along with a lower redox potential of Sec compared to Cys provide principal differences between these amino acids and account for the fact that Cys cannot functionally replace Sec in known selenoproteins. Indeed, comparison of kinetic parameters for several selenoproteins with their natural Cys-containing counterparts in other organisms revealed that selenoproteins are superior catalysts due to greatly increased $k_{cat}$ [6,12].

The Sec/Cys pairs in homologous sequences may be viewed as an evolutionary advancement of inserting Sec in place of Cys in positions that require redox function and that are dependent on strong reactivity and full ionization of an amino acid side chain. On the other hand, it appears that incorporation of Sec is metabolically expensive due to the scarcity of selenium in the environment and the necessity to maintain a multi-component Sec incorporation system for insertion of just a few amino acids. This suggests that in addition to an evolutionary advantage of utilizing Sec in protein, organisms may also be under selective pressure to inactivate the Sec insertion system. This paradigm may help to explain the unusual distribution of selenoproteins in living organisms, i.e., the presence of selenoprotein genes in all domains of life, but also their absence in certain representatives of these domains.

**Mammalian selenoprotein gene signature**
Observations of Sec/Cys pairs in homologous sequences as well as obligate requirement for the presence of TGA codons that encode Sec (Figure 1) and SECIS elements (Figure 2) that dictate Sec insertion in selenoprotein genes may be formally defined as a set of criteria that distinguish selenoprotein genes from other genes and may also help to identify new selenoproteins. We designated these criteria as Mammalian Selenoprotein Gene Signature (MSGS). According to MSGS, a protein gene contains a Sec codon if the following are satisfied:

1) *Sec is conserved and is flanked by homologous sequences*, i.e., conservation of Sec-encoding TGA and of sufficiently long Sec-flanking amino acid sequences within the gene for this protein between different mammals.

2) *The SECIS element is conserved*, i.e., conservation of the SECIS element sequence and structure in the 3'-UTRs of mammalian genes for this protein.

3) *There are distinct Cys- and/or Sec-containing homologs*, i.e., occurrence of genes (usually in lower eukaryotes) that contain a Cys codon in place of TGA in the gene for this protein (or occurrence of distinct homologous genes that conserve TGA) and meet an additional criterion of encoding sufficiently long homologous sequences on both sides of the Cys/Sec (or Sec/Sec) pair.

Application of MSGS may allow identification of new selenoproteins and testing whether a predicted gene encodes a selenoprotein. According to MSGS, a new mammalian selenoprotein gene should contain a conserved SECIS element in the 3'-UTR of its mRNA and have a conserved Sec (encoded by TGA) and Sec-flanking regions in the protein sequence in other mammals. In addition, a new protein will also have eukaryotic homologs that contain a Cys residue in place of Sec, or alternatively, distinct homologs that conserve Sec will be present in a eukaryotic genome.

Although extensive sequence information is now available for genomes of a great variety of organisms, including humans, identification of selenoproteins has been difficult. The lack of a common amino acid motif in selenoprotein sequences and in particular in Sec-flanking regions, and the low conservation of a SECIS element nucleotide sequence, do not permit a straightforward identification of selenoprotein genes in nucleotide sequence databases. Typically, the finding of a new selenoprotein gene required experimental verification of Sec incorporation into a protein, using amino acid analysis of an isolated protein or metabolic $^{75}$Se-labeling of cells expressing the protein. These techniques are time consuming and may require significant amounts of isolated proteins. There are several examples (when these analyses were not performed) of misinterpretation of TGA codons as Sec codons instead of stop signals [41] and stop signals instead of Sec codons [13,42,43].

The MSGS criteria are specific enough to identify all known selenoproteins in nucleotide sequence databases. Moreover, searches of non-redundant databases of Genbank did not reveal any exceptions, i.e., proteins that satisfy all MSGS criteria but do not contain Sec. While relying on a formal MSGS definition of a selenoprotein gene, one may potentially overlook putative atypical selenoproteins (those that have unusual SECIS elements or do not have Cys- or Sec-containing homologs), all currently known selenoproteins can be specifically described by MSGS.

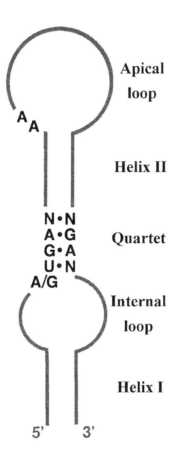

**Figure 2.** Eukaryotic SECIS element. Structural features of the stem-loop structure are indicated on the right. Conserved nucleotides in the Quartet, a purine preceding the Quartet and the AA motif in the apical loop are indicated. See text for further discussion.

The MSGS criteria were initially formulated for mammalian sequences. However, they appear to be sufficient to recognize selenoproteins in all eukaryotes. Indeed, selenoproteins that satisfy the MSGS criteria and are homologs of known selenoproteins could be found in the trematode, *Schistosoma mansoni*, the fly, *Drosophila melanogaster*, the nematode, *Caenorhabditis elegans*, the slime mold, *Dictyostelium discoideum*, the parazitic protozoa, *Leishmania major*, the ascidian, *Halocynthia roretzi* and many other eukaryotic organisms (Kryukov and Gladyshev, unpublished).

**SECISearch**
Since the only common feature (besides Sec-encoding UGA codons) in eukaryotic selenoprotein genes is the presence of SECIS elements in 3'-UTR regions (Figure 2), a practical approach to identify selenoprotein genes is to identify SECIS elements. However, low sequence conservation in these stem-loop structures makes this approach challenging. Nevertheless, two groups recently applied this strategy to identify new selenoproteins in mammalian EST databases [32,33]. This was possible because SECIS elements, while having low primary sequence conservation, strictly conserve their secondary structure and the free energy characteristics. Accurate computational description of SECIS elements allowed selecting true SECIS elements among millions of candidate sequences [32]. Moreover, this approach is sufficient to analyze entire eukaryotic genomes. A recent analysis of the *Drosophila* genome using SECISearch, a program that is capable of recognition of the absolute majority of known selenoproteins, revealed only three selenoproteins (SPS2, G-rich and BthD) in fruit flies. The presence of such a small number of selenoprotein genes in a genome provides a new powerful tool to characterize selenoprotein functions that are dependent on dietary selenium. Indeed, recent analysis demonstrated that selenium deficiency shortens, while dietary supplementation normalizes the life span of fruit flies in a process that likely involves one or more of the three identified selenoproteins [34].

An alternative strategy to search for selenoprotein genes may include homology analyses of Sec-flanking sequences. The idea behind this approach is that if UGA encodes Sec in a protein, downstream sequences would be homologous in its orthologs, whereas if UGA is a stop signal, such homology would be lacking. This strategy has not yet been used in database searches. However, it is used as a component of MSGS criteria.

The use of described bioinformatics tools (SECISearch, MSGS criteria) and the development of new computational approaches will likely lead to identification of the majority of selenoproteins in mammalian genomes, including humans, in the near future. The information on human selenoprotein sequences may provide avenues for new exciting research to link functions of all selenoproteins to biological and health-related properties of dietary selenium.

**Aknowledgements:** Supported by NIH grant GM61603.

## References

1. JT Rotruck, AL Pope, HE Ganther, AB Swanson, DG Hafeman, WG Hoekstra 1973 *Science* 179:588
2. JE Cone, RM Del Rio, JN Davis, TC Stadtman 1976 *Proc Natl Acad Sci USA* 73:2659
3. I Chambers, J Frampton, P Goldfarb, N Affara, W McBain, PR Harrison 1986 *EMBO J* 5:1221
4. F Zinoni, A Birkmann, TC Stadtman, A Bock 1986 Proc Natl Acad Sci USA 83:4650
5. GE Garcia, TC Stadtman 1991 *J Bacteriol* 173:2093
6. DL Hatfield, VN Gladyshev, J Park, SI Park, HS Chittum, HJ Baek, BA Carlson, ES Yang, ME Moustafa, BJ Lee 1999 *Comp Nat Prod Chem* 4:353
7. MJ Berry, L Banu, YY Chen, SJ Mandel, JD Kieffer, JW Harney, PR Larsen 1991 *Nature* 353:273
8. VN Gladyshev, SV Khangulov, MJ Axley, TC Stadtman 1994 *Proc Natl Acad Sci USA* 91:7708
9. JC Boyington, VN Gladyshev, SV Khangulov, TC Stadtman, PD Sun 1997 *Science* 275:1305
10. MJ Axley, A Bock, TC Stadtman 1991 *Proc Natl Acad Sci USA* 88, 8450
11. E Garcin, X Vernede, EC Hatchikian, A Volbeda, M Frey, JC Fontecilla-Camps 1999 *Structure Fold Des* 7:557
12. TC Stadtman 1996 *Annu Rev Biochem* 65:83
13. MJ Guimaraes, D Peterson, A Vicari, BG Cocks, NG Copeland, DJ Gilbert, NA Jenkins, DA Ferrick, RA Kastelein, JF Bazan, A Zlotnik 1996 *Proc Natl Acad Sci USA* 93:15086
14. M Wagner, D Sonntag, R Grimm, A Pich, C Eckerskorn, B Sohling, JR Andreesen 1999 *Eur J Biochem* 260:38
15. JR Andreesen, M Wagner, D Sonntag, M Kohlstock, C Harms, T Gursinsky, J Jager, T Parther, U Kabisch, A Grantzdorffer, A Pich, B Sohling 1999 *Biofactors* 10:263
16. VN Gladyshev, SV Khangulov, TC Stadtman 1996 *Biochemistry* 35:212
17. T Schrader, A Rienhofer, JR Andreesen 1999 *Eur J Biochem* 264:862
18. O Meyer, L Gremer, R Ferner, M Ferner, H Dobbek, M Gnida, W Meyer-Klaucke, R Huber 2000 *Biol Chem* 381:865
19. R Wilting, S Schorling, BC Persson, A Bock 1997 *J Mol Biol* 266:637
20. R Brigelius-Flohe 1999 Free Radic Biol Med 27:951
21. F Ursini, S Heim, M Kiess, M Maiorino, A Roveri, J Wissing, L Flohe 1999 *Science* 285:1393
22. ES Arner, A Holmgren 2000 *Eur J Biochem* 267:6102
23. QA Sun, L Kirnarsky, S Sherman, VN Gladyshev, submitted.
24. B Rozell, HA Hansson, M Luthman, A Holmgren 1985 *Eur J Cell Biol* 38:79
25. SR Lee, JR Kim, KS Kwon, HW Yoon, RL Levine, A Ginsburg, SG Rhee 1999 *J Biol Chem* 274:4722
26. QA Sun, F Zappacosta, VM Factor, PJ Wirth, DL Hatfield, VN Gladyshev 2000 *J Biol Chem*, in press
27. J Kohrle 1999 *Biochimie* 81:527
28. GM Lacourciere 1999 *Biofactors* 10:237
29. RF Burk RF, KE Hill 1999 *Bioessays* 21:231
30. GV Kryukov, VN Gladyshev 2000 *Genes Cells* 5:1049

31. SC Vendeland, MA Beilstein, JY Yeh, W Ream, PD Whanger 1995 *Proc Natl Acad Sci USA* 92:8749
32. GV Kryukov, VM Kryukov, VN Gladyshev 1999 *J Biol Chem* 274:33888
33. A Lescure, D Gautheret, P Carbon, A Krol 1999 *J Biol Chem* 274:38147
34. FJ Martin-Romero, GV Kryukov, AV Lobanov, BJ Lee, VN Gladyshev, DL Hatfield, submitted
35. KV Korotkov, E Kumaraswamy, J Zhou, DL Hatfield, VN Gladyshev, submitted
36. E Kumaraswamy, A Malykh, KV Korotkov, S Kozyavkin, Y Hu, SY Kwon, ME Moustafa, BA Carlson, MJ Berry, BJ Lee, DL Hatfield, AM Diamond, VN Gladyshev 2000 *J Biol Chem* 275:35540
37. A Böck, K Forchhammer, J Heider, W Leinfelder, G Sawers, B Veprek, F Zinoni 1991 *Mol Microbiol* 5:515
38. JF Atkins, A Böck, S Matsufuji, RF Gesteland. In: *The RNA World*, ed., RF Gesteland, TR Cech and JF Atkins, Cold Spring Harbor Laboratory Press, Cold Spring Harbor, New York, 1999 pp 637
39. VN Gladyshev, GV Kryukov *BioFactors*, in press
40. XM Xu, X Zhou, BA Carlson, LK Kim, TL Huh, BJ Lee, DL Hatfield 1999 *FEBS Lett* 454:16
41. L Cataldo, K Baig, R Oko, MA Mastrangelo, KC Kleene 1996 *Mol Reprod Dev* 45:320
42. PY Gasdaska, JR Gasdaska, S Cochran, G Powis 1995 *FEBS Lett* 373:5
43. VN Gladyshev, KT Jeang, TC Stadtman 1996 *Proc Natl Acad Sci USA* 93:6146

# Chapter 10. Bacterial selenoenzymes and mechanisms of action

Thressa C. Stadtman

*National Heart, Lung, and Blood Institute, National Institutes of Health, Bethesda, MD 20892, USA*

**Summary:** There are two types of bacterial selenoenzymes known at present: those that contain selenium in the form of selenocysteine residues in the polypeptide chains and those that instead contain a catalytically essential selenium in a dissociable cofactor form. Among the latter are the molybdopterin containing enzymes, nicotinic acid hydroxylase, xanthine dehydrogenase and the recently discovered purine hydroxylase. The crystal structure of a related enzyme, carbon monooxide dehydrogenase, has been determined. This enzyme, unlike the other three, is present in various aerobic bacteria. *Escherichia coli* formate dehydrogenase is a selenium-dependent enzyme that contains selenocysteine, molybdenum, a molybdopterin-guanine dinucleotide and an iron-sulfur center. The crystal structures of the oxidized and reduced enzymes have been reported. This enzyme differs from classical molybdenum hydroxylases in that oxygen from water is not incorporated in the product. Clostridial glycine reductase is a selenocysteine-dependent enzyme that reduces glycine to acetylphosphate and ammonia. The reaction mechanism of this enzyme involves the cleavage of a carbon-nitrogen bond of the Schiff base derivative of the glycine substrate and represents a new role of the ionized selenol group of the selenocysteine residue in metabolism. Selenium-carboxymethylselenocysteine is formed as the enzyme-bound intermediate on the selenoprotein A subunit of the enzyme complex in this reaction. Reductive cleavage of the selenoether, transfer to a cysteine residue of another subunit forming an acetylthioester and reaction with phosphate results in the formation of acetylphosphate.

## Introduction

Selenium that occurs in most of the known selenoproteins and selenoenzymes of bacterial and animal origin is present as selenocysteine residues that correspond in position to TGA codons in the respective genes. This UGA-directed highly specific incorporation of selenocysteine in proteins is to be contrasted to the occasional non-specific substitution of selenomethionine and selenocysteine for methionine and cysteine, respectively, that occurs in many biological systems. The amounts of these selenoamino acids that are incorporated non-specifically generally reflect the amount of selenium in the diet or in the culture medium and can be a function of the amount that enters sulfur metabolic pathways. In higher plants and in *Saccharomyces cerevisiae*, for example, the non-specific occurrence of selenium in proteins appears to predominate and few, if any, specific selenoenzymes have been identified.

In 1957, a report by Pinsent [1] that formate dehydrogenase activity in *Escherichia coli* depended on selenium and molybdenum in the growth medium laid the groundwork for recognition of the importance of selenium in prokaryotic metabolism. However, more than fifteen years passed before selenium was shown to be an essential constituent of an enzyme, in this case clostridial glycine reductase [2]. There was a lapse of three more years until the selenium moiety was identified as a selenocysteine residue in the selenoprotein A component of the enzyme complex [3]. Using methodology developed in these studies, selenocysteine was shown to be present in the mammalian enzyme glutathione peroxidase. In addition to the glycine reductase complex, several other selenocysteine-containing enzymes now are known to be present

**Table 1.** Selenium dependent enzymes in bacteria

**A. Selenium occurs in a selenocysteine residue**

Clostridial glycine reductase

Certain formate dehydrogenases

Some hydrogenases of methanogens

Formylmethanofuran dehydrogenase

Selenophosphate synthetase of

*Methanococcus jannashii* and *Haemophilus influenzae*

Clostridial proline reductase

**B. Selenium occurs in a readily dissociable form**

Nicotinic acid hydroxylase

Xanthine dehydrogenase in clostridia

Carbon monoxide dehydrogenase in various aerobic bacteria

Purine hydroxylase in clostridia

in prokaryotes (Table 1). Also, there is a growing list of enzymes that lack selenocysteine and instead contain selenium in the form of a dissociable cofactor (Table 1). Initial clues that selenium in enzymes such as nicotinic acid hydroxylase [4] and xanthine dehydrogenase might have a role in catalysis came from observations that activities of these enzymes in various anaerobic bacteria were considerably elevated when growth media were supplemented with selenium [5,6]. Evidence that the selenium in nicotinic acid hydroxylase was present in a form that was dissociated by heat, by treatment with chaotropic agents and by alkylation under conditions that clearly distinguished it from selenocysteine came from studies on purified enzyme preparations carried out by Dilworth [7,8]. Treatment of the enzyme during turnover conditions with a variety of alkylating agents caused immediate inactivation and this was accompanied by quantitative liberation of bound $^{75}$Se-labeled selenium as the corresponding dialkyl selenide. These observations, together with the absence of a $^{75}$Se-labeled selenocysteine residue in the protein clearly showed the occurrence of a different type of selenium cofactor. Quantitation of molybdenum, FAD, a "pterin-like" cofactor together with iron and acid labile sulfide present in purified enzyme preparations confirmed the similarity of nicotinic acid hydroxylase to xanthine oxidase and other known molybdenum hydroxylases [9]. Although there was considerable variation in selenium content, as determined by atomic absorption of different enzyme preparations, correspondence of hydroxylase activity and the amount of selenium present indicated the importance of the selenium cofactor.

**Properties of selenium-dependent hydroxylases**
The enzymes in this group contain molybdenum (or in some cases tungsten) which is present in a protein-bound molybdopterin type of cofactor. In at least one of the enzymes, nicotinic acid hydroxylase of *Clostridium barkeri*, the selenium of the enzyme bound dissociable cofactor is coordinated to the molybdenum [10] and this selenium is essential for oxidation of nicotinic acid by the enzyme [11]. In several members of the family of molybdopterin hydroxylases, in particular the selenium-containing forms listed in Table 1, molybdenum (or in some cases tungsten) is coordinated to a dinucleotide derivative of the pterin [12]. This may be a dinucleotide of cytosine, guanine, adenine or hypoxanthine. A molybdopterin-cytosine dinucleotide, tentatively identified in nicotinic acid hydroxylase, has been identified in carbon monoxide dehydrogenase, an enzyme present in several aerobic bacteria [13,14]. The crystal structure of this enzyme has been determined [13,14]. To date no eukaryotic molybdohydroxylases have been shown to contain selenium.

**Carbon monoxide dehydrogenase**
The reaction catalyzed by CO dehydrogenase is $CO + H_2O = CO_2 + 2 e + 2 H^+$. The crystal structure of CO dehydrogenase from the aerobic organism, *Oligotropha carboxydovorans*, was determined at 2.2 Å [13]. The enzyme is

a dimer of heterotrimers with two separate catalytic domains, one on each of the trimers. A large 88.7 kDa subunit contained one molybdopterin-cytosine dinucleotide with 1 Mo per subunit, a medium 30.2 kDa flavoprotein subunit contained 1 FAD, and a small 17.8 kDa iron sulfur protein contained two [2Fe-2S] centers. The crystal structure of CO dehydrogenase from *Hydrogenophaga pseudoflava*, determined at 2.25 A also has been reported [14]. Some molecules of both enzymes contained selenium bonded to the sulfur of a cysteine residue in the large subunit, Cys-388 of *O. carboxydovorans* or Cys-385 of *H. pseudoflava*. This was modeled as a Cys-S-Se⁻ and termed S-selanylcysteine. In *H. pseudoflava* a modified arginine residue, γ-hydroxy-Arg-384, of unknown function preceded Cys-385. Although CO dehydrogenase is termed a selenium-dependent enzyme and isolated enzyme preparations are reported to contain 0.2 - 0.5 equivalent of selenium per mole, earlier experiments [15] showed that activation of the enzyme was achieved to the same extent by aerobic treatment with selenite or by anaerobic treatment with sulfide plus dithionite. In the recent review by Meyer et al. [16], results of EXAFS studies and additional information on the groups bonded to Mo in active forms of enzyme and enzyme inactivated by cyanide treatment together with the crystallographic data are reevaluated. A possible model of the selenium adduct on the enzyme is considered to be a bridging selenotrisulfide attached to the Mo [Mo = S-Se-S-Cys-388] in which the sulfur attached to Mo is the cyanolyzable sulfido group. The role of a selenium-atom in catalysis therefore needs to be clarified to support the claim of specificity of activation. In fact, strictly aerobic bacteria are not generally known to contain selenium-specific moieties. In some forms of CO dehydrogenase that appear to be fully active, the selenium content is always less than stoichiometric and in the enzyme from the thermophilic bacterium *S. thermoautotrophicus* that exhibits a high specific catalytic activity, there is a complete absence of selenium. Thus, the authors state that the function of selenium in catalysis remains to be elucidated for CO dehydrogenase. For detailed information as to atomic distances and structural changes at the Mo-site of various forms of the enzyme, the reader is referred to the Meyer et al. review in *Biological Chemistry* [16].

**Clostridial xanthine dehydrogenase and purine hydroxylase**
*Clostridium purinolyticum* [17] served as the source of a new molybdopterin-hydroxylase, purine hydroxylase [18] that utilizes purine, 2-hydroxy purine and hypoxanthine (2,6- dihydroxypurine) as substrates. The latter is the intermediate involved in conversion of purine to xanthine. A separate enzyme, xanthine dehydrogenase, was detected in side fractions during purification of purine hydroxylase and both enzymes were obtained in an essentially homogeneous form [18]. Purine hydroxylase, a protein of about 590 kDa, contains four dissimilar subunits and the native protein appears to be a $\alpha_4\beta_4\gamma_4\delta_4$

structure. One large subunit of 54 kDa, two medium 42 and 34 kDa subunits and one small 20 kDa subunit were detected. Xanthine dehydrogenase, by contrast, consists of three different subunits, a large 80 kDa, a medium 35 kDa and a small 16 kDa. The xanthine dehydrogenase from *Eubacterium barkeri* was reported [19] to contain subunits of similar sizes. The amino terminal sequences of all of the subunits of these enzymes from *C. purinolyticum* were reported [18]. When purine hydroxylase, labeled with $^{75}$Se, was treated with cyanide, the radioactivity was liberated and enzyme activity was lost. In this respect, the dependence of activity on the presence of selenium resembles the nicotinic acid hydroxylase that contains an essential selenium coordinated to the molybdenum atom. Regarding the various enzymes that contain an essential selenium in the putative form of R-S-Se-, there are a number of interesting questions that remain to be answered. What is the origin of the selenium atom under physiological conditions, at what stage of formation of the enzyme is the selenium atom introduced and is the source a selenium-delivery protein such as is the case for selenophosphate synthesis by selenophosphate synthetase?

## Selenium-dependent formate dehydrogenases

*Escherichia coli* formate dehydrogenase H (FDH$_H$) is an example of a molybdopterin-selenoprotein that contains selenium in the form of a selenocysteine residue within the polypeptide chain instead of a dissociable selenium cofactor [20,21]. Moreover, this enzyme is not a classical Mo-hydroxylase in that it does not catalyze incorporation of oxygen from water into product. Instead FDH$_H$ oxidizes formate directly to carbon dioxide and both oxygen atoms in the product are derived from the substrate [22]. Only after equilibration with solvent to form bicarbonate is there appearance of oxygen from water in the formate oxidation product. FDH$_H$ is the terminal member of the respiratory pathway when *E. coli* is grown anaerobically on glucose. A selenocysteine residue (Cys-140) is essential for full activity of the enzyme. When this is replaced by cysteine, enzyme activity is only about 0.3% that of wild type enzyme and a selenocysteine/serine mutant is inactive catalytically [23]. Wild type enzyme after reduction with formate is very oxygen labile suggesting sensitivity of the Mo(IV) and or Mo(V) reduced states. Inactivation of the enzyme by reaction with alkylation reagents requires prior reduction of enzyme by formate indicating that only in the reduced form is the selenium available as the ionized selenol of the selenocysteine residue.

A series of *E. coli* mutants that were unable to synthesize FDH$_F$ and were isolated by Marie-Andree Mandrand-Berthelot [24], proved to be invaluable in elucidating the pathway of specific selenium-incorporation in selenocysteine in selenoenzymes. Thus, a gene that encoded an 80 kDa protein, formate dehydrogenase, surprisingly contained a TGA codon in the open reading frame [20]. Experiments showed that readthrough of this codon required selenium for translation of the complete 80kDa protein [25], and later the peptide sequence analysis confirmed that the selenocysteine in the protein at position 140 corresponded to the UGA codon in the message [21]. The elegant studies of August Bock and his associates have unraveled the sequence of events in *E. coli* and related species that are required for the complicated steps in this overall process [see Chapter 2].

Crystallographic analysis of *E. coli* FDH$_H$ [26] confirmed that the selenium of the selenocysteine residue is coordinated to the Mo atom and the latter is bound to two molybdopterin-guanine dinucleotides present on the protein. Reduction of the enzyme by formate converts Mo(VI) to Mo(IV) and electrons are transferred to the single FeS center on the enzyme. The suggestion that protonation of the selenoenzyme residue is involved in the reaction is countered by results of EPR studies [22] suggesting that a His residue is the proton acceptor and by EXAFS results [27] suggesting a selenosulfide involving one of the thiolene sulfurs bonded to the Mo atom accepts the liberated proton. Clearly additional studies are required to clarify this issue.

**Clostridial glycine reductase**

The reductive deamination of glycine, eventually to acetate and ammonia, is catalyzed by several anaerobic bacteria through the actions of glycine reductase and acetate kinase. This energy conserving process involves the synthesis of an enzyme-bound acetyl thiolester that is cleaved by phosphate to give acetyl phosphate. In a final reaction step acetylphosphate and ADP are converted to ATP by the widely distributed enzyme, acetate kinase. Glycine reductase from *Clostridium sticklandii* was one of the first selenium-dependent enzymes to be identified [2] and the selenium moiety in this enzyme was the first to be identified as selenocysteine [3].

The role of the ionized selenol group in this enzyme appears to facilitate cleavage of a carbon-nitrogen bond in an initial Schiff base intermediate formed from the carbonyl group of a pyruvate and the substrate glycine [28, 29]. Displacement of the nitrogen of glycine by the ionized selenol results in transfer of the carboxymethyl group to selenium forming a selenoether derivative of the selenocysteine residue in the selenoprotein A subunit of glycine reductase. Reductive cleavage of the selenoether followed by transfer to a cysteine residue of the protein C component of glycine reductase gives rise to a Cys-S-acetylthiolester which upon reaction with phosphate is converted to acetyl phosphate [30, 31]. The gene sequences for the selenoprotein A components of glycine reductase from *Clostridium purinolyticum* and *C. sticklandii* contained TGA codons corresponding to selenocysteine in the proteins [32,33]. Although the full-length gene product was expressed in *E. coli*, the SECIS sequence of the clostridial enzyme was not recognized and the resulting protein was active only to the extent of non-specific incorporation of selenocysteine via mischarged Cys-tRNA. In the absence of selenium in the medium a full length protein product that was completely inactive catalytically was produced. Cloning and sequencing of the selenoprotein A component and also partial sequencing of the other protein components of glycine reductase from related clostridial species has been reported by Andreesen and associates [34,35].

The presence of an additional TGA codon in the gene encoding the protein B component that forms the Schiff base derivative with substrate suggests the presence of another selenocysteine residue, but there seems to be no information concerning its function in the reaction. In view of the fact that there is preliminary evidence that not all selenocysteine residues in proteins are necessarily involved directly in catalysis, it is premature to speculate as to the role of selenocysteine in the protein B component. For further details on glycine reductase, the reader is referred to various current reviews on the subject of selenoenzymes.

**References**

1. J Pinsent 1954 *Biochem J* 57:10
2. DC Turner, TC Stadtman 1973 *Archiv Biochem Biophys* 154:366
3. JE Cone, R Martin del Rio, JN Davis, TC Stadtman 1976 *Proc Nat Acad Sci USA* 73:2659
4. D Imhoff, JR Andreesen 1979 *FEMS Microbiol* 5:155
5. R Wagner, JR Andreesen 1979 *Arch Microbiol* 121:255
6. GL Dilworth 1982 *Arch Biochem Biophys* 219:30
7. GL Dilworth 1983 *Arch Biochem* 221:565
8. JS Holcenberg, ER Stadtman 1969 *J Biol Chem* 244:1194
9. R Hille 1996 *Chem Reviews* 96:2757
10. VN Gladyshev, SV Khangulov, TC Stadtman 1994 *Proc Nat Acad Sci USA* 91:232
11. VN Gladyshev, SV Khangulov, TC Stadtman 1996 *Biochemistry* 35:212
12. C Kisker, H Schindelin, DC Rees 1997 *Ann Rev Biochem* 66:233
13. H Dobbek, L Gremer, O Meyer, R Huber 1999 *Proc Nat Acad Sci USA* 96:8884
14. P Hanzelmann H Dobbek, L Gremer, R Huber, O Meyer 2000 *J Mol Biol* 301:1221
15. Meyer, KV Rajagopalan 1984 *J Biol Chem* 259:5612
16. O Meyer, L Gremer, R Ferner, M Ferner H Dobbek M Gnida, W. Meyer-Klaucke, R Huber 2000 *Biol Chem* 381:865
17. P Durre, W Andersch, JR Andreesen 1981 *Int J Syst Bacteriol* 31:184
18. WT Self, TC Stadtman 2000 *Proc Nat Acad Sci USA* 97:208
19. T Schrader, A Reinhofer, JR Andreesen 1999 *Eur J Biochem* 264:8652
20. F Zinoni, A Birkmann, TC Stadtman, A Bock 1986 *Proc Nat Acad Sci USA* 83:4630
21. TC Stadtman, JN Davis, W-M Ching, F Zinoni A Bock 1991 *BioFactors* 3:21
22. SV Khangulov, VN Gladyshev, GC Dismukes, TC Stadtman 1998 *Biochemistry* 37:3518
23. MJ Axley, A Bock, TC Stadtman 1991 *Proc Nat Acad Sci USA* 88:8450
24. BA Haddock, M-A Mandrand-Berthelot 1982 *Biochem Soc Trans* 10:478
25. F Zinoni, A Birkmann, W Leinfelder, A Bock 1987 *Proc Nat Acad Sci USA* 84:3156
26. JC Boyington, VN Gladyshev, SV Khangulov, TC Stadtman, PD Sun 1997 *Science* 275:1305

27. GN George, CM Colangelo, J Doug, RA Scott, SV Khangulov, VN Gladyshev, TC Stadtman 1998 *J Am Chem Soc* 120:1267
27. H Tanaka, TC Stadtman 1979 *J Biol Chem* 254:447
28. RA Arkowitz, RH Abeles 1990 *J Am Chem Soc* 112:870
29. RA Arkowitz, RH Abeles 1989 *Biochemistry* 28:4639
30. TC Stadtman 1989 *Proc Nat Acad Sci USA* 86:7853
31. GE Garcia, TC Stadtman 1991 *J Bacteriology* 173:2093
32. GE Garcia, TC Stadtman 1992 *J Bacteriology* 174:7080
33. M Lubbers, JR Andreesen 1993 *Eur J Biochem* 217:791
34. S Kreimer, JR Andreesen 1995 *Eur J Biochem* 234:192

# Chapter 11. Selenoprotein P

Kristina E. Hill and Raymond F. Burk

*Department of Medicine, Vanderbilt University School of Medicine, Nashville, TN 37232, USA*

**Summary:** Selenoprotein P is the major plasma selenoprotein. It is a heparin binding protein and associates with endothelial cells throughout the body. Selenoprotein P contains multiple selenocysteine residues and is glycosylated. It is present in isoforms, at least one of which appears to be generated by termination at a UGA codon in the open reading frame. Evidence that selenoprotein P defends against oxidative injury and that it transports selenium to the brain has been presented. Selenium deficiency causes plasma levels of selenoprotein P to decline in rats and in humans.

## Introduction
In the early 1970s two groups reported that a plasma protein rapidly incorporated $^{75}$Se following $^{75}$Se-selenite administration to the rat [1,2]. Later, after glutathione peroxidase had been demonstrated to be a selenoprotein, Herrman showed that the rapidly labeling plasma protein was distinct from glutathione peroxidase [3]. In 1982, two groups characterized the protein further, confirming that it was not glutathione peroxidase [4,5] and showing that it incorporated selenium more efficiently than did glutathione peroxidase [4]. Once its selenium had been demonstrated to be selenocysteine, the protein was accepted as the second selenoprotein to be identified in animals and the name selenoprotein P was given to it because of its plasma location [5].

## Purification and characterization in plasma
In spite of the fact that selenoprotein P is a plasma protein, its purification was difficult to achieve. Efforts using conventional chromatographic techniques only produced preparations of rat selenoprotein P that were partially pure [5]. Such a preparation was eventually used to generate monoclonal antibodies, and selenoprotein P was purified to homogeneity in 1987 from rat plasma using a monoclonal antibody column [6]. Several groups have subsequently reported purifying human selenoprotein P using a variety of methods [7-11].

Selenoprotein P is the major plasma selenoprotein. Immunoprecipitation of it demonstrated that it accounts for over 60% of the selenium in rat plasma [12]. If 15 µg selenium/100 g body weight is used as the reference selenium content of the rat, selenoprotein P in plasma accounts for 7% of total body selenium. Its peptide concentration in rat plasma is 30 µg/ml [12]. It is also the major selenoprotein in human plasma [13], but its peptide concentration is much lower than in rat plasma and probably around 5 µg/ml.

The half-lives of selenoprotein P and plasma glutathione peroxidase (GSHPx-3) in rat plasma have been compared [14]. Selenoprotein P turned over with a half-life of 3-4 h while GSHPx-3 had a half-life of 12 h. Using the plasma concentration and the half-life, the turnover of selenoprotein P was calculated to be approximately 4% of that of albumin. Thus, a large amount of this protein is made in the organism and it contains a significant fraction of whole body selenium.

**Cloning and sequencing**

The selenoprotein P gene has been cloned and sequenced for several species. The first to be reported was the cDNA sequence for the rat gene [15]. Others that have been sequenced are human [16], mouse [17], bovine [18], and zebrafish [19]. Two distinct genes have been identified and sequenced for the zebrafish [19]. Selenoprotein P nucleotide sequences from all species contain multiple in-frame UGA codons in the open reading frame of the gene. The two zebrafish genes contain the most and the fewest in-frame UGA codons, with the long form containing 17 UGAs and the short form containing a single UGA. The rat, mouse, and human cDNAs contain 10 UGAs while the bovine cDNA contains 12. Thus, the predicted protein for each species contains multiple selenocysteine residues with the exception of the short form in the zebrafish which is expected to contain a single selenocysteine residue. Consistent with the presence of selenoprotein P in plasma, a nucleotide sequence coding for a signal peptide is present in each gene.

Alignment of the amino acid sequences deduced from the nucleotide sequences shows significant conservation from species to species (Figure 1). Among mammalian sequences, there is a 60-80% identity of their implied amino acid sequences. As mentioned previously, the zebrafish has a short form of selenoprotein P that contains a single UGA codon and a long form that contains 17 UGA codons. The zebrafish long form is similar in length to the mammalian proteins; however, the overall identity is only 35%. There are several regions where the identity is significantly higher. In particular, the sequence flanking the first selenocysteine residue has a 66% identity from residue 28 to residue 71. In this segment is a thioredoxin like motif that is conserved in all species. It consists of a selenocysteine residue and a cysteine residue that are separated by two amino acids.

## Selenoprotein P

```
residue     1
rat         ESQGQSPA-CKQAPPWNIGDQNPMLNSEGTVTVVALLQASUYLCLLQASRLEDLRIKLENQ
mouse       ESQGQSSA-CYKAPEWYIGDQNPMLNSEGKVTVVALLQASUYLCLLQASRLEDLRIKLESQ
human       ESQDQSSL-CKQPPAWSIRDQDPMLNSNGSVTVVALLQASUYLCIIEASKLEDLRVKLKKE
bovine      ESQGQSSY-CKQPPPWSIKDQDPMLNSYGSVTVVALLQASUYLCILQASRLEDLRVKLEKE
zebrafish1  ESETEGAR-CKLPPEWKVGDVEPMKNALGQVTVVAYLQASULFCLEQASKLNDLLLKLENQ
zebrafish2  EKESNGSRICKPAPQWEIDGKTPMKELLGNVVVVALLKASUHFCLTQAARLGDLRDKLANG

            61
rat         GYFNISYIVVNHQGSPSQLKHAHLKKQVSDHIAVYRQDEHQTDVWTLLNGNKDDFLIYDR
mouse       GYFNISYIVVNHQGSPSQLKHSHLKKQVSEHIAVYRQEEDGIDVWTLLNGNKDDFLIYDR
human       GYSNISYIVVNHQGISSRLKYTHLKNKVSEHIPVYQQEENQTDVWTLLNGSKDDFLIYDR
bovine      GYSNISYVVVNHQGISSRLKYVHLKNKVSEHIPVYQQEENQPDVWTLLNGNKDDFLIYDR
zebrafish1  GYPNIAYMVVNNREERSQRL-HHLLQERLLNITLYAQDLSQPDAWQAVNAEKDDILVYDR
zebrafish2  GLTNISFMVVNEQDSQSRAMYWELKRRTAQDIPVYQQSPLQNDVWEILEGDKDDFLVYDR

            121
rat         CGRLVYHLGLPYSFLTFPYVEEAIKIAYCEKRCGNCSFTSLEDEAFCKNVSSATASKTTE
mouse       CGRLVYHLGLPYSFLTFPYVEEAIKIAYCEERCGNCNLTSLEDEDFCKTVTSATANKTAE
human       CGRLVYHLGLPFSFLTFPYVEEAIKIAYCEKKCGNCSLTTLKDEDFCKRVSLATVDKTVE
bovine      CGRLVYHLGLPFSFLTFTYVEDSIKTVYCEDKCGNCSLSRPQDEDVCKRVFLATKEKTAE
zebrafish1  CGRLTYHLSLPYTILSHPHVEEAIKHTYCDRICGECSLESSAQLEECKKATEEVNKPVEE
zebrafish2  CGYLTFHIVLPFSFLHYPYIEAAIRATYHKNMC-NCSLNANFSISESSDSTKNDPAGENN

            181
rat         PSE-EHNHHKH---------------HDKHGHEHLGSSKPSENQQPGALDVETSLPPSGL
mouse       PSE-AHSHHKH---------------HNKHGQEHLGSSKPSENQQPG--PSETTLPPSGL
human       TPS-PHYHHEH---------------HHNHGHQHLGSSELSENQQPGAPNAPTHPAPPGL
bovine      ASQ-RHHHPHPHSHPHPHPHPHPHPHHGHQLHENAHLSESPKPDTPDTPENPPTSGL
zebrafish1  EPRQDHGHHEHGHHE-HQGEAERHRHGHHHP--------------------------H
zebrafish2  QRPNSTEPVTAAHHHHHQ------HEPHHHHHNPYPNSHKKSGDSDVTGKPKEPPHHSQE

            225
rat         HHHHHHHKHKGQHRQGHLESUDMGA-SEGLQLSLAQRKLURRGCINQLLCKLSEESGAATS
mouse       HHHH---RHRGQHRQGHLESUDTTA-SEGLHLSLAQRKLURRGCINQLLCKLSKESEAAPS
human       HHHH---KHKGQHRQGHPENRDMPA-SEDLQDL--QKKLCRKRCINQLLUQFPKYSESALS
bovine      HHHHH--RHKGPQRQGHSDNCDTPVGSESLQPSLPQKKLURKCINQLLUQFPKYSESALS
zebrafish1  HHHHHHRGQQQVDVDQQVLSQVDFGQVAVETPMMKRPUAKHSRUKVQYSUQQGADSPVA--
zebrafish2  HVHNHR

            285
rat         SCCCHCRHLIFEK--SGSAITUQCAENLPSLCSUQGLFAEEKVIES-CQCRSPPAA-UH-S
mouse       SCCCHCRHLIFEK--SGSAIAUQCAENLPSLCSUQGLFAEEKVTES-CQCRSPPAA-UQ-N
human       SUCCHCRHLIFEK--TGSAITUQCKENLPSLCSUQGLRAEENITES-CQURLPPAA-UQIS
bovine      SCCCHCRHLVFEK--TGSAITUQCTEKLPSLCSUQGLLAEENVIES-UQURLPPAA-UQAA
zebrafish1  SUCUHURQLFGGEGNGRVAGLUHCDEPLPASUPUQGLKEQDNHIKETUQURPAPPAEUELS

            341
rat         QH-VSPTEASPNUSUNNKTKKUKUNLN
mouse       QP-MNPMEANPNUSUDNQTRKUKUHSN
human       QQ-LIPTEASASURUKNQAKKUEUPSN
bovine      GQQLNPTEASTKUSUKNKAKMUKUPSN
zebrafish1  Q---------PTUVUPAGDATUGURKK
```

**Figure 1.** Alignment of selenoprotein P sequences predicted from rat, mouse, human, bovine, and zebrafish genes. The sequences shown are the mature peptide after removal of the signal peptide. The residue numbers shown correspond to the rat sequence. Selenocysteine residues, U, and cysteine residues, C, are shaded. Alignment gaps have been inserted for conservation of maximum identity. See text for discussion.

When considered as a group, the selenocysteine and cysteine residues are highly conserved from species to species, including the long form from zebrafish. The last third of the protein contains the highest concentration of selenocysteine and cysteine residues. In the C-terminal 100 amino acid residues, there are 17 selenocysteine/cysteine residues and their positions are conserved. The conservation among species of the positions of the selenocysteine and cysteine residues suggests that these residues may be critical to the function of the protein.

Genomic clones of human, mouse and bovine selenoprotein P have been obtained and completely or partially sequenced [20-23]. The gene has 5 exons with 4 intervening introns in all species. The exon/intron junctions in the gene conform to the GT/AG rule. Analysis of the promoter region of the human gene by one group revealed the presence of a TATA box, a CAAT box, two AP1 (activator protein-1) sites, an SP1 (stimulating protein-1) site and a GAS (interferon-γ activation site) motif [21]. Other motifs identified in the promoter region of the human gene included C/EBP (CCAAT/enhancer binding protein), HNF-1 (HNF=hepatic nuclear factor), HNF-3, and a TRE-like (TRE=TPA response element) motif [20]. The promoter region of the mouse gene was found to contain multiple motifs including a TATA-box, HNF3β, GATA-1 (GATA-binding factor-1), BRN-2 (BRN=POU factor Brain) and SRY (sex-determining region Y gene product) motifs [22]. The promoter region of the bovine gene contained additional motifs [23]. Of particular interest in the bovine promotor are γ-IRE (γ-interferon responsive element) and MRE2 (metal responsive element) consensus sequences. The consensus motifs identified in the selenoprotein P promoter regions from the three species suggest a possible role for the protein in inflammatory conditions and possible tissue specific responses.

Transfection of promoter deletion constructs made from a 1.8 kb fragment of the 5' flanking region of the human genomic clone revealed that the TATA box and SP1 site were necessary for selenoprotein P transcription [21]. Characterization of this 1.8-kb fragment of the promoter region showed that interleukin 1β, tumor necrosis factor α, and interferon-γ repressed promoter activity in transfected HepG2 cells. Thus, the human selenoprotein P gene promoter is responsive to cytokines and repression of selenoprotein P expression may occur during acute phase reaction.

**Expression**

Selenoprotein P is widely expressed. Northern analysis has shown the mRNA to be present in rat liver, heart, brain, kidney, testis and muscle [24, 25]. In the mouse, selenoprotein P mRNA has been shown to be present in placenta and uterus as well as liver, heart, brain, kidney, testis, and muscle [26]. Several regions of the rat, mouse, and bovine brain were shown to contain selenoprotein P mRNA by in situ hybridization [18,22]. Selenoprotein P mRNA has been detected in human liver, kidney, and intestine [16, 27]. Studies using cultured cells have shown that astrocytes, myocytes, hepatocytes and Leydig cells from the testis express the protein [25,27,28].

In general, dietary selenium restriction results in decreased expression of all selenoproteins. Even moderate selenium deficiency leads to a decline in liver glutathione peroxidase (GSHPx-1). Other selenoproteins, including selenoprotein P, are more resistant to selenium deficiency and decline after liver glutathione peroxidase has fallen [29,30].

Selenoprotein P mRNA has been shown to be lower in colorectal adenomas than in normal colon mucosa [31]. It was undetectable in a rat renal cell carcinoma model [32]. The underlying reason for the decrease in selenoprotein P expression by these neoplastic tissues is not known.

A possible role for selenoprotein P in metal detoxification has been suggested. However, studies with cultured bovine kidney cell lines showed no induction of selenoprotein P mRNA following incubation with cadmium or zinc [23].

Expression of recombinant selenoprotein P has been reported [33]. The recombinant protein was detected in cell media from a transiently transfected human epithelial kidney cell line, HEK293. It was necessary to modify the Kozak leader sequence that precedes the initiation codon to improve expression of recombinant protein. Overexpression of the recombinant protein was not observed in HepG2 and CHO cells, which have high endogenous expression of the protein. Among the multiple factors that could affect expression of recombinant selenoprotein P, the readthrough of the UGA codon and insertion of selenocysteine appear to be the most critical.

**Primary structure, glycosylation, and isoforms**
Although only one rat selenoprotein P mRNA was detected by northern analysis, studies of the purified rat protein indicated that two or more forms of it were present [12,34]. Glycosylation of the protein was also detected. Therefore, it was deemed necessary to characterize the purified protein preparation.

The heparin binding nature of selenoprotein P was used to separate its isoforms. Binding of proteins to heparin is mediated by positively charged amino acids, usually lysine and arginine. However, histidine can also participate in heparin binding if it is positively charged. It was reasoned that selenoprotein P binding to heparin might be mediated, at least in part, by histidine residues because of their abundance in the protein (28 histidines out of a total of 366 amino acid residues). The pKa of histidine is 7 so the pKa of histidine in a polypeptide would be expected to be near the physiological pH range.

When purified selenoprotein P was passed over a heparin Sepharose column at pH 7.0 it bound to the column [34]. Most of the protein was eluted in three peaks when a pH gradient from 7.0 to 8.5 was applied. This strongly suggested that histidine was involved in the heparin binding. The second peak that eluted yielded a single protein band on SDS/PAGE that had an Mr of 45 kDa [34]. The third peak yielded a single protein band of Mr 57 kDa. These two preparations of apparently pure isoforms were characterized further by amino acid analysis, amino terminal sequencing, and carboxy terminal sequencing [35]. The data obtained on study of the shorter isoform indicated that it corresponded to a polypeptide that was predicted by the selenoprotein P cDNA (figure 1) but that terminated at the serine just before the second selenocysteine. Thus, this isoform appeared to use the second UGA in the mRNA as a stop codon instead of as a selenocysteine codon. The data obtained on study of the longer isoform (third peak) indicated that it was a full length polypeptide. Its termination was at the UAA of the mRNA. All 10 UGAs were read through, presumably with incorporation of selenocysteines. These results suggested that the mRNA for selenoprotein P could be read in different ways. At least one UGA in the open reading frame could serve as a stop to yield the short isoform or could be read through, designating incorporation of selenocysteine to yield the full length isoform.

The first peak off the heparin Sepharose column yielded two protein bands on SDS/PAGE [34]. One band was approximately 57 kDa and the other was approximately 45 kDa. Because these apparent isoforms could not be separated from each other, they were not further analyzed in that study. They may represent isoforms that terminate at other UGAs [36]. Thus, two isoforms have been characterized and it is suspected that two more exist.

Efforts were made to verify the amino acid sequence of selenoprotein P that was predicted by the cDNA. Protease digestion of purified selenoprotein P was carried out and the resulting peptides were separated by HPLC [35]. Edman chemistry was used to sequence the peptides and 80% of the predicted amino acid sequence was confirmed. As expected, the signal peptide was not detected in the mature protein. All peptides detected could be traced to the cDNA sequence (Figure 1). Thus, there was no indication that alternative mRNAs were present, and the results were consistent with all isoforms being produced from a single sequence.

Digestion of selenoprotein P with glycosidases indicated that it contained carbohydrate [12]. This was supported by the fact that it migrated at a greater apparent mass on SDS/PAGE than its predicted peptide weight. Five N-glycosylation sites were identified at asparagines 64, 155, 169, 351, 356. During sequencing, the first three sites were confirmed to contain carbohydrate because the asparagines were not detected when their peptides were sequenced [35]. Thus, selenoprotein P is a glycoprotein with at least three sites of N-glycosylation.

**Selenium content**
Selenium analysis of the first purified preparations of selenoprotein P indicated that the protein contained more than one selenium atom per molecule [12]. The selenium content was estimated to be $7.5 \pm 1.0$ atoms of selenium per molecule. This was based on peptide analyses and used the full length peptide weight as the denominator. At the time, it was not known why the selenium content was not 10 atoms per protein molecule as predicted by the cDNA. It seems likely that the original results were the result of analyzing a mixture of isoforms, each of which contained a different number of selenocysteine residues. Now that isoforms are known, characterization of each of them will be necessary to determine whether they contain selenocysteines at all the positions corresponding to the UGAs in the mRNA.

Selenoprotein P is the only selenoprotein known to have more than one selenium atom per polypeptide chain. One selenocysteine is located in the N-terminal end of the protein where it is paired with a cysteine in a thioredoxin like motif (see Figure 1). The other selenocysteines are in the last one-third of the sequence. Most of the selenocysteines must be internal to the protein because they were detected as reduced in the purified protein [12]. Had they been on the surface of the molecule, they would have been oxidized during purification. A great deal remains to be learned about the selenium in selenoprotein P.

**Heparin binding and localization**
Although selenoprotein P was originally described as a plasma protein, it also binds to cells. An immunohistochemical study showed that capillaries in the brain and in the kidney glomerulus were lined with selenoprotein P [37]. Arterial endothelial cells and hepatic sinusoidal endothelial cells also bound the protein.

Selenoprotein P appears to be expressed by all tissues that have been studied as indicated above. These findings are compatible with a need for selenoprotein P in the interstitial space of tissues. Plasma proteins have limited access to this compartment so plasma selenoprotein P produced by the liver might not reach the interstitium reliably. Thus, secretion of the protein by cells in the tissues would ensure that it was available in their interstitial spaces. The large amount of selenoprotein P in plasma is available to bind to endothelial cells and to leave the vasculature at sites of increased permeability such as areas of inflammation. These results indicate that selenoprotein P is likely to be present in almost all extracellular locations and they allow the prediction that it functions in the extracellular space.

Selenoprotein P is a heparin binding protein and it seems likely that its association with cells is through binding to heparan sulfate proteoglycans. The heparin binding site was sought by modifying the rat protein with compounds that bind to histidine, lysine, and arginine residues [38]. The study suggested that histidines and/or lysines were responsible for the heparin binding. Another report has demonstrated that human selenoprotein P binds to heparin with two affinities—one high and the other low [39]. Thus, the location of selenoprotein P in the extracellular space may depend on its heparin binding properties.

**Function**

Two major functions have been proposed for selenoprotein P. One is that it serves in oxidant defense in the extracellular space and the other is that it serves as a means of transport of selenium from the liver to other tissues. Both these potential functions can be supported and are discussed here.

**Diquat model**

Over 20 years ago the observation was made that selenium deficient mice were much more sensitive to injury by the redox cycler paraquat than were control mice [40]. Based on this work, our group developed a rat model in which low dose diquat administration led to massive lipid peroxidation and liver necrosis within several hours [41]. Neither control nor vitamin E deficient rats manifested this form of injury, even at higher doses of diquat. Injection of 50 µg of selenium 10 h before the diquat was given prevented the injury even though neither liver nor plasma glutathione peroxidase had risen significantly. It was postulated that a selenium dependent factor other than glutathione peroxidase was protecting against the oxidative injury caused by diquat.

In 1995, our group presented a study of diquat administration in selenium deficient rats that took advantage of the $F_2$ isoprostanes as indices of lipid peroxidation [42]. That study showed that only a small amount of lipid peroxidation took place in the liver, even though massive liver necrosis occurred. Because of that result, we postulated that the diquat induced damage was taking place in a small, but vital, compartment of the liver such as the plasma membrane. That study also showed that the protection against lipid peroxidation and liver necrosis afforded by selenium injection correlated with the appearance of selenoprotein P, but not with the appearance of glutathione peroxidase.

After selenoprotein P was shown to bind to the hepatic sinusoidal endothelial cells [37], it was suspected that the diquat injury to the liver might begin in them. Recently, an electron microscopy study has demonstrated that administration of diquat to the selenium deficient rat leads to centrilobular endothelial cell injury [43]. Within 20 minutes of administration, centrilobular endothelial cells had nuclear changes and occasional disruption of cell to cell contacts. By 60 minutes, endothelial cells were missing from the centrilobular regions and hepatocytes were exposed to the blood. By 120 minutes, hepatocytes were undergoing necrosis. Thus, the centrilobular endothelial cells are injured and detach before hepatocyte necrosis occurs. Loss of endothelial cells is presumably the primary cause of the hepatic necrosis.

These results suggest that selenoprotein P protects against the oxidative liver injury caused by diquat. Firstly, the protection associated with injection of selenium correlates with the appearance of selenoprotein P and not with the appearance of glutathione peroxidase [41,42]. Secondly, selenoprotein P associates with endothelial cells in the liver [37] and they are the primary site of injury by diquat [43].

A hypothesis of the selenoprotein P effect in protecting against diquat is presented in Figure 2.

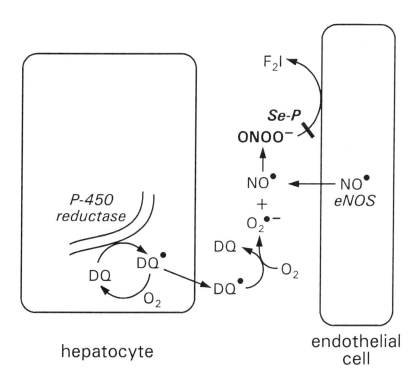

**Figure 2.** Model of liver injury by diquat (DQ) in selenium deficient rats. Selenoprotein P (Se-P) is postulated to protect the centrilobular endothelial cell against peroxynitrite($ONOO^-$) attack. $F_2I$ is $F_2$ isoprostanes.

Diquat accepts an electron from reductases such as cytochrome P450 reductase to become a free radical. It donates the electron to molecular oxygen, but under low oxygen conditions has been shown to diffuse out of the cell and react with extracellular molecules [44]. The oxygen tension is very low in the centrilobular region of the liver and we speculate that the diquat radical can reach the interstitial space in that environment where it can react with molecular oxygen to form superoxide. Superoxide could then react with nitric oxide to form peroxynitrite. Because selenium compounds and selenoproteins have been shown to be effective scavengers of peroxynitrite [45,46], we postulate that selenoprotein P bound to the endothelial cell protects it against this oxidant. This protection would be weakened in selenium deficiency with the decrease in selenoprotein P concentration. No direct proof of this scheme is available but it is compatible with the studies that have been reported.

**Brain uptake**
The plasma location of selenoprotein P and its rapid incorporation of injected selenium suggested to some investigators that it might be a selenium transport protein [47]. Our group carried out a study to evaluate this possibility [14]. $^{75}$Se-labeled selenoprotein P and GSHPx-3 were prepared and injected intravenously into selenium deficient and control rats. Uptake by tissues was compared at times up to 24 h. It was reasoned that $^{75}$Se would be taken up to a greater extent from a transport form in selenium deficient tissues than in selenium replete tissues. The only tissue in which this expectation was fulfilled was the brain. Within minutes of $^{75}$Se-selenoprotein P injection, $^{75}$Se accumulation by the brain was greater in the selenium deficient rats than in the control rats. Such a differential was not seen in other tissues, nor was it seen in any tissue, including the brain, when the source of the $^{75}$Se was GSHPx-3. Thus, it was concluded that a mechanism exists whereby the brain is able to accumulate selenium present in selenoprotein P in a rather specific manner.

The brain has long been known to retain its selenium efficiently in selenium deficiency compared with other tissues [48]. This finding allows the hypothesis that receptors for selenoprotein P are located at the blood brain barrier that allow the brain to take up the protein or its selenium. The study further suggests that other tissues do not have such receptors and thus cannot compete with the brain.

**Peroxidase and peroxynitrite reductase activities**
A group in Japan has reported detection of phospholipid hydroperoxide glutathione peroxidase activity of human selenoprotein P [10]. GSHPx-3 has been reported to have a similar activity [49]. Whether these selenoproteins function as glutathione peroxidases in vivo is not known and further work is

needed on how reducing equivalents could be supplied to such extracellular proteins.

A German group has presented work showing that human selenoprotein P is capable of protecting plasma proteins from modification with nitrotyrosine as a result of incubation with peroxynitrite [50]. This is a promising preliminary observation and warrants further study.

**Selenoprotein P in humans**
Human selenoprotein P has been purified and studied by several groups [7-11, 51]. The preparations obtained vary somewhat but most groups have presented evidence that isoforms are present. Antibodies have been generated to the protein and used for immunoassays.

Selenoprotein P has been measured in a number of population groups [13, 52-54]. Of special interest are the Chinese living in selenium deficient regions of that country. One study of men living in a selenium deficient region found that plasma selenoprotein P concentrations were 23% of those in men from an area that had been supplemented with selenium [13]. Administration of 200 µg selenium per day for 14 days to the men with low selenium status caused a rise in the selenoprotein P concentration to 88% of the control. Plasma glutathione peroxidase activity and selenium content correlated well with the values of selenoprotein P, showing that selenoprotein P can serve as an indicator of selenium status.

Patients with cirrhosis have depressed selenoprotein P levels that correlate with the severity of their illness [55]. The reason for the low selenoprotein P does not appear to be selenium deficiency because plasma glutathione peroxidase, which originates in the kidney, is not depressed. Rather, it seems likely that impairment of selenoprotein P synthesis by the diseased liver is the cause for the low plasma selenoprotein P concentration.

**Conclusions and future directions**
In addition to its status as a selenoprotein, selenoprotein P has several features that are highly unusual. It is the only selenoprotein that has been established to have more than one selenium atom. It is the only known selenoprotein that uses a UGA codon in the open reading frame of its mRNA alternatively for termination and for insertion of selenocysteine. It has highly unusual pH dependent heparin binding properties. These features need further characterization.

Selenoproteins usually have redox functions. The presence of so many selenium atoms in a single polypeptide suggests that selenoprotein P has an unusual redox function. Although glutathione peroxidase and peroxynitrite reductase activities have been reported for human selenoprotein P, the structure of the protein is more complicated than that of other enzymes that carry out these activities. Therefore, the biochemical function of selenoprotein P needs much more characterization.

Selenoprotein P has been implicated as protecting against oxidative injury of endothelial cells and as supporting brain selenium accumulation. Better characterization of these putative functions is needed, as is consideration of other functions.

This chapter demonstrates that selenoprotein P is now accepted as an important component of selenium metabolism and function in animals. It also indicates that a great deal remains to be learned about selenoprotein P and that the field of research on it is wide open.

**Acknowlegement:** The authors' research on selenoprotein P is supported by NIH ES 02497.

**References**

1. KR Millar 1972 *NZJ Agric Res* 15:547
2. RF Burk 1973 *Proc Soc Exp Biol Med* 143:719
3. JK Herrman 1977 *Biochim Biophys Acta* 500:61
4. RF Burk, PE Gregory 1982 *Arch Biochem Biophys* 213:73
5. MA Motsenbocker, AL Tappel 1982 *Biochim Biophys Acta* 704:253
6. J-G Yang, J Morrison-Plummer, RF Burk 1987 *J Biol Chem* 262:13372
7. B Åkesson, T Bellew, RF Burk 1994 *Biochim Biophys Acta* 1204:243
8. R Daher, F Van Lente 1994 *Clin Chem* 40:62
9. B Eberle, HJ Haas 1995 *J Trace Elements Med Biol* 9:55
10. Y Saito, T Hayashi, A Tanaka, et al 1999 *J Biol Chem* 274:2866
11. U Sidenius, O Farver, O Jøns, et al 1999 *J Chromatogr B* 735:85
12. R Read, T Bellew, J-G Yang, et al 1990 *J Biol Chem* 265:17899
13. KE Hill, Y Xia, B Åkesson, et al 1996 *J Nutr* 126:138
14. RF Burk, KE Hill, R Read, et al 1991 *Am J Physiol* 261:E26
15. KE Hill, RS Lloyd, J-G Yang, et al 1991 *J Biol Chm* 266:10050
16. KE Hill, RS Lloyd, and RF Burk 1993 *Proc Natl Acad Sci* 90:537
17. P Steinert, M Ahrens, G Gross, et al 1997 *Biofactors* 6:311
18. K Saijoh, N Saito, MJ Lee, et al 1995 *Mol Brain Res* 30:301
19. GN Kryukov, VN Gladyshev 2000 *Genes Cells* 5 in press
20. Y Yasui, K Hasada, J-G Yang, et al 1996 *Gene* 175:260
21. I Dreher, TC Jakobs, J Köhrle 1997 *J Biol Chem* 272:29364
22. P Steinert, D Bächner, L. Flohé 1998 *Biol Chem* 379:683
23. M Fujii, K Saijoh, T Kobayashi, et al 1997 *Gene* 199:211
24. RF Burk, KE Hill 1994 *J Nut*r 124:1891
25. X Yang, KE Hill, MJ Maguire, et al 2000 *Biochim Biophys Acta* 1474:390
26. JW Kasik, EJ Rice 1995 *Placenta* 16:67
27. I Dreher, C Schmutzler, F Jakob, et al 1997 *J Trace Elem Med Biol* 11:83
28. M Koga, H Tanaka, K Yomogida, et al 1998 *Biol Reproduction* 58:261
29. J-G Yang, KE Hill, RF Burk 1989 *J Nutr* 119:1010
30. KE Hill, PR Lyons, RF Burk 1992 *Biochem Biophys Res Commun* 185:260
31. H Mörk, OH Al-Taie K Bähr, et al 2000 *Nutr Cancer* 37:108
32. T Tanaka, S Kondo, Y Iwasa, et al 2000 *Am J Pathol* 156:2149
33. RM Tujebajeva, JW Harney, MJ Berry 2000 *J Biol Chem* 275:6288
34. HS Chittum, S Himeno, KE Hill, et al 1996 *Arch Biochem Biophys* 325:124
35. S Himeno, HS Chittum, RF Burk 1996 *J Biol Chem* 271:15769

36. S Ma, KE Hill, RM Caprioli, et al *Characterization of rat selenoprotein P by mass spectrometry.* in *7th International Symposium on Selenium in Biology and Medicine.* 2000. Venice, Italy.
37. RF Burk, KE Hill, ME Boeglin, et al 1997 *Histochem Cell Biol* 108:11
38. RJ Hondal, KE Hill, RF Burk 1999 *FASEB J* 13:A875
39. GE Arteel, S Franken, J Kappler, et al 2000 *Biol Chem* 381:265
40. JS Bus, SD Aust, JE Gibson 1974 *Biochem Biophys Res Commun* 58:749
41. RF Burk, RA Lawrence, JM Lane 1980 *J Clin Invest* 65:1024
42. RF Burk, KE Hill, JA Awad, et al 1995 *Hepatology* 21:561
43. JB Atkinson, KE Hill, RF Burk 2001 *Lab Invest* in press
44. MC Cornu, GA Moore, Y Nakagawa, et al 1993 *Biochem Pharmacol* 46:1333
45. S Padmaja, GL Squadrito, JN Lemercier, et al 1996 *Free Radic Biol Med* 21:317
46. H Sies, VS Sharov, LO Klotz, et al 1997 *J Biol Chem* 272:27812
47. MA Motsenbocker, AL Tappel 1982 *Biochim Biophys Acta* 719:147
48. D Behne, W Wolters 1983 *J Nutr* 113:456
49. Y Yamamoto, K Takahashi 1993 *Arch Biochem Biophys* 305:541
50. GE Arteel, V Mostert, H Oubrahim, et al 1998 *Biol Chem* 379:1201
51. V Mostert, I Lombeck, J Abel 1998 *Arch Biochem Biophys* 357:326
52. MEK Persson-Moschos, L Stavenow, B Åkesson, et al 2000 *Nutr Cancer* 36:19
53. AJ Duffield, CD Thomson, KE Hill, et al 1999 *Am J Clin Nutr* 70:896
54. Y Xia, P Ha, KE Hill, et al 2000 *J Trace Elem Exp Med* 13:333
55. RF Burk, DS Early, KE Hill, et al 1998 *Hepatology* 27:794

# Chapter 12. Selenoprotein W: A muscle protein in search of a function

L. Walt Ream, William R. Vorachek and Philip D. Whanger

*Departments of Microbiology and Environmental and Molecular Toxicology, Oregon State University, Corvallis, OR 97331, USA*

**Summary:** Dietary selenium deficiency causes white muscle disease, which results in calcification of skeletal and cardiac muscle in sheep, and is associated with a cardiomyopathy called Keshan disease in humans. This prompted us to search for selenoproteins abundant in skeletal and cardiac muscle, which led to the discovery of selenoprotein W (SelW). This small (87 amino acid) protein contains one selenocysteine residue and exists in four forms. One type has glutathione bound to a specific cysteine residue, which suggests that SelW has an antioxidant function, as several other selenoproteins do. Another SelW form contains an unknown 41 MW moiety, SelW is also present with both adducts or with none. SelW contains a putative calcium-binding domain near its carboxy terminus, although there is no evidence for a direct association between SelW and calcium. SelW is highly conserved (83% identical) among the six mammals studied to date (human, monkey, rat, mouse, sheep, and pig) suggesting that SelW plays an important role in the health of skeletal and cardiac muscle. SelW is abundant in skeletal and cardiac muscle and brain in primates and sheep, but it is not abundant in rodent heart muscle. In contrast to primates and sheep, rodent cardiac muscle does not deteriorate upon selenium depletion. Dietary selenium affects SelW mRNA and protein accumulation, but selenium levels do not affect the transcription rate of the SelW gene. Instead, adequate selenium is required to stabilize SelW mRNA, and selenium depletion reduces the half life of SelW mRNA. Although the metabolic function of SelW remains unknown, selenium deficiency eliminates SelW from skeletal and cardiac muscle, suggesting a link between SelW and healthy muscle tissues.

## Background
In 1958, Schubert and colleagues [1] at Oregon State University first showed that selenium deficiency causes white muscle disease, a myopathy that results in calcification of skeletal and cardiac muscles [2]. The prior year Schwartz and Foltz discovered that selenium supplements prevent liver

necrosis was observed in rats fed a diet low in vitamin E and sulfur-containing amino acids [3], and this was the reason selenium was studied in connection with white muscle disease. Until that time, the only known significance of selenium was its toxicity. These remarkable discoveries initiated a long and challenging effort to change the public perception of selenium. A newly discovered enzyme, glutathione peroxidase (GPx), was reported [4] the same year selenium was shown to prevent liver necrosis in rats. However, it took 15 years to discover that GPx contained selenium [5]. The discovery that GPx was a selenoenzyme stimulated intensive research on selenium biochemistry, and a number of prokaryotic and eukaryotic selenoproteins are now known and discussed in this book by others. A review on selenoprotein W (SelW) has been written with a more detailed background, problems with purification, rationale for isolation of this selenoprotein and related information [6]. This review will concentrate on more recent results.

## Purification and physical characteristics of selenoprotein W

The discovery that white muscle disease is caused by selenium deficiency prompted a search for muscle selenoproteins, which may be necessary to maintain healthy muscle tissue. Accordingly, lambs were given radioactive selenium, and radiolabeled muscle proteins were examined [7,8]. This led to the discovery of SelW, which is abundant in muscle from selenium-supplemented lambs, but absent or barely detectable in muscle from selenium-deficient lambs with white muscle disease [9]. Although initial studies were conducted with lambs [7,10,11], later studies were performed with rats [12]. The demonstration that selenium was present in SelW as selenocysteine indicated that we were dealing with a selenoprotein [11].

SelW was purified by ammonium sulfate fractionation, Sephadex G-50 gel filtration, cation exchange chromatography on CM-Sephadex, and reverse phase high pressure liquid chromatography using C-18 Vydac columns [12]. Four forms of the protein were separated by the cation exchange and reverse phase chromatography steps. Molecular weights of the four proteins, as determined by matrix-assisted laser desorption/ionization time of flight mass spectrometry (MALDI), revealed masses of 9550, 9590, 9858 and 9898. The 9550 form does not contain adducts, whereas the 9590 form contains an unknown 41 MW moiety. The 9858 form contains reduced glutathione bound to a cysteine residue at position 37 [14]. The reduced glutathione can only be removed by dithiothreitol at elevated temperatures, suggesting that SelW must be denatured to remove the glutathione adduct. The highest molecular weight form contains both glutathione and the unknown 41 MW moiety.

## Selenoprotein W amino acid sequences are highly conserved

The peptide sequence of the first 60 amino acids of rat SelW formed the basis for further work. Rat skeletal muscle SelW cDNA was isolated using degenerate polymerase chain reaction (PCR) primers based on reverse translation of the partial peptide sequence [15]. Reverse transcription-coupled PCR product synthesized from rat muscle mRNA template was used to screen a muscle cDNA library prepared from selenium-supplemented rats. The cDNA sequence confirmed the partial protein sequence, including a selenocysteine residue encoded by UGA at position 13, and allowed us to predict the rest of the amino acid sequence [15]. RNA folding algorithms predict a stem-loop structure in the 3' untranslated region of the SelW mRNA that constitutes a form II selenocysteine insertion sequence (SECIS) element similar to those identified in other selenoprotein mRNAs [16, 17]. Thus, SelW is an authentic selenoprotein encoded by a typical mRNA with a SECIS element that allows interpretation of an in-frame UGA as a selenocysteine codon.

Mouse, sheep and monkey skeletal muscle cDNA libraries were constructed, and a human skeletal muscle cDNA library was obtained from a commercial source [16]. Porcine cDNA was isolated from smooth muscle. Probes prepared from a cloned rat SelW cDNA were used to screen these cDNA libraries. In all species the selenocysteine codon (UGA) lies at position 13. UGA is also the gene termination codon in the rodent, pig and sheep SelW genes. Only one other example of a eukaryotic mRNA that uses UGA for both termination and selenocysteine incorporation has been reported: the mRNA encoding GPx of *Schistosoma mansoni* [17]. The SelW SECIS element contains an unusually long stem, which brings the SECIS structure close to the UGA termination codon [16]. This close proximity may prevent the translational apparatus from interpreting the UGA termination codon as a selenocysteine codon. In contrast, UAA is the stop codon in primates. The sheep, pig and primate open reading frames each contain 87 codons, one less than the rodent SelW genes. The nucleotide sequences of the coding regions are about 80% identical among these six species.

SelW is highly conserved among the six mammalian species studied to date: 83% of the amino acid residues are identical in all six species (Figure 1) [16]. Rat and mouse SelW proteins are identical, as are human and monkey SelW. The sheep, pig, and primate SelW are one amino acid shorter than the rodent proteins. The selenocysteine residue at position 13, and the cysteine residues at positions 10 and 37 are conserved in the proteins from all six species. These residues may be essential for either catalysis or maintenance of tertiary structure. For example, peptide mapping studies showed that glutathione binds selenoprotein W at cysteine 37 [14]. The high degree of sequence conservation between species, including residues that are likely critical for SelW function (e.g., SeCys 13 and Cys 37), suggests that

SelW plays an important role in mammals.

SelW contains a motif (residues 60-71) that strongly resembles calcium-binding domains found in such proteins as calmodulin and calbindin [18] (Figure 2). Because selenium deficiency results in soft tissue calcification, it is attractive to speculate that SelW may bind calcium; however, no direct evidence exists for an interaction between SelW and calcium. If SelW is involved in regulation of calcium metabolism, it would help to explain the link between selenium and this element in selenium deficiency diseases.

|         | 1 |   |   |   |   |   |   |   |   |   |   |   |   |   | 15 |
|---------|---|---|---|---|---|---|---|---|---|---|---|---|---|---|----|
| Pig     | M | G | V | A | V | R | V | V | Y | C | G | A | SeC | G | Y |
| Primate |   | A | L |   |   |   |   |   |   |   |   |   |   |   |    |
| Rodent  |   | A | L |   |   |   |   |   |   |   |   |   |   |   |    |
| Sheep   |   | A |   | V |   |   |   |   |   |   |   |   |   |   |    |

|         | 16 |   |   |   |   |   |   |   |   |   |   |   |   |   | 30 |
|---------|----|---|---|---|---|---|---|---|---|---|---|---|---|---|----|
| Pig     | K  | S | K | Y | L | Q | L | K | K | K | L | E | D | E | F  |
| Primate |    |   |   |   |   |   |   |   |   |   |   |   |   |   |    |
| Rodent  |    | P |   |   |   |   |   |   |   |   | E |   | H |   |    |
| Sheep   |    | P |   |   |   |   |   |   |   |   |   |   |   |   |    |

|         | 31 |   |   |   |   |   |   |   |   |   |   |   |   |   | 45 |
|---------|----|---|---|---|---|---|---|---|---|---|---|---|---|---|----|
| Pig     | P  | G | R | L | D | I | C | G | E | G | T | P | Q | V | T  |
| Primate |    |   |   |   |   |   |   |   |   |   |   |   |   | A |    |
| Rodent  |    |   | C |   |   |   |   |   |   |   |   |   |   |   |    |
| Sheep   |    |   |   |   |   |   |   |   |   |   |   |   |   |   |    |

|         | 46 |   |   |   |   |   |   |   |   |   |   |   |   |   | 60 |
|---------|----|---|---|---|---|---|---|---|---|---|---|---|---|---|----|
| Pig     | G  | F | F | E | V | L | V | A | G | K | L | V | H | S | K  |
| Primate |    |   |   |   |   | M |   |   |   |   |   | I |   |   |    |
| Rodent  |    |   |   |   |   | T |   |   |   |   |   |   |   |   |    |
| Sheep   | R  |   |   |   |   | F |   |   |   |   |   |   |   |   |    |

|         | 61 |   |   |   |   |   |   |   |   |   |   |   |   |   | 75 |
|---------|----|---|---|---|---|---|---|---|---|---|---|---|---|---|----|
| Pig     | K  | G | G | D | G | Y | V | D | T | E | S | K | F | L | K  |
| Primate |    | K |   |   |   |   |   |   |   |   |   |   |   |   |    |
| Rodent  |    | R |   |   |   |   |   |   |   |   |   |   | R |   |    |
| Sheep   |    |   |   |   |   |   |   |   |   |   |   |   |   |   |    |

|         | 76 |   |   |   |   |   |   |   |   |   |   |   | 88 |
|---------|----|---|---|---|---|---|---|---|---|---|---|---|----|
| Pig     | L  | V | A | A | I | K | A | A | L | A | Q | G | *  |
| Primate |    |   |   |   |   |   |   |   |   |   |   |   | *  |
| Rodent  |    |   | T |   |   |   |   |   |   |   |   | C | Q  |
| Sheep   |    |   |   |   |   |   |   |   |   |   |   | A | *  |

**Figure 1.** Selenoprotein W deduced amino acid sequences in six species. Only the amino acid residues that differ from the pig sequence are shown for primate, rodent, and sheep SelW.

## Selenoprotein W gene structure

The genomic sequences of selenoprotein W genes from rat and mouse show that each gene contains six exons and five introns. The rat SelW gene extends over 4931 base pairs (bp) from the beginning of exon 1 to the end of exon 6. Transcription of the rat SelW gene begins 51 or 52 bp upstream of the ATG initiation codon [19]. The first 400 bp upstream of the coding

## Selenoprotein W: A muscle protein in search of a function

region contain sequences that promote strong transcription in both muscle and brain cells.

**Selenoprotein W structure and antibody production**

Primate SelW adopts a different tertiary structure than SelW from rodents or sheep. Polyclonal antibodies raised against a synthetic peptide antigen react differently with rodent and primate SelW, indicating that these proteins present different epitopes despite their highly conserved primary sequences (87.5% identical; Figure 1).

SelW contains two hydrophilic regions: amino acids 13 to 31 and 51 to 69. Rabbits immunized with a synthetic peptide containing residues 13 to 31 from rat SelW produced polyclonal antibodies that detected SelW in tissues from rodents and sheep but not primates [21]. Because antibodies raised against the rodent-derived peptide do not recognize primate SelW, we compared the sequences of the rodent and primate peptides. Primate and rodent SelW differ at three positions in this region: Ser (primate) or Pro (rodent) at residue 17, Lys or Glu at residue 24, and Asp or His at residue 28 (Figure 1). Although each substitution is nonconservative and thus may significantly alter an epitope, the change at position 17 from Pro to Ser appears responsible for the failure of primate SelW to cross react with antibodies raised against the peptide based on the rodent SelW sequence. Both sheep and primate SelW have Lys and Asp at residues 24 and 28, repectively, but rodent and sheep SelW, which cross react, each have Pro at position 17, in contrast to primate SelW. Thus, the proline at position 17 in rodent SelW is postulated to alter its antigenicity and likely causes structural changes due to proline-induced bending.

|  |  |  |  |  | Gly |  |  |  |  |  |  |
|---|---|---|---|---|---|---|---|---|---|---|---|
|  |  |  | Lys |  |  |  |  |  |  |  |  |
| SelW: | Lys | Lys | Arg | Gly | Asp | Gly | Tyr | Val | Asp | Thr | Glu Ser |
| Consensus: | Asp | Lys | Asp | Gly | Asp | Gly | Tyr | Ile | Asp | Phe | Glu Glu |
|  |  | Asn |  | Asn |  |  |  | Val | Glu | Thr | Asp |
|  |  |  |  |  |  |  |  |  |  |  | Asn |

**Figure 2.** Possible calcium-binding domain in selenoprotein W. Consensus sequence derived from 19 Ca-binding sites in human troponin C, sea urchin Ca-binding protein, *Schizosaccharomyces pombe* calmodulin, and human calbindin. Alternative residues found in SelW from some species are shown above the sequence, and alternative residues found in other Ca-binding proteins are shown below the consensus sequence.

To obtain antibodies specific for primate SelW, we converted the selenocysteine codon of the human SelW coding sequence to a cysteine codon, added six histidine codons at the 3' end of the gene, and fused the gene to a strong promoter in a prokaryotic expression vector [21]. Histidine-tagged primate SelW expressed in *Escherichia coli* was purified by chromatography over nickel-agarose and reverse phase HPLC. Rabbit polyclonal antibodies raised against this protein were used in western blots to determine the distribution of SelW in monkey and human tissues.

### Tissue distribution of selenoprotein W
Western blots of proteins extracted from tissues of rats fed a commercial (selenium-adequate) chow revealed the highest amounts of SelW in skeletal muscle and brain [20]. Lower levels accumulated in spleen and testis, but SelW was not detected in liver, kidney, intestinal mucosa, lungs, heart, plasma, or erythrocytes. In rhesus monkey and human tissues, SelW levels were highest in skeletal muscle, heart, and brain. In contrast to rat tissues, we detected SelW in all human and monkey tissues examined, including tongue, spleen, kidney, testis, ovary, and liver, which had the lowest level of SelW [21]. The tissue distribution of SelW in sheep and primates is similar: cardiac muscle contains high levels of SelW in these species, whereas rodents lack SelW in heart tissue. Selenium depletion is associated with cardiomyopathy in humans and sheep, but not rats, suggesting that SelW may be important for cardiac muscle health in humans and sheep, but not rodents.

### Dietary selenium affects selenoprotein W levels
The influence of deficient (no added selenium), adequate (0.1 mg selenium/kg), or excessive (4 mg selenium/kg) levels of dietary selenium on SelW content was measured in various rat tissues. In rats fed a selenium-deficient diet, SelW protein is undetectable in skeletal muscle. Rats fed this diet supplemented with 0.1 mg selenium/kg accumulate SelW in muscle, but SelW levels are much higher in muscle of rats fed a diet with 4 mg/kg selenium [22]. Thus, the levels of SelW in rat skeletal muscle depend upon selenium intake. Regardless of the dietary selenium level, SelW does not accumulate in several other tissues, including liver, thyroid, pancreas, pituitary, and eyes. In contrast, SelW levels fluctuate in response to dietary selenium in heart, lung, prostate, esophagus, small intestine, tongue, skin, and diaphragm as well as skeletal muscle. In other tissues such as kidney and seminal vesicles, SelW accumulates only in rats fed a diet with excess selenium.

The influence of dietary selenium on SelW levels was examined in sheep fed either a low selenium diet (0.02 mg selenium/kg) or a diet supplemented with selenium [23]. After 10.5 weeks, selenium-deficient sheep contained less SelW in skeletal muscle, heart, tongue, lung, spleen, kidney, and liver than sheep fed a selenium-supplemented diet. In contrast, SelW content in brain did not differ between selenium-deficient and selenium-supplemented sheep. Both selenium content and levels of the selenoenzyme glutathione peroxidase decreased significantly in brain from selenium-deficient sheep, indicating that brain tissue retained SelW preferentially during selenium deprivation. In selenium-supplemented sheep, skeletal muscle and heart contained similar levels of SelW; this differs from rats, which have very little SelW in cardiac muscle. Thus, the tissues that contain the highest levels of SelW in selenium-supplemented sheep, skeletal and cardiac muscle, are the ones that deteriorate in selenium-deficient sheep with white muscle disease.

SelW levels were determined in rhesus monkey and human tissues [21]. SelW levels were highest in cardiac and skeletal muscle and lowest in liver. In contrast to rat, SelW was found in all tissues examined (muscle, tongue, heart, brain, spleen, kidney, liver, testis and ovary). Thus, the tissue distribution of SelW in primates is similar to sheep, but differs from rats with respect to cardiac muscle. In order to obtain information on the effects of dietary selenium on SelW levels in human tissues, we examined samples from spontaneously aborted fetuses from women living in selenium-deficient (Xichang), adequate (Shijiazuang) or excessive (Enhi) regions of China [24]. Tissues from six selenium-adequate adults killed in accidents were also examined. Tissue selenium content correlated with regional selenium status: fetuses from the selenium-deficient region exhibited lower selenium levels in skeletal muscle, brain, and heart than fetuses from regions with adequate or excessive selenium. Thus, fetal SelW levels in skeletal and cardiac muscle closely reflected selenium status. SelW content in skeletal muscle was usually higher in selenium-adequate adults than in selenium-adequate fetuses, even though fetal muscle contains higher selenium levels than adult muscle. This suggests that selenium content is not the sole factor governing the SelW level in a particular tissue, and SelW levels may vary during normal development.

Although SelW content in rat brain was unaffected by selenium status when the entire organ was examined as a whole [25], SelW levels in specific regions within the brain responded differently to dietary selenium [26]. In rats, SelW levels in the cortex and cerebellum were not significantly affected by dietary selenium, but selenium increased SelW levels in the thalamus [26]. These results indicate that different regions of the brain should be examined separately to assess the effects of dietary selenium on selenoprotein levels.

**Selenoprotein W mRNA accumulation depends on selenium**
One method selenium regulates SelW levels is through the alteration of mRNA levels. Northern blots indicated that SelW mRNA accumulation increased in muscle four [15] to six [27] fold in rats fed excess selenium compared to selenium-deficient rats. Thus, dietary selenium affects SelW mRNA accumulation as well as SelW protein levels. Tissue specific

accumulation of SelW mRNA also correlates with SelW protein levels, as expected. For example, in primates SelW mRNA levels were highest in cardiac and skeletal muscle, which is the same pattern observed for SelW protein accumulation [21]. Hence, SelW protein levels correlate with SelW mRNA accumulation with respect to tissue specificity and response to dietary selenium.

## Selenium levels affect selenoprotein W mRNA stability

To examine the influence of selenium on SelW transcription rate and mRNA levels, rat L8 muscle cells were cultured in serum-free medium with and without selenium [28]. We used northern blots to measure mRNA accumulation and nuclear run-on assays to determine SelW transcription rates. Selenium depletion caused a drop in SelW mRNA levels, whereas selenium supplementation maintained SelW mRNA at the original level (Figure 3). Nuclear run on analysis showed that the selenium content of the medium did not affect the transcription rate of SelW. Therefore, selenium did not affect SelW transcription, but instead is required for stabilization of SelW mRNA, which is also true for another selenoprotein, cellular GPx [29].

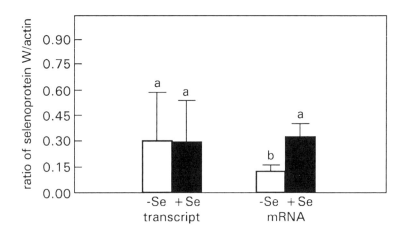

**Figure 3.** Newly synthesized (transcript) and steady-state (mRNA) levels of SelW mRNA in rat L8 muscle cells cultured with (hatched bars) or without (open bars) selenium in the medium. a and b are significantly different ($P < 0.05$).

Decay of SelW mRNA was examined in cells, grown in serum-free medium with or without selenium, in which further RNA synthesis was blocked with alpha-amanitin. When cells were grown in low selenium medium, SelW mRNA levels decreased more rapidly than in cells grown in selenium-supplemented medium (Figure 4). The half-lives of SelW mRNAs were 56 hours in the absence of selenium, but 120 hours in the presence of selenium. This two fold greater half-life in the presence of selenium further suggests that selenium is required to stabilize SelW mRNA.

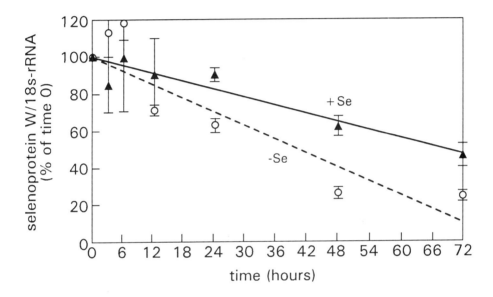

**Figure 4.** Decay of SelW mRNA with (closed triangles) and without (open circles) selenium. See text for discussion

## What is metabolic function of selenoprotein W?

In selenium-supplemented lambs, SelW accumulates to the highest levels in muscle and heart, which are the tissues affected in white muscle disease, and selenium depletion eliminates or drastically reduces SelW from these tissues. Thus, SelW may be involved in skeletal and cardiac muscle metabolism [15,20,23]. Many of the selenoenzymes identified thus far are involved in redox reactions, and SelW may also have antioxidant functions. This possibility is strengthened by the discovery that glutathione is bound to the major species of SelW [13,14]. A study empolying rat glial cells was undertaken to elucidate the possible antioxidant function of SelW [30,31]. Using inducible LacSwitch expression vectors, SelW was overexpressed 22-fold relative to control cells. Glial cells that contained elevated levels of SelW survived treatment with an oxidizing agent (2,2'-Azobis (2-amidinopropane) dihydrochloride) at a somewhat higher rate than control cells, suggesting that SelW may have an antioxidant function.

**Acknowledgements:** Public Health Service Research Grants DK 38306, DK 38341 and DK 54226 supported the research reported in this communication.

## References

1. OH Muth, JE Oldfield, LF Remmert, JR Schubert 1958 *Science* 128:1090
2. JR Schubert, OH Muth, JE Oldfield, LF Remmert 1961 *Fed Proc* 20:689
3. K Schwarz, CM Foltz 1957 *J Amer Chem Soc* 79:3292
4. GC Mills 1957 *J Biol Chem* 229:189
5. JT Rotruck AL Pope, HE Ganther, AB Swanson, DG Hafeman, WG Hoekstra 1973 *Science* 179:588
6. PD Whanger 2000 *Cellular Molecular Life Science* 57:00
7. ND Pedersen, PD Whanger, PH Weswig, OH Muth 1972 *Bioinorgan Chem* 2:33
8. RS Black, MJ Tripp, PD Whanger, PH Weswig 1978 *Bioinorgan Chem* 8:161
9. PD Whanger, SC Vendeland, MA Beilstein 1993 In *Trace Elements in Man and Animals* TEMA 8; M. Anke, D. Meissner, CL Mills, (Eds); pp 119-126, Verlag Media Touristik, Gersdorf, Germany
10. PD Whanger, ND Pedersen, PH Weswig 1972 *Fed Proc* 31:691
11. MA Beilstein, MJ Tripp, PD Whanger 1981 *J Inorgan Biochem* 15:339
12. SC Vendeland, MA Beilstein, CL Chen, ON Jensen, E. Barofsky, PD Whanger 1993 *J Biol Chem* 268:17103
13. MA Beilstein, SC Vendeland, E Barofsky, ON Jensen, PD Whanger 1996 *J Inorgan Biochem* 61: 117-124
14. Q-P Gu, MA Beilstein, E Barofsky, LW Ream, PD Whanger 1999 *Arch Biochem Biophys* 361:23
15. SC Vendeland, MA Beilstein, J-Y Yeh, WL Ream, PD Whanger 1995 *Proc Natl Sci USA* 92:8749
16. Q-P Gu, MA Beilstein, SC Vendeland, W Lugade, LW Ream, PD Whanger 1997 *Gene* 193:187.
17. C Roche, DL Williams, J Khalife, T LePresle, A Capron, RJ Pierce 1994 *Gene* 138:149.
18. T Takeda, M Yamamota 1987 *Proc Natl Acad Sci USA* 84:3580
19. PD Whanger, SC Vendeland, Q-P Gu, MA Beilstein, LW Ream 1997 *Biomedical Environmental Sciences* 10:190
20. J-Y Yeh, MA Beilstein, JS Andrews, PD Whanger 1995 *The FASEB J* 9:392
21. Q-P Gu, Y Sun, LW Ream, PD Whanger 2000 *Molec Cell Biochem* 204:49
22. Y Sun, P-C Ha, JA Butler, B-R Ou, J-Y Yeh, PD Whanger 1998 *J Nutr Biochem* 9:23
23. J-Y Yeh, Q-P Gu, MA Beilstein, NE Forsberg, PD Whanger 1997 *J Nutr* 127:394
24. JA Butler, Y Xia, Y Zhou, Y Sun, PD Whanger 1999 *The FASEB J* 13:A248
25. Y Sun, JA Butler, NE Forsberg, PD Whanger 1999 *Nutr Neuroscience* 2:227
26. Y Sun, JA Butler, PD Whanger 2000 *J Nutr Biochem* in press
27. J-Y Yeh, SC Vendeland, Q-P Gu, JA Butler, B-R Ou, PD Whanger 1997 *J Nutr* 127: 2165.
28. Q-P Gu 1997 Ph.D. thesis, Oregon State University, Corvallis, OR
29. MJ Christensen, KW Burgener 1992 *J Nutr* 122:1620
30. Y Sun, Q-P Gu, PD Whanger 1998 *The FASEB J* 12:A824
31. Y Sun, Q-P Gu, PD Whanger 2000 *J Inorganic Biochemistry* in press

# Chapter 13. The 15 kDa selenoprotein (Sep15): functional studies and a role in cancer etiology

Vadim N. Gladyshev

*Department of Biochemistry, University of Nebraska, Lincoln, NE 68588, USA*

Alan M. Diamond

*Department of Human Nutrition, University of Illinois at Chicago, Chicago, IL 60612, USA*

Dolph L. Hatfield

*Section on the Molecular Biology of Selenium, Basic Research Laboratory, National Cancer Institute, National Institutes of Health, Bethesda, MD 20892, USA*

**Summary:** The 15 kDa selenoprotein (Sep15) is one of several recently identified selenoproteins. It contains a single selenocysteine residue in the middle of a 162-amino acid open reading frame and has no detectable homology to known proteins. The human Sep15 gene spans 51 kb, has 5 exons and is located on chromosome 1 at position p31. The gene contains two single nucleotide polymorphisms in the 3'-untranslated region (3'-UTR) including one in the SECIS element, that are distributed differently between Caucasians and African Americans. Sep15 localizes to the endoplasmic reticulum where it is tightly bound to UDP-glucose:glycoprotein glucosyltransferase, a protein involved in the quality control of protein folding. Sep15 may be involved in the chemopreventive effect of dietary selenium. This hypothesis is based on its differential expression in normal and malignant tissues, the distribution and functional consequences of natural polymorphisms within its gene, and the location of the Sep15 gene in a region that is often altered in a variety of cancers.

**Introduction**
The number of selenoproteins identified thus far in vertebrates stands at 22 (see Chapter 9). The function of only a few of these proteins is known. Since there is mounting evidence that at least some of the beneficial effects of selenium on health, including its anticarcinogenic properties, are mediated through selenoproteins, it is important to identify, characterize and determine functions for as many selenoproteins as possible.

One of these proteins for which the function is not known is the 15 kDa selenoprotein (Sep15). Sep15 was discovered about three years ago [1] when it was purified from human T cells as a protein that was strongly radiolabeled when cells were grown in the presence of $^{75}$Se. The Sep15 cDNA sequence was deduced from the analysis of the EST database following sequencing of several tryptic peptides obtained from the isolated selenoprotein. The new open reading frame (ORF) contained an in-frame TGA codon (Figure 1) that was predicted to encode selenocysteine (Sec). This conclusion was also based on the presence of selenium in the protein and the fact that sequences downstream of the in-frame TGA matched those predicted to encode tryptic peptides in the isolated human T-cell Sep15.

Analysis of human Sep15 sequence using the EST database revealed that this selenoprotein was expressed in a variety of human cell types and identified its orthologs in other mammals. However, Sep15 sequences had no homology to known proteins making this protein a difficult target for functional characterization. Recent biochemical and genetic studies [2-4] significantly advanced our understanding of the genetics and biochemistry of Sep15. These studies identified a binding partner for Sep15 that implicated this selenoprotein in the quality control of protein folding [4] and revealed unexpected genetic factors that implicated Sep15 in the dietary effect of selenium in cancer prevention [2,3]. These studies are discussed in this chapter.

**The human Sep15 gene**
The complete sequence of the human Sep15 gene has been recently determined [2]. The gene spans a region of 51 kilobase pairs on chromosome 1 and consists of five exons and four introns (see Figure 1). The 5'-untranslated region (5'-UTR) and the first 27 amino acid residues of a putative signal peptide are included in the first exon. This peptide is not present in Sep15 isolated from human T cells [1] consistent with posttranslational processing of the signal peptide. Such gene organization, when the coding sequence for an amino terminal signal peptide is located in a separate exon, has been observed for many other mammalian proteins. The largest exon in the human Sep15 gene is exon 5 which encodes C-terminal sequences and also includes the 3'-UTR. The TGA codon that dictates Sec incorporation into the resulting Sep15 gene product is located in exon 3. The site of initiation of transcription in the Sep15 gene was determined by primer extension and was found to lack a TATA box that is often found upstream of RNA polymerase II transcribed mammalian genes [2].

Two polymorphic sites, which are located in the 3'-UTR at positions 811 and 1125 in human cDNA, have been identified [1-3] and their locations in the Sep15 mRNA sequence are shown in Figure 1. In all DNA samples examined to date, $C^{811}$ is found associated with $G^{1125}$ and $T^{811}$ is associated with $A^{1125}$ suggesting the presence of two alleles. Interestingly, the polymorphism at 1125 site is located in the apical loop of selenocysteine insertion sequence (SECIS) element.

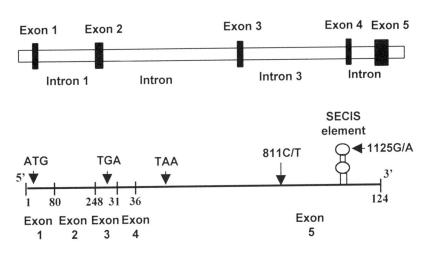

Figure 1. Structural organization of the human 15 kDa selenoprotein gene. In the upper panel, the exon-intron organization and relative sizes of exons and introns in the 51-kb human Sep15 gene are shown. Horizontal lines correspond to introns and flanking regions, and the closed squares to exons. In the lower panel, the organization of the human cDNA sequence obtained by splicing of the Sep15 gene is shown. The relative positions of the ATG initiation and the TGA Sec codons, the TAA termination signal and the detected polymorphisms (811C/T and 1125G/A) are shown. The location of 1125G/A polymorphism in the apical loop of the SECIS element is also indicated. The long horizontal line corresponds to the Sep15 cDNA, and short vertical lines correspond to exon-exon junctions. Numbers under junction sites correspond to last nucleotides in preceding exons.

## Amino acid sequences

Coding sequences within Sep15 genes do not show detectable homology to known proteins, nor do they have clear sequence motifs indicative of a protein structure, cofactor composition or function (Figure 2). However, Sep15 genes were found in a wide variety of animals, including rats, mice, zebrafish, fruit flies and nematodes (Figure 2) [1,2]. The position of Sec is conserved in the Sep15 vertebrate sequences, whereas in insects and nematodes, Sec is replaced with Cys (Figure 2). Homology analyses also indicate that vertebrate Sep15 sequences are highly conserved. For example, mature human and mouse Sep15 differ by only three amino acid residues.

```
Rat     MAAGQGGWLRPALGL RLLLATAFQAVSALG AEFSSEACRELGFSS 45
Mouse   MAAGQGGWLRPALGL RLLLATAFQAVSALG AEFASEACRELGFSS 45
Human   MAAGPSGCLVPAFGL RLLLATVLQAVSAFG AEFSSEACRELGFSS 45
Bovine  MAARRDGWLGPAFGL RLLLATVLQTVSALG AEFSSESCRELGFSS 45
Ciona   MVGPKRRKNGAIISK LLQLFLLGMIHSHAE ASLSAQECADLGFSS 45
Shrimp  -------MMVDPGGI GLFIAATLLSIVEAV QELSTEECFAVGLNK 38
Droso   --------------- --------------- ---------------  0
C.eleg  ---------MWVIFL LLAAVVSPMFGEVEE YKIDVEECKAAGFNP 36

Rat     N-LLCSSCDLLGQFN LLPLDPVCRGCCQEE AQFETKKLYAGAILE 89
Mouse   N-LLCSSCDLLGQFN LLPLDPVCRGCCQEE AQFETKKLYAGAILE 89
Human   N-LLCSSCDLLGQFN LLQLDPDCRGCCQEE AQFETKKLYAGAILE 89
Bovine  N-LLCSSCDLLGQFN LLQLDPDCRGCCQEE AQFETKKLYAGAILE 89
Ciona   E-LMCGSCSLLPKFN LTMLEDDCKKCCQSE VEEDTAKRFHSAILE 89
Shrimp  ANLLCSSCDTLKEFN LDVLEANCRGCCNVD DVNATPTKYPRAILE 83
Droso   ---MCSSCEKLDDFG LDTIKPQCKQCCTLD QQPAAQRTYAKAILE 42
C.eleg  ETLKCGLCERLSDYH LETLLTDCLQCCIKE -EEFKHEKYPTAILE 80

Rat     VCGUKLGRFPQVQAF VRSDKPKLFR-GLQI KYVRGSDPVLKLLDD 133
Mouse   VCGUKLGRFPQVQAF VRSDKPKLFR-GLQI KYVRGSDPVLKLLDD 133
Human   VCGUKLGRFPQVQAF VRSDKPKLFR-GLQI KYVRGSDPVLKLLDD 133
Bovine  VCGUKLGRFPQVQAF VRSDKPKLFK-GLQI KYVRGSDPVLKLLDD 133
Ciona   VCGUKIGRYPQVQAF VKGEKSRAFS-NLKI KYVRGADPVIKLLNE 133
Shrimp  VCGURLGAFPQVQAF VKSDRPAAFP-NLTI KYVRGADPIIKLMDE 127
Droso   VCTCKFRAYPQIQAF IQSGRPAKFP-NLQI KYVRGLDPVVKLLDA  86
C.eleg  VCECNLARFPQVQAF VHKDMARQFGGKVKV KHVRGVRPQVALKDA 125

Rat     NGNIAEELSILKWNT DSVEEFLSEKLERI- ---------------162
Mouse   NGNIAEELSILKWNT DSVEEFLSEKLERI- ---------------162
Human   NGNIAEELSILKWNT DSVEEFLSEKLERI- ---------------162
Bovine  SGNIAEELSILKWNT DSVEEFLSEKLERI- ---------------162
Ciona   DEQVQDTLSITKWNT DSVEEFLNEKLIRV- ---------------162
Shrimp  DGDVMETLAIDKWNT DSXEEFLNTYLILPG QRGRSR---------163
Droso   SGKVQETLSITKWNT DTVEEFFETHLAKDG AGKNSYSVVEDADGD 132
C.eleg  DFKTKEVLSVEKWDT DTLIDFFNQWLE--- ---------------153

Rat     ----------- 162
Mouse   ----------- 162
Human   ----------- 162
Bovine  ----------- 162
Ciona   ----------- 162
Shrimp  ----------- 163
Droso   DDEDYLRTNRI 143
C.eleg  ----------- 153
```

**Figure 2.** Amino acid alignment of animal Sep15 sequences. Residues identical in all sequences are highlighted. U represents selenocysteine and is shown by a closed circle above the sequence. Accession numbers for human and mouse Sep15 sequences are AF051894 and AF288740, respectively. Rat, bovine, *Ciona intestinals* (Ciona) and *Penaeus vannamei* (shrimp) sequences were assembled from EST sequences. The *Drosophila* (Droso) Sep15 sequence (accession number AE003523) was predicted by the *Drosophila* genome project. This sequence appears to lack the first exon containing the N-terminal signal peptide. The *C. elegans* (C.eleg) sequence (accession number AC024872) was predicted by the *C. elegans* Genome Project and corrected to remove an additional intron.

## Pattern of expression

Following the initial EST analyses [1], expression of the Sep15 gene was examined in a number of human and mouse tissues by northern analysis and immunoblot assays [2]. Highest levels of gene expression were observed in prostate, liver, kidney, testes and brain, while lower levels were found in lung, spleen and skeletal muscle.

The high expression of Sep15 in the prostate may be relevant to the recent observation of an abundant $^{75}$Se-labeled selenoprotein that migrated as ~15-16 kDa species on SDS PAGE gels [5]. In selenium deficiency, expression of this protein was conserved, in contrast to that of another selenoprotein, glutathione peroxidase 1. However, ~15-16 kDa protein has not been isolated or its gene sequenced, making it difficult to compare this protein and Sep15. Mammalian cells contain a number of selenoproteins with predicted molecular masses similar to that of mature Sep15 [6]. Thus, the rat prostate 15 kDa selenoprotein detected by Behne and colleagues [5] is either Sep15, one of the selenoproteins identified recently by searching mammalian genomes with various bioinformatics tools (Chapter 9), or a protein yet to be characterized.

## Association of Sep15 and UGTR

Initial purification of Sep15 from human T cells employed reversed-phase chromatography as a final isolation step. Thus, Sep15 was isolated in a denatured state [1]. However, this protein migrated as ~160 kDa species on native gels. This observation suggested that Sep15 is either composed of multiple identical subunits or strongly binds another protein to form the 160 kDa complex. Taking advantage of the high expression level of Sep15 in the prostate, the protein was isolated in the native state from this organ by combination of conventional chromatography and HPLC [4].

Native Sep15 was found to occur in a complex with a 150 kDa protein. The latter protein was then identified by methods of protein microchemistry as UDP-glucose:glycoprotein glucosyltransferase (UGTR), an endoplasmic reticulum-resident protein [7]. The function of UGTR is to glucosylate misfolded proteins, thus retaining them in the endoplasmic reticulum until they are correctly folded or transferring them to degradation pathways [8]. The Sep15 gene was fused to a green florescent protein in order that the expressed fusion protein could be tracked intracellularly. Consistent with its association with UGTR, Sep15 was localized to the endoplasmic reticulum [4]. Furthermore, the N-terminal signal peptide, which was cleaved in the mature protein, was found to be essential for its translocation. The C-terminal sequence of Sep15 was not involved in retaining this protein in the endoplasmic reticulum, and most likely, Sep15 is retained in this compartment by its association with UGTR. Sep15 is the first known endoplasmic reticulum-resident selenoprotein and its complex with UGTR suggests a role of selenium in the quality control of protein folding.

**Influence of the $A^{1125}/G^{1125}$ polymorphism on selenoprotein expression**
The fact that the $A^{1125}/G^{1125}$ polymorphic site occurs in the apical loop of the SECIS element suggested that it might differentially influence the level of Sep15 translation. This possibility prompted us to study the ability of polymorphisms in the Sep15 gene to support selenoprotein expression, as well as to study the allelic frequencies within the human population and their association with malignancy.

Initially, the ability of the nucleotide positions comprising the $A^{1125}/G^{1125}$ polymorphism to influence selenoprotein expression was examined [2,3]. The Sep15 3'-UTR, encoding one or the other polymorphism, was attached to the deiodinase 1 (D1) gene in an expression vector. The levels of D1 in the chimeric constructs were compared to that of wild type D1 [2]. The level obtained with the $A^{1125}$ construct was approximately 75% of that observed with wild type and the level with the $G^{1125}$ construct was approximately 50% of that of wild type. These data suggested that the identity of the base at position 1125 ultimately may influence Sep15 levels.

In contrast to the 1125 site, the base at position 811 did not influence the level of Sep15 expression. That is, a $T^{811}/G^{1125}$ construct yielded approximately the same level of expression as $C^{811}/G^{1125}$ and a $C^{811}/A^{1125}$ construct yielded approximately the same level of expression as $T^{811}/A^{1125}$ [2,3].

To further examine the consequences of the polymorphic nucleotide positions with regard to their ability to influence the translation of mRNAs containing an in-frame UGA codon, a construct was employed that permitted us to test SECIS function by monitoring expression levels of full size and truncated proteins. Again, the G form resulted in lower expression of the reported construct. Thus, the data examining SECIS element efficiency as a function of the two polymorphic positions in Sep15 clearly demonstrated that the identity of the base at position 1125 influences the level of UGA readthrough. However, in addition to influencing selenoprotein expression at concentrations of selenium provided by 10% fetal bovine serum in cell culture media, these polymorphisms differentially responded to additional selenium supplementation of culture media [2,3]. Although the $G^{1125}$ allelic form resulted in lower SECIS efficiency compared to the $A^{1125}$ form in the absence of selenium supplementation, it was more responsive to increasing selenium levels than the $A^{1125}$ form. The possibility that these properties may have a role in cancer prevention is discussed below.

## Possible role of Sep15 in cancer etiology

The finding that the Sep15 gene occurs in humans as two allelic forms allowed us to determine the frequencies at which these alleles are present in the human population. Over 700 samples were analyzed by the PCR amplification of the polymorphic region followed by diagnostic restriction enzyme digestion [3]. This analysis revealed surprising differences in allele frequencies between Caucasians and African Americans. In addition, this study revealed differences in allele frequencies obtained from tumors of breast or head and neck origin when compared to cancer-free individuals among African Americans [3]. Furthermore, loss of heterozygosity (LOH) at the Sep15 gene locus was detected when DNA from circulating lymphocytes of an African American patient with a supraglotis tumor was compared to the DNA obtained from that tumor [3]. Development of cancer is often accompanied by the LOH at the sites that encode protective genes that are also called tumor suppressor genes. Thus, genetic changes at the Sep15 locus suggest that the loss of the Sep15 gene (or possibly a gene in the vicinity of the Sep15 gene) contributes to the development of certain cancers.

The human Sep15 gene is located on chromosome 1 at position p31. This region is commonly mutated or deleted in human cancers and the presence of a tumor suppressor gene on 1p31 has been suggested, but its identity has not been established [9,10]. It is possible that the loss of one copy of Sep15 gene may result in a decrease in the biosynthesis of this selenoprotein gene product in malignant tissues relative to the corresponding normal tissues.

The observation that natural polymorphisms found in the 3'-UTR of the Sep15 gene may influence its translational efficiency in response to selenium levels may also be relevant to the putative role of the protein in cancer. It is possible that individuals representing different combinations of these two haplotypes may express different amounts of Sep15 and, in addition, they may differentially respond to changes in dietary selenium (i.e., differentially control translational levels of Sep15 in response to changes in selenium levels).

The fact that prostate in both humans and mice expresses high levels of Sep15 may also be relevant to the link between Sep15 and cancer. Human selenium supplementation trials revealed that dietary selenium can reduce the incidence of prostate cancer [11] and it is likely that the greatest protection is provided to those individuals with lower selenium intake. In addition, epidemiological data have indicated a statistically significant inverse correlation between selenium in the diet and prostate cancer [12]. However, the mechanism of cancer prevention by selenium is poorly characterized [13] and no selenoprotein has been implicated in such protection. The recent data on Sep15 raise the possibility that this protein may function in the prevention of cancer and possibly serve as an agent by which selenium supplementation exerts its chemopreventive effect.

Although maximal protection against carcinogen-induced cancers in rodents has been achieved by providing dietary selenium in amounts exceeding levels that are necessary for maximal expression of GPx1 and GPx3 [13; and see Chapters 8, 14 and 23], these studies should not necessarily be viewed as proof that selenoproteins are not involved in cancer prevention. It is possible that distinct selenoproteins exhibit different expression patterns at high concentrations of selenium or when individuals are under environmental or genetic stress, such as those that are predisposed to the risk of developing certain cancers. Consistent with this notion was the observation that liver tumors of TGFα/*c-myc* double transgenic mice (a transgenic model of hepatocarcinoma) that were maintained on a selenium-sufficient diet had reduced levels of Sep15 compared to adjacent hepatic tissues [2]. In addition, Sep15 was essentially undetectable in a mouse prostate cancer cell line, while this protein was abundant in normal mouse prostate [2]. If lower levels of the Sep15 predispose an individual to malignancy, then the observation of differences in SECIS element function between the two naturally occurring alleles in the human population may indicate a segment of the population who are either at greater risk of cancer or whom might benefit from selenium supplementation. These observations, together with the fact that alterations in the region on chromosome 1 where the Sep15 gene is located is often associated with cancer progression suggest a possible role of Sep15 in cancer etiology.

## Concluding remarks

Sep15 was discovered by virtue of the presence of selenium in the protein and the relative abundance of Sep15 compared to other selenoproteins. However, characterization of its gene sequences failed to detect significant homology to known proteins. Recent studies on the selenoprotein have suggested insights into its function and a role in human health. Its identification as a binding partner to the UGTR protein implicates Sep15 in the control of protein folding in the endoplasmic reticulum. In addition, several biochemical and genetic observations have suggested that Sep15 may be involved in cancer development and may mediate, at least in part, the chemoprotective effect of selenium. While these data hold great promise, further studies are necessary to determine the relevance of Sep15 to both protein folding and cancer prevention.

**Acknowledgements:** We thank members of our laboratories for contributions they made in characterization of Sep15 and its gene. This study was supported by a grant from the Cancer Research Foundation of America (to VNG).

## References

1. VN Gladyshev, KT Jeang, JC Wootton, DL Hatfield 1998 *J Biol Chem* 273:8910

2. E Kumaraswamy, A Malykh, KV Korotkov, S Kozyavkin, Y Hu, SY Kwon, ME Moustafa, BA Carlson, MJ Berry, BJ Lee, DL Hatfield, AM Diamond, VN Gladyshev 2000 *J Biol Chem* 275:35540
3. Y Hu, KV Korotkov, R Mehta, DL Hatfield, C Rotimi, A Luke, TE Prewitt, RS Cooper, W Stock, EE Vokes, ME Dolan, VN Gladyshev, AM Diamond 2001 *Cancer Res* in press
4. KV Korotkov, E Kumaraswamy, Y Zhou, DL Hatfield, VN Gladyshev, submitted
5. M Kalcklosch, A Kyriakopoulos, C Hammel, D Behne 1995 *Biochem Biophys Res Commun* 217:162
6. GV Kryukov, VM Kryukov, VN Gladyshev 1999 *J Biol Chem* 274:33888
7. C Ritter, A Helenius 2000 *Nat Struct Biol* 7:278
8. AJ Parodi 2000 *Annu Rev Biochem* 69:69
9. S Avigad, H Benyaminy, Y Tamir, D Luria, I Yaniv, J Stein, B Stark, R Zaizov 1997 *Eur J Cancer* 33:1983
10. S Mathew, VV Murty, GJ Bosl, RS Chaganti 1994 *Cancer Res* 54:6265
11. LC Clark, GF Combs, BW Turnbull, EH Slate, DK Chalker, J Chow, LS Davis, RA Glover, GF Graham, EG Gross, A Krongrad, JL Lesher, HK Park BB Sanders, CL Smith, JR Taylor 1996 *J Am Med Assoc* 276:1957
12. K Yoshizawa, WC Willett, SJ Morris, MJ Stampfer, D Spiegelman, EB Rimm, E Giovannucci 1998 *J Natl Cancer Inst* 90:1219
13. HE Ganther 1999 *Carcinogenesis* 20:1657

# Chapter 14. Selenoproteins of the glutathione system

Leopold Flohé

*Department of Biochemistry, Technical University of Braunschweig, Mascheroder Weg 1, D-38124 Braunschweig, Germany*

Regina Brigelius-Flohé

*Department of Vitamins and Atherosclerosis, German Institute for Research of Nutrition (DIfE), Arthur-Scheunert-Allee 114-116, D-14558 Potsdam, Germany*

**Summary:** The protein family of glutathione peroxidases (GPx) is spread over the entire living kingdom. Four distinct molecular clades characterized by an active site selenocysteine residue coexist in vertebrates. Such selenoperoxidases have sporadically been detected in lower animals, while GPx homologs having the active site selenocysteine replaced by cysteine are more common and appear to be the only representatives of the family in plants and bacteria.

Distinct biological roles of the individual GPx types are inferred from differences in catalytic efficiency, substrate specificity, tissue distribution, subcellular compartmentalization and selenium-dependency of biosynthesis. All selenocysteine-containing GPx types are highly efficient in the reduction of hydroperoxides having apparent rate constants of this partial reaction, $k'_1$, near $10^7$ $M^{-1}sec^{-1}$, while the cysteine homologs, with a $k'_1$ near $10^4$ $M^{-1}sec^{-1}$, are poor peroxidases. The scope of accepted hydroperoxides appears to increase from the cytosolic GPx (cGPx) and the gastrointestinal type (GI-GPx) over the extracellular one (pGPx) to phospholipid hydroperoxide GPx (PHGPx), while the specificity for glutathione (GSH) declines in this order. Compelling evidence defines cGPx as an emergency device to detoxify $H_2O_2$ and soluble hydroperoxides in pathological conditions. Being dispensible for survival, as evidenced by inverse genetics, cGPx is nevertheless essential for the maintenance of hydroperoxide homeostasis in oxidatively unchallenged conditions, as demonstrated by mimicking the development of Keshan disease in mice lacking the cGPx gene, i.e, cGPx (-/-) mice. The biological role of pGPx is still elusive, as is that of GI-GPx. Being unable to substitute for cGPx in challenged (-/-) mice, the functions of these enzymes have to be sought in local regulation of peroxide-dependent processes. PHGPx appears to be a key player in silencing leukotriene biosynthesis, in dampening cytokine-dependent NFκB activation, and in sperm differentiation. The

## Introduction

Glutathione (GSH; γ-glutamyl-cysteinyl-glycine) had been recognized to maintain the intracellular redox-balance long before its role as a substrate of selenium-containing peroxidases was discovered [1]. In the late 1950s and early 1960s, the relevance of adequate GSH levels to cellular integrity was corroborated by clinical observations indicating that redox-cycling drugs or pro-oxidant xenobiotics may cause hemolytic anemia in certain subjects who otherwise appeared healthy. The underlying defects, deficiencies in glucose-6-phosphate dehydrogenase [2], 6-phosphogluconate dehydrogenase [3], glutathione reductase [4,5], glutathione synthetase [6,7], or γ-glutamyl-cysteine synthetase [8], share a common feature in that reduced glutathione can not be regenerated at adequate rates when the red blood cells are challenged oxidatively. Nevertheless, the obvious interpretation that GSH might be essential for the reduction of hydroperoxides was not generally accepted, even though G. C. Mills had discovered an enzymatic activity in red blood cells that appeared to protect hemoglobin from oxidative denaturation by reducing $H_2O_2$ at the expense of GSH [9]. The reluctancy to accept that this "GPx" was relevant to $H_2O_2$ detoxification persisted due to the firm belief that the extreme efficiencies of heme-containing peroxidases, notably catalase, could not be accomplished by any other type of enzyme [10]. For quite some time, this dogma was seriously challenged by only a few publications. For example, Cohen and Hochstein [11] demonstrated that $H_2O_2$ did not denature hemoglobin in red blood cells despite the abundance of catalase, unless GSH decreased to almost undetectable levels. GPx deficiency was also reported to be associated with hemolytic disorders [12]. Furthermore, Flohé and Zimmermann [13,14] found catalase to be less efficient than GPx in preventing peroxidative destruction of mitochondrial membranes and Sies et al. [15] provided compelling evidence for a pivotal role of GPx in cytosolic hydroperoxide metabolism of perfused rat liver. Finally, the kinetic analysis of isolated bovine GPx unequivocally revealed a molar efficiency higher than that of catalase [16,17].

The early history of GPx is discussed herein because it led to the first identification of a mammalian selenoprotein. Hoekstra's group in Wisconsin had observed that red blood cells of selenium-deficient rats resembled those of patients with glucose-6-phosphate dehydrogenase deficiency. That is, these cells lysed upon $H_2O_2$ exposure more readily than normal ones. All enzymes constituting the pentose-phosphate shunt, however, were normal and the GSH levels were rather elevated [18]. Unlike in most cases of clinical drug-induced hemolysis, the utilization rather than the regeneration of GSH appeared impaired. When the selenium-deficient cells were finally analyzed for GPx, the activity was found severely depressed and could be re-expressed by supplementing the rats with selenium. Moreover, when the rats were supplemented with $^{75}$Se, the radioactivity co-chromatographed with a protein fraction displaying GPx activity. Intrigued by the first report on these findings [19], Flohé and co-workers subjected a bovine GPx preparation, that had survived several chromatographic purifications [20], and could even be crystallized [21], to neutron activation analysis. They found precisely four gramatoms of selenium per mole of the homotetrameric enzyme [22]. This stoichiometric selenium content was then soon confirmed for GPx from sheep [23], rat [24], and man [25].

For more than a decade, GPx, which we now call the classical or cytosolic GPx (cGPx; GPx-1), remained the only known mammalian selenoprotein. Accordingly, most of the selenium deficiency syndromes in higher animals were tentatively attributed to the role of selenium in antioxidant defense. This proposal can no longer be upheld, since selenium has been discovered in several distinct families of enzymes and in a growing number of proteins of still undefined function [26,27]. It may even be questioned whether all peroxidases of the GPx family are exclusively engaged in hydroperoxide detoxification. In fact, many recent observations suggest that the structural divergence within this peroxidase family is associated with a pronounced functional specialization.

**Basic characteristics of the GPx family**
Glutathione peroxidase activity may be associated with a variety of proteins that may or may not be related phylogenetically. The reaction of GSH with ROOH has been reported to be also catalysed by GSH-S-transferases [28], selenoprotein P [29] and a peroxiredoxin [30] and none of these proteins have any significant sequence homology with cytosolic GPx. The latter enzyme is a member of an old protein family that is distributed throughout the entire living kingdom. The family is clearly split into several molecular clades (Figure 1) which may coexist within a species. They may or may not contain selenium and it has to be questioned whether the homology of the GPx selenium-free proteins justifies the assumption of an efficient peroxidase activity. Neither can a glutathione peroxidase activity be uncritically attributed to any of the related proteins.

One of the common features of the GPx family is a strictly conserved catalytic triad composed of (seleno)cysteine, glutamine and tryptophan residues. This triad is made up by distant loops of the subunit. Selenocysteine is present in the four prototypes of GPx prevailing in, yet not restricted to, vertebrates: 1) in cGPx [31,32]; 2) in the extracellular GPx (pGPx, GPx-3) discovered by Takahashi and coworkers [33,34]; 3) in phospholipid hydroperoxide GPx (PHGPx, GPx-4) identified by Ursini et al. [35-38]; and 4) in the gastro-intestinal GPx (GI-GPx, GPx-2) first described

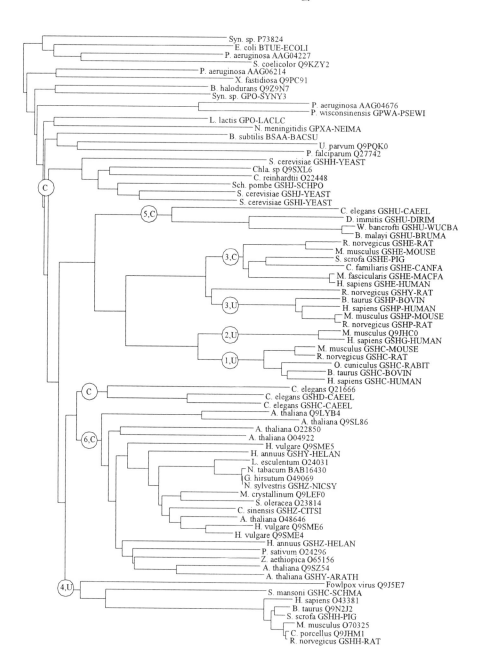

**Figure 1.** Dendrogram of the GPx superfamily. Branches designated U contain selenocysteine, and those marked with C comprise cysteine homologs. 1, cGPx of mammals; 2, GI-GPx of mammals; 3, pGPx of mammals (3,U) and their cysteine homologs (3,C); 4, mammalian PHGPx and related selenoproteins of *Schistosoma mansoni* and fowlpox virus; 5, cuticular cysteine homologs in helmints; and 6, plant GPx homologs; the remaining ones are cysteine homologs in prokarya and lower eukarya.

*Selenoproteins of the glutathione system* 161

by Chu et al. [39]. Cysteine-containing congeners are also seen in vertebrates where they appear to be a side branch of pGPx [40], and appear to be abundant in plants, lower fungi, protozoa and bacteria (see Figure 1). As already deduced from the x-ray analysis of cGPx [41], the selenol in the triad is hydrogen-bonded to the imino nitrogen of the tryptophan residue and to the amido nitrogen of the glutamine residue (Figure 2). Thereby the selenol is dissociated and polarized to facilitate the reaction with the hydroperoxide substrate, which is the first step of the catalytic cycle. In this microenvironment rate constants for the oxidation of the selenocysteine by $H_2O_2$ near $10^7$ $M^{-1}$ $sec^{-1}$ are consistently achieved [16,17,42,43]. A cysteine residue within this triad also appears largely ionized, as is suggested by an equally fast reaction with iodoacetate [44]. Nevertheless, the reaction of such activated cysteines with hydroperoxides proved to be three orders of magnitude slower, as demonstrated with muteins of cGPx [45] and PHGPx [44] and with a GPx homolog of *Plasmodium falciparum* [46]. The validity of the proposed activation of selenium or sulfur, respectively, has been established by systematic site-directed mutagenesis of the residues constituting the catalytic triad [44].

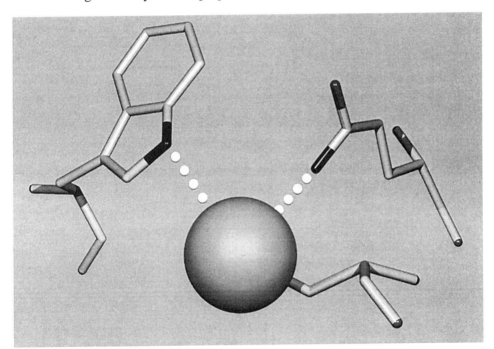

**Figure 2.** Molecular model of the catalytic triad of bovine cGPx [41,61] composed of selenocysteine (U52), glutamine (Q87), and tryptophan (W165). Nitrogen atoms are marked in dark, the selenium-atom is shown as a ball, possible hydrogen bridges are indicated as lines of four small dots. This triad is conserved with either selenocysteine or cysteine in the entire GPx family.

The product of the first step of catalysis is an intermediate in which the selenium (or sulfur) is oxidized to a selenenic (or sulfenic) acid derivative (E-SeOH). Chemical proof of such an intermediate is still lacking, yet stoichiometric considerations do not allow the assumption of a different oxidation state. It can not be excluded, however, that the highly reactive selenenic acid intermediate stabilizes itself by forming a bond to one of the nearby nitrogens with elimination of water. Thereby, an intermediate similar to the GPx mimic ebselen could be formed. Irrespective of the precise structure of the first intermediate, a selenadisulfide (disulfide) bond between the enzyme and the thiol substrate can be formed in the consecutive step. This may happen, as frequently proposed [47], by a reaction of RSH, typically of GSH, with the SeOH group in the oxidized enzyme or, as demonstrated for the ebselen catalysis [48], by reductive cleavage of a Se-N bond by GSH. Reaction of the resulting second intermediate (E-Se-SG) with another thiol then regenerates the ground-state reduced enzyme. The existence of the second intermediate could be verified by reacting cGPx with [$^{35}$S]GSH and isolation of a product containing $^{35}$S and Se in a 1/1 stoichiometry [49,50]. The intermediate, either formed from reduced enzyme, GSH and ROOH or enzyme plus GSSG, can also be detected by electrochemical methods [51,52]. Further, dead-end inhibition is observed with compounds being substituted at the C-atom next to the thiol group (e.g., penicillamine and mercaptosuccinate), presumably because the attack of the intermediate E-Se-SR by GSH is sterically hindered [47].

Needless to state that the catalysis, as outlined, reflects a typical "enzyme substitution mechanism". Accordingly, ping-pong kinetics were observed whenever a member of the GPx family was kinetically analyzed [16,17,42,43,46]. Usually, the kinetics of glutathione peroxidases can be described by the Dalziel equation:

$$\frac{[E_o]}{V} = \frac{\varphi_1}{[ROOH]} + \frac{\varphi_2}{[GSH]} \qquad (1)$$

which is a simplified version of the general Dalziel equation for a three-substrate ping-pong mechanism [53]:

$$\frac{[E_o]}{V} = \varphi_0 + \frac{\varphi_1}{[A]} + \frac{\varphi_2}{[B]} + \frac{\varphi_3}{[C]} \qquad (2)$$

Equation (1) describes a two-substrate reaction, which is formally justified, since substrates B and C are identical, i.e., GSH. The empirical coefficient $\varphi_2$, therefore, characterizes all GSH-dependent steps. Its physical meaning is accordingly complex, but can be defined as the reciprocal value of $k_2'$ which is the net forward rate constant of the reductive part of the catalytic cycle. $\varphi_1$ is the reciprocal net rate constant for the oxidation of the enzyme by hydroperoxide, $k'_1$. $k'_1$ approximates the real forward rate constant $k_1$, because the oxidation of the enzyme by ROOH can be assumed to be irreversible. It thus characterizes the peroxidatic efficiency. As a rule, the term $\varphi_o$ is zero for GPx-type enzymes which implies that $K_m$ and $V_{max}$ values are infinite. This observation has often been discussed to indicate low affinities of GPx-type enzymes for their substrates. It may be stressed here that this conclusion is not justified. In case of cGPx, the infinite $K_m$ values simply mean that the formations of enzyme-substrate complexes are slower than the reactions within such complexes, which is by no means surprising with regard to the chemical reactivity of the functional groups involved in the catalysis.

The empirically accessible kinetic coefficients $\varphi_1$ and $\varphi_2$ (or the pertinent k' values) can be used to calculate the substrate turnover by GPx-type enzymes for conditions which are not easily analyzed experimentally. Such estimates reveal that under physiological conditions GPx turnover is primarily determined by the $k_1$ for a particular hydroperoxide and the enzyme concentration. The cellular GSH content is irrelevant in this case, unless it drops to less than 10% of that normally seen in cells. In other words, the ratio of rate constants and physiological substrate concentrations complies with an in vivo situation where the fully reduced enzyme is primed to scavenge the hydroperoxide molecule.

**Molecular and functional divergence**
Thus far, all members of the GPx family display glutathione peroxidase activity. This does not necessarily imply, however, that this activity in each case represents the main or physiological function. Furthermore, recent data do not support the assumption that GPx enzymes coexisting in an organism might simply substitute for each other.

While cGPx, GI-GPx and pGPx are tetrameric, PHGPx-type enzymes are monomeric. The subunit interaction sites in cGPx and pGPx are known from x-ray crystallography [41,54] and are not present in mammalian PHGPx [55] and in a non-selenium homolog of *P. falciparum* [46], which is also monomeric. In the tetrameric enzyme, the active site selenium is located in a valley formed by interacting subunits, whereas in the monomeric enzyme (PHGPx), the active site should be more accessible, which according to molecular modelling has a spheric shape with the active site selenium exposed in the center of a flat lipophilic depression. This peculiarity may explain the broad specificity of PHGPx for hydroperoxides comprising, apart

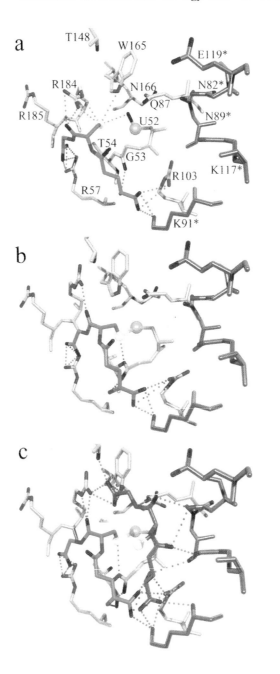

**Figure 3**. Molecular models [see 61] based on x-ray data of bovine cGPx [41] demonstrating the specific interactions of catalytic intermediates with GSH. Blue color, nitrogen; red color, oxygen; and grey and dark grey color, carbons of first and second subunit; and yellow and orange balls, sulfur and selenium, respectively. a shows the oxidized enzyme (E-SeOH) complexed with GSH; b, an intermediate with a covalently bound glutathionyl residue (E-Se-SG); and c, a complex with a second GSH (E-Se-SG·GSH).

from H$_2$O$_2$, phosphatidylcholine hydroperoxide and cholesterolester hydroperoxides even when these are integrated into lipoproteins or biomembranes [56]. In contrast, the specifity of cGPx, and likely that of GI-GPx, is restricted to H$_2$O$_2$ and soluble organic hydroperoxides [57,58]. pGPx, which also acting on hydroperoxides of phosphatides, but not on cholesterolesters, adopts an intermediate specificity [56,59].

In addition, pronounced differences in donor substrate specificity within the GPx family have been reported. In fact, a strict specificity for GSH was only documented for cGPx. Based on activities with a large number of cysteine derivatives and other low molecular weight thiols, both carboxylic groups of GSH were implicated in substrate binding [60]. The structure of cGPx shows the active site selenocysteine surrounded by four arginine residues at a suitable distance for electrostatic interaction with the substrate carboxylates. The precise mode of substrate binding could unfortunately not be unambiguously verified due to low substrate occupancy in crystallized cGPx [41]. Molecular modelling [61] predicts that the glycine carboxylate of GSH is bound to Arg57 and its γ-glutamate tail is attracted by the positive charges of Arg103 and Lys91' of an adjacent subunit (Figure 3a). By this orientation, which is further supported by various hydrogen bridges, the sulfur of GSH is directed into an ideal position to react with the oxidized selenocysteine. In the resulting intermediate, the orientation of the now covalently bound glutathionyl residue is essentially the same, but the interaction with Arg184 and Trp165 appears to be weakened (Figure 3b). The main driving forces for binding the second molecule of GSH appear to be the excess positive charges provided by Arg103, Lys103' and Arg184 [Arg103 and Lys 91' and by Arg184], while the nearby Arg185 is not easily brought into contact with the substrate (Figure 3c).

The basic residues implicated in binding of GSH are strictly conserved in cGPx of various species. GI-GPx comes closest to cGPx only having Lys91 replaced by glutamine and Arg185 by threonine. In pGPx, only Arg103 and Arg185 are conserved. In PHGPx, none of the charged residues nor any functional substitute are retained. These characteristics suggest a decreasing specificity for GSH in the order cGPx > GI-GPx > pGPx >> PHGPx. Several observations underscore the relevance of this deduction: 1) The $k'_2$ values, which reflect the speed of complex formation with GSH rather than catalytic efficiency, are at least one order of magnitude higher for cGPx [16] than for pGPx [43] and PHGPx [42]; 2) pGPx was found to be equally active with GSH, thioredoxin and glutaredoxin [62]; 3) PHGPx, apart from being active with various dithiols [63], tends to act on various protein thiols when the predominant cellular thiol, GSH, becomes limiting [64-66]; and 4) the *P. falciparum* GPx homolog, which apart from having selenocysteine replaced by cysteine resembles mammalian PHGPx, clearly prefers thioredoxins over GSH [46].

**Tissue distribution and subcellular location**
The tissue-specific expression of glutathione peroxidases has recently been reviewed extensively [67] and is only briefly summarized and up-dated herein.

The tissue distribution of cGPx largely parallels oxidative metabolism. It is high in liver, kidney, lung, red blood cells and placenta, but is also detected in other tissues. Within cells, it is present as a soluble enzyme of the cytosol and the mitochondrial matrix space. In rat liver, 75% of the enzyme is located in the cytosol and 25% in mitochondria, while the amounts in the remaining compartments are negligible [68]. The mitochondrial and the cytosolic fractions of cGPx are derived from the same gene, since in cGPx (-/-) mice, the activity is abrogated in both the cytosol and mitochondria [69].

Levels of cGPx are consistently higher in female than in male rats [70] and appear to be upregulated by estrogens [71], a phenomenon still waiting for a molecular explanation. The expression of cGPx may be stimulated by oxidative stress to some extent (reviewed in [72]) and an oxygen-responsive element was identified in the 5' flanking region of the cGPx gene [73].

A typical leader sequence classifies pGPx as an extracellular enzyme. Accordingly, pGPx is only detected in extracellular body fluids such as blood plasma, chamber water of the eye, or amniotic fluid where it is found without the leader sequence as a glycosylated protein [34]. Closely related androgen-responsive non-selenium homologs of pGPx, also initially synthesized with a leader sequence, are present in seminal fluid without the leader sequence [74].

GI-GPx is primarily found in the epithelium lining of the gastrointestinal tract from the esophagus to the rectum [75,76]. It is, however, also present in the human liver cell line HepG2 [75] and has recently been detected in a mammary cancer cell line [77]. In human ileum, GI-GPx is higher in the crypts than in the villi and particularly enriched in Panet cells at the bottom of the crypts. It is considered to be a cytosolic protein. It usually appears to be diffusely spread over the whole cell body, when visualized by immune histochemistry [78]. In human colon crypts, however, it forms a cap-like structure on the apical site of the nucleus suggesting an association with the Golgi system. It does, however, not co-localize with typical markers of cis and trans Golgi such as βCOP and TGN38 [78].

The distribution of PHGPx is unusual in several respects. High activities are found in endocrine tissues in general, and particularly in postpuberal testis. PHGPx also appears to account for most of the GPx activity of brain. PHGPx activities, however, do not consistently match PHGPx protein measured by immunological techniques [79], which indicates the occurrence of enzymatically inactive enzyme forms, e.g., in liver [79] and in spermatozoa [65]. The subcellular distribution is equally puzzling and varies largely between cell types and status of differentiation. PHGPx has been described as a cytosolic protein, but also to be associated with nuclei [64], nuclear proteins [64], mitochondria [80-83] and as a constituent of the mitochondrial capsule of sperm [65 and Chapter 22). Also, the PHGPx gene contains two potential start codons that can mutually be used in a tissue-specific manner. Use of the first start codon results in a pre-protein containing a leader sequence targeting the protein to mitochondria [82,83]. Upon import, the leader sequence is cleaved and the mitochondrial PHGPx becomes indistinguishable from the cytosolic form, which is synthesized if the second ATG codon is used [83]. Factors determining the alternative use of the start codons and tissue-specific expression have not yet been identified.

**Response to selenium**
Glutathione peroxidases respond differently to selenium deficiency. They occupy extreme positions in the "hierarchy of selenoproteins". cGPx is commonly considered to rank lowest in this hierarchy, which means, it rapidly declines upon selenium deprivation and is resynthesized with considerable delay upon selenium repletion [72]. The extracellular GPx behaves similar to cGPx [84]. PHGPx, however, has long been considered as the example of a selenoprotein ranking high in the hierarchy since it is fairly resistant to selenium deficiency and is rapidly synthesized when selenium is replenished [85]. More recently, GI-GPx was thought to compete with the deiodinases as the most resistant selenoprotein to selenium deficiency within the hierarchy [86,87]. The relative position of selenoproteins within this hierarchy is believed to reflect their relative biological importance. The selenium-responsiveness of the four GPx types, however, conflicts with this assumption. GI-GPx and cGPx reflect the two extremes on the scale with being responsive to selenium. The genes for both these selenoproteins have been knocked out without creating any overt pathological effects [88-90].

The molecular mechanisms leading to differential synthesis of selenoproteins under selenium restriction are far from being clear. The wide range of selenium responsiveness of the closely related glutathione peroxidases offers a unique opportunity to scrutinize mechanistic concepts presently discussed [72,91]. As a rule, the ranking of selenoproteins parallels the stability of the pertinent mRNA. The ranking of the four GPx types fully supports the view that mRNA stability is the most important factor determining GPx gene expression. The mRNAs of cGPx and pGPx are degraded rapidly, PHGPx mRNA remains fairly stable and GI-GPx mRNA tends to increase during selenium deficiency [86,93-95]. The question of how selenium specifically stabilizes the individual mRNAs, however, remains unanswered. The instability of cGPx mRNA has been attributed to "cytoplasmatic nonsense-mediated decay", a phenomenon describing the elimination of mRNA species having a stop codon located at a certain distance from a pre-mRNA splice site [91]. The process of detecting the in-frame UGA codon in the cGPx mRNA, which in the absence of bioavailable selenium means stop or nonsense, is believed to compete with selenium-dependent selenocysteine incorporation at the ribosome. While this mechanism may well explain the instability of cGPx mRNA, it is hard to understand why the mRNAs of PHGPx and GI-GPx that have their UGA in a homologous position as cGPx mRNA remain stable or are increased during selenium deficiency. Differences in translational efficiencies can not account for the different stabilities either. In reporter gene constructs, the 3' flanking region of cGPx mRNA harboring the SECIS element proved to be more efficient in incorporating selenocysteine than that of GI-GPx mRNA, when tested under selenium-adequate conditions [86]. The maximum SECIS efficiencies, however, do not correlate with the selenium-responsiveness of SECIS efficiencies. Interestingly, the reporter gene-linked 3'UTR of GI-GPx, in contrast to that of cGPx, enabled a comparatively low, but largely selenium-independent selenocysteine incorporation. This might indicate that the GI-GPx mRNA is stabilized by a factor binding to its SECIS in a selenium-independent manner or having such high affinity to selenium that it remains saturated with selenium in experimental selenium deficiency. A likely candidate for this stabilizing factor appeared to be the eukaryotic homolog of SelB, which in bacteria binds to both, the SECIS and the selenocysteyl-loaded tRNA$^{(ser)sec}$ [96]. After the eukaryotic SelB has been characterized [97,98], it is evident that the situation must be more complex; the tRNA$^{(ser)sec}$ binds specifically to eukaryotic SelB [97a,97b,98], but not to SBP2, which in eukaryotes is the protein recognizing the SECIS element [99]. Eukaryotic SelB apparently does not interact directly with the SECIS or the 3'-UTR, but in vivo forms a supercomplex with tRNA$^{(ser)sec}$, mRNA and SBP2 on the ribosome [97a,97b]. It thus remains an appealing hypothesis that the differential stabilities of selenoprotein mRNAs and, in consequence, the selenium responsiveness of selenoprotein biosynthesis result from different affinities of SBP2 to the individual SECIS elements. These affinities could well be modulated by loading of SelB with tRNA$^{(ser)sec}$ in the translation complex. Yet additional factors binding to the mRNAs in a sequence specific manner or modulating transcription in a selenium-dependent manner cannot at present be ruled out.

**Mutual support, complementation or distinct functions?**
A group of co-existing enzymes, like the glutathione peroxidases, that have identical catalytic mechanisms and similar efficiencies and specificities may be suspected to play the same role in biology of guaranteeing optimum performance by means of redundancy. Complementary distribution would suggest regional responsibilities without questioning the principle function. Yet what is the evidence that all types of glutathione peroxidases are

providing more or less the same role in a living organism?

The cGPx knock-out mouse tells us that cGPx is little else than an emergency device to cope with oxidative challenge exerted by hydroperoxides. Unchallenged cGPx (-/-) mice developed normally and even grew faster and tolerated elevated oxygen tension [88]. The lack of any overt phenotype in these mice is not surprising in light of the clinical phenotype associated with patients harboring deficiencies in GSH regeneration. Such patients are normal as long as they are unchallenged with hyperoxides and the related genetic defects were accordingly rated as "non-diseases" [100]. It is surprising that the complete lack of cGPx in mice is tolerated as a more or less pronounced impairment of GSH regeneration due to decreased efficiencies or faster aging of mutated enzymes in patients. Like patients with glucose-6-phosphate dehydrogenase deficiency, the cGPx (-/-) mice are highly susceptible to oxidative damage, as has been shown by exposure to redox-cycling herbicides [89,101,102] and bacterial lipopolysaccharides that trigger an oxidative burst in phagocytes [103].

The cGPx (-/-) mouse demonstrates that a seemingly normal life without cGPx is threatened by certain environmental hazards. Most importantly, the cGPx (-/-) mouse proved to be a model for the classical Chinese selenium-deficiency syndrome of Keshan disease [104]. When these mice were exposed to a non-virulent Coxsackie strain, the virus rapidly mutated into a virulent form [105], as had been observed previously with selenium-deficient mice [106,107]. Based on these studies, the debate whether Keshan disease is caused by selenium-deficiency itself or a viral infection [108] can likely be settled. A decrease in cGPx, the enzyme responding fastest to selenium deprivation, results in elevated steady-state-levels of hydroperoxides. Elevated hydroperoxides in turn accelerate the mutation rate of usually benign viral strains and allow virulent mutants to become dominant. This interpretation is supported by repetition of in vivo experiments in a tissue culture system [109], which lacks a potentially disturbed immune surveillance that was implicated in the development of pathogenic viral strains. A decreased non-specific immune response due to the lack of cGPx is not supported by experimental data in any case. Overexpression of cGPx in tissue culture cells reduces TNFα-induced NFκB activation [110], as does selenium-supplementation [111,112], and enhances the spread of HIV, probably due to suppressed apoptosis [113].

The data obtained with cGPx (-/-) mice and cells overexpressing cGPx support the anticipated role of cGPx as an antioxidant enzyme. The experiments that oxidatively challenged cGPx (-/-) mice were all performed with selenium-deprived and selenium-adequate animals, and, as a rule, the selenium-status did not significantly alter the results. This surprising outcome allows two alternative interpretations. Either the selenium deficiency was not sufficient to decrease the other selenoproteins to any relevant degree or, more likely, none of the remaining selenoproteins can efficiently substitute for cGPx in countering a systemic oxidative challenge. The latter conclusion may be considered as provocative in view of the presence of three more selenoprotein peroxidases and of the metabolic link of thioredoxin reductases to peroxiredoxin-type peroxidases. The emerging data, however, appear to justify a reconsideration of the biological roles of the various GPx types in any case.

Plasma GPx has a limited chance to counter a serious challenge by hydroperoxides, since micromolar concentrations of extracellular GSH or the tiny amounts of extracellular reduced thioredoxin would soon be consumed in absence of any regenerating system. Up to now, the biological role of plasma GPx remains speculative. It is implicated in the reduction of lipid hydroperoxides in low density lipoproteins (LDL) [114,115] and thus might be relevant to Steinberg's ideas on atherogenesis [116]. In view of the low steady-state level of peroxidized lipoproteins, the limited reduction capacity of pGPx may suffice and could be sustained by export of reduced GSH [117]. The substrate specificity of pGPx, however, is not ideal for the reduction of peroxidized LDL, since it does not reduce peroxidized cholesterol esters [114]. The enzyme could reduce soluble lipid hydroperoxides and might also regulate eicosanoid metabolism, which have been implicated in the activation of cyclooxygenase, the key enzyme of prostaglandine synthesis [118]. Similarly, the activation of other lipoxygenases, which typically remain dormant in the absence of any hydroperoxides, could be prevented by pGPx, as occurs in vitro by cGPx [119]. This hypothesis, however, suffers from the lack of knowledge about the relevance of extracellular lipid hydroperoxides to the activation of intracellular lipoxygenases.

More plausibly, pGPx, which is a highly efficient peroxidase, could be regarded as a redox buffer that is required to discriminate between irrelevant and serious inflammatory stimuli [67]. Whenever $\cdot O_2^-$ release by NADPH oxidase is triggered by any kind of irrelevant irritation, the elicited "microburst", by peroxide-mediated amplification mechanisms, could turn into a full-blown inflammatory response, if $H_2O_2$ were not scavenged instantly. On the other hand, the capacity to scavenge $H_2O_2$ has to be overcome if, as in case of a serious infection, a real host defense reaction is required. The puzzling situation of having a fast acting peroxidase in an environment with almost no reducing substrate may have evolved to make the most economic use of the oxidative host defense machinery. This defense machinery is indispensable for survival in a hostile environment, but self-destructive when overreacting. Clearly, pGPx, which is sometimes designated as an orphan enzyme due to lack of regenerating substrates and of general interest, deserves to be further investigated.

GI-GPx, because of its location in the intestine, has been proposed to prevent systemic access of food-born peroxides, which could pre-exist in food that is generated by the intestinal flora or by the mucosa when they metabolize xenobiotics [120]. However, experimental evidence in support of this concept is lacking. The GI-GPx gene was recently knocked out in mice. Reportedly, GI-GPx (-/-) mice appear to have a normal phenotype [121], but mice deficient in both cGPx and GI-GPx showed retarded growth after weaning and morphological alterations of the intestinal mucosa such as shortened villi and loss of goblet cells [121]. Additional histological details of these mice have not yet been made available nor have any experimental challenges to environmental stress been reported.

Pathological changes that occur in the double knockout, but not in either of the single knockout mice, suggest a mutual complementation of GI-GPx and cGPx. This does not necessarily imply that the two enzymes complement each other in antioxidant defense or in the protection against alimentary peroxides. They could complement each other in regulating cellular processes subject to redox regulation, e.g., proliferation rate or apoptosis. Inhibition of apoptosis has been documented in the overexpression of both cGPx [113] and PHGPx [122-124], which may serve a potential role common to all selenoperoxidases. In the intestinal epithelium, a steep gradient of GI-GPx content is present which declines from the proliferating stem cells in the lower crypts to the luminally exposed cells gradually undergoing apoptosis [78]. If this delicate balance of events is indeed regulated by GI-GPx, the absence of both selenoperoxidases could unmask a "disregulation" that is compensated for by cGPx in the GI-GPx knockout. The intriguingly high concentration of GI-GPx in Panet cells, which are involved in mucosal immunity, and not absorption, and its association in colon cells with vesicular structures [78] also suggest highly specialized functions that remain to be elucidated. Cautiously stated, the proposal that GI-GPx only lines the intestine to protect the organism against peroxides hypothetically derived from the largely anaerobic gut content appears naive.

PHGPx was discovered and characterized as an enzyme preventing progression of lipid peroxidation in biomembranes due to its unique ability to reduce hydroperoxo-groups in complex lipids [35]. In this context, it is synergistically supported by tocopherols, which reduce lipid peroxy radicals to lipid hydroperoxides. The latter, if not reduced by PHGPx, would re-initiate free radical-mediated lipid peroxidation by Haber-Weiss or Fenton-type chemistry [125]. Over the years, however, growing evidence indicated that protection of biomembranes against unspecific oxidative damage is only one and possibly not the most important role of PHGPx.

With other glutathione peroxidases, PHGPx shares the ability to reduce lipoxygenases [126,127], to inhibit apoptosis [122,123], and to suppress cytokine-induced NFκB activation [128]. In at least two cases, PHGPx was shown to be the biologically relevant regulator. In the first case, selenium-deficient rat basophilic leukemia cells as well as whole animals overproduce 5-lipoxygenase products comprising the potent pro-inflammatory leukotrienes. This phenomenon can be abrogated by short-term re-supplementation of selenium that normalizes the level of PHGPx, but not that of cGPx [126]. Similarly, leukotriene biosynthesis is suppressed in transformed rat basophilic leukemia cells selectively overexpressing PHGPx [123]. Within cells, therefore, PHGPx appears to be the principal selenoperoxidase in charge of reducing 5-lipoxygenase. Secondly, interleukin-1, as an example of a very potent pro-inflammatory cytokine, exerts its transcriptional activity by means of a complex phosphorylation cascade ultimately leading to activation and nuclear translocation of NFκB. This process has for long been recognized to be redox-regulated and affected by selenium status [reviewed in 129,130]. A moderate overexpression of PHGPx in the human ECV cell line completely abrogated interlukin-1 induced NFκB activation, while a huge variation of cGPx activity achieved by deprivation and resupplementation of selenium in pertinent control cells did not [128]. NFκB activation induced by hydroperoxides is similarly suppressed by overexpression of PHGPx in rabbit aortic smooth muscle cells [124]. PHGPx can thus be regarded as a key regulator of intracellular events initiating inflammatory responses and in this context cannot be substituted by the more abundant cGPx. How the peculiar specificity of PHGPx in regulating such process is achieved is unknown. It may be due to its preference for hydrophobic lipids or to subcellular microcompartmentalization. The observation that PHGPx is sometimes found oxidatively cross-linked to other proteins [64,65] suggests another possibility. That is, PHGPx, through reaction of its oxidized selenium with accessible thiols in particular proteins, could also act as a peroxide-dependent thiol-modifying agent.

Despite ongoing efforts, a knockout mouse, which could further shed light on the role of PHGPx, has not yet been obtained. Chimeric mice having more than 50% PHGPx (+/-) cells only produced PHGPx(+/+) offspring [131]. The relevance of this finding to the role of PHGPx in male fertility is discussed in a separate chapter of this book (see Chapter 22).

**Clinical relevance**

Both cGPx and pGPx are widely used to monitor the selenium status in man, other animals, and in tissue culture. Since both enzymes respond rapidly to a shortage in bioavailable selenium and are easily measured in small blood samples, they may be considered the biomarkers of choice. The enzyme activities not only indicate whether a particular selenium compound is absorbed and taken up by cells, but also to what extent it is metabolized for use in selenoprotein biosynthesis. Monitoring GPx activities is therefore superior to any chemical determination of selenium if the nutritional value of food or supplements is to be evaluated. pGPx activities are preferentially used to monitor short-term variations of selenium status, while cGPx activity of erythrocytes, due to their life time of 120 days and lack of de novo protein synthesis, indicate average selenium supply over the preceeding 4 months. Other selenoproteins have sporadically been recommended to monitor selenium status, which is justified if these also rank low in the hierarchy of selenoproteins. Determination of less selenium-responsive selenoproteins is, of course, required to evaluate their role in experimentally severe selenium deficiency.

Severe selenium deficiency, as observed in some rural areas of China where the soil is low in selenium, is associated with Keshan and Kashin-Beck disease [27,47,104,108 and Chapter 18]. Selenium deficiency is not a problem in Western countries with access to a variety of foods, even though the soil in the particular region may be lacking in selenium. Severe selenium-deficiency has also been recognized to be associated with myopathies and cardiac disturbances [see Chapter 18], but such abnormalities were almost exclusively seen in highly industrialized countries in patients maintained on total parenteral nutrition diets unsupplemented with selenium [47]. These diets, however, are no longer used.

Slight to moderate acute selenium deficiency, preferentially affecting the levels of pGPx and cGPx, does not appear to be uncommon. The pathophysiological consequences may be deduced from experiments with cGPx (-/-) mice and from clinically well documented deficiencies in GSH regeneration such as favism. A predisposition to hemolytic crises upon exposure to pro-oxidant drugs or xenobiotics can be anticipated and related oxidative damage in metabolically active parenchymal tissue can not be ruled out. Selenium deficiency should further aggravate systemic infections, which, in the initial phase at least, are associated with massive and ubiquituous $H_2O_2$ production by phagocytes. In fact, low selenium and GPx activities are almost consistently observed in patients with septic syndroms [132,133] and selenium supplementation is being considered in these conditions [134].

The relevance of long lasting cGPx or pGPx depression due to minor selenium deficiency to the development and progression of chronic diseases remains a matter of ongoing debate. The Finish Mobile Clinic Health Examination cohort study began in 1968 and enrolled a total of 58,000 men and women to examine the relationship of serum selenium levels and health. This study revealed a reduced risk of seronegative, but not of seropositve, rheumatoid athritis at high serum selenium levels, a trend towards reduced risk of asthma and, only in women, a reduced risk of coronary heart disease. No relevant benefit of selenium in stroke, diabetes, Alzheimer's and Parkinson's diseases was observed. Associated variations in GPx activities, thus, can only have a minor impact on such chronic diseases [135].

The relevance of a disturbed peroxide metabolism to the initiation of malignancies remains an attractive hypothesis. It was supported by early epidemiolical studies showing an inverse relationship of cancer mortality and blood selenium levels [136], an increased incidence of certain cancer types in patients with a calculated low selenium intake [137], and by animal studies on the prevention of chemical carcinogenesis dating back to 1949 [138]. As has been recently reviewed by Ganther [139], however, experimental and clinical cancer prevention by selenium is consistently achieved only with higher dosages of selenium than are required to optimize the levels of glutathione peroxidases or other known selenoproteins. Thus, the hypothesis that glutathione peroxidases, by metabolizing genotoxic hydroperoxides, might contribute to cancer prevention also remains elusive.

**Acknowledgements:** The preparation of this article was supported by the Deutsche Forschungsgemeinschaft (grants Fl61/12-1 and Bri778/5-1). The Figures were kindly prepared by Drs. H.-J. Hecht and K.-D. Aumann, GBF, Braunschweig, Germany.

**References**

1. S Colowick, A Lazarow, E Racker, DR Schwarz, E Stadtman, H Waelsch (Eds) 1954 *Glutathione* Academic Press, New York
2. PE Carson, CL Flanagan, CE Ickes, AS Alving 1965 *Science* 124:484
3. C Lausecker, P Heidt, D Fischer, H Hartleyb, GW Löhr 1965 *Arch Franc Pediatr* 22:789
4. PE Carson, GJ Brewer, C Ickes 1961 *J Lab Clin. Med* 58:804
5. GW Löhr, HD Waller 1962 *Med Klin* 57:1521
6. P Boivin, C Galand 1965 *Nouv Rev Franc Hemat* 5:707
7. P Boivin, C Galand, R André, J Debray 1966 *Nouv Rev Franc Hemat* 6:859
8. PN Konrad, FH Richards, FH Valentine, DE Paglia 1972 *New Engl J Med* 286:557
9. GC Mills 1957 *J Biol Chem* 229:189
10. P Nicholls 1972 *Biochim Biophys Acta* 279:306
11. G Cohen, P Hochstein 1963 *Biochemistry* 2:1420
12. TF Necheles 1974 *Glutathione* L Flohé, HC Benöhr, H Sies, HD Waller, A Wendel (Eds) Georg Thieme Publishers, Stuttgart, p 173
13. L Flohé, R Zimmermann 1970 *Biochim Biophys Acta* 223: 210
14. L Flohé, R Zimmermann 1974 *Glutathione* L Flohé, HC Benöhr, H Sies, HD Waller, A Wendel (Eds) Georg Thieme Publishers, Stuttgart, p 245
15. H Sies, C Gerstenecker, H Menzel, L Flohé 1972 *FEBS Letters* 27:171
16. L Flohé, G Loschen, WA Günzler, E Eichele 1972 *Z Physiol Chem* 353:987
17. WA Günzler, H Vergin, I Müller, L Flohé 1972 *Z Physiol Chem* 353:1001
18. JT Rotruck, AL Pope, HE Ganther, WG Hoekstra 1972 *J Nutr* 102:689
19. JT Rotruck, WG Hoekstra, AL Pope, H Ganther, A Swanson, D Hafeman 1972 *Federation Proc* 31:691Abs
20. L Flohé, B Eisele, A Wendel 1971 *Z Physiol Chem* 352:151
21. L Flohé, WA Günzler 1974 *Glutathione* L Flohé, HC Benöhr, H Sies, HD Waller, A Wendel (Eds) Georg Thieme Publishers, Stuttgart, p 132
22. L Flohé, WA Günzler, HH Schock 1973 *FEBS Letters* 32:132
23. SH Oh, HE Ganther, WG Hoekstra 1974 *Biochemistry* 13:1825
24. W Nakamura, S Hosoda, K Hayashi 1974 *Biochim Biophys Acta* 358:251
25. YC Awasthi, E Beutler, SK Srivastava 1975 *J Biol Chem* 250:5144

26. L Flohé, JR Andreesen, R Brigelius-Flohé, M Maiorino, F Ursini 2000 *IUBMB Life* 49:411
27. J Köhrle, R Brigelius-Flohé A Böck, R Gärtner, O Meyer, L Flohé 2000 *Biol Chem* 381:849
28. RF Burk, RA Lawrence 1978 "Non-selenium-dependent glutathione peroxidase" *Functions of glutathione in liver and kidney* H Sies and A Wendel (Eds) Springer Verlag, Berlin, p 114
29. Y Saito, T Hayashi, A Tanaka, Y Watanabe, M Suzuki, E Saito. K Takahashi 1999 *J Biol Chem* 274:2866
30. J-W Chen, C Dodia, SI Feinstein, MK Jain, AB Fisher 2000 *J Biol Chem.* 275:28421
31. JW Forstrom, JJ Zakowski, AL Tappel 1978 *Biochemistry* 17:2639
32. WA Günzler, GJ Steffens, A Grossmann, SM Kim, F Ötting, A Wendel, L Flohé 1984 *Hoppe Seylers Z Physiol Chem* 365:195
33. K Takahashi, N Avissar, J Whitin. H Cohen 1987 *Arch Biochem Biophys* 256:677
34. K Takahashi, M Akasaka, Y Yamamoto, C Kobayashi, J Mizoguchi, J Koyama 1990 *J Biochem (Tokyo)* 108:145
35. F Ursini, M Maiorino, M Valente, L Ferri. C Gregolin 1982 *Biochim Biophys Acta* 710:197
36. R Schuckelt, R Brigelius-Flohé, M Maiorino, A Roveri, J Reumkens, W Strassburger, F Ursini, B Wolf, L Flohé 1991 *Free Radic Res Commun* 14:343
37. R Brigelius-Flohé, KD Aumann, H Blöcker, G Gross, M Kiess, KD Klöppel, M Maiorino, A Roveri, R Schuckelt, F Ursini, E Wingender, L Flohé 1994 *J Biol Chem* 269:7342
38. RA Sunde, JA Dyer, TV Moran, JK Evenson, M Sugimoto 1993 *Biochem Biophys Res Commun* 193:905
39. FF Chu, JH Doroshow, RS Esworthy 1993 *J Biol Chem* 268:2571
40. NB Ghyselinck, JP Dufaure 1990 *Nucleic Acids Res* 18:7144
41. O Epp, R Ladenstein, A Wendel 1983 *Eur J Biochem* 133:51
42. F Ursini, M Maiorino. C Gregolin 1985 *Biochim Biophys Acta* 839:62
43. RS Esworthy, FF Chu, P Geiger, AW Girotti, JH Doroshow 1993 *Arch Biochem Biophys* 307:29
44. M Maiorino, KD Aumann, R Brigelius-Flohé, D Doria, J van den Heuvel, J McCarthy, A Roveri, F Ursini, L Flohé 1995 *Biol Chem Hoppe Seyler* 376:651
45. C Rocher, JL Lalanne, J Chaudiere 1992 *Eur J Biochem* 205:955
46. H Sztajer, B Gamain, K-D Aumann, C Slominanny, K Becker-Brandenburg, R Brigelius-Flohé, L Flohé 2001 *J Biol Chem* in press
47. L Flohé 1989 *Glutathione: Chemical, Biochemical and Medical Aspects - Part A* D Dolphin, R Poulson, O Aramovic (Eds). John Wiley & Sons Inc., New York, p 643
48. MJ Parnham, E Graf 1987 *Biochem Pharmacol* 36:3095
49. RJ Kraus, JR Prohaska, HE Ganther 1980 *Biochim Biophys Acta* 615:19
50. HE Ganther, RJ Kraus, SJ Foster 1984 *Methods Enzymol* 107:593
51. C Lehmann, U. Wollenberger, R Brigelius-Flohé, FW Scheller 1998 *J Electroanal Chem* 455:259
52. C Lehmann, U. Wollenberger, R Brigelius-Flohé, FW Scheller 2000 *J Electroanalysis* in press
53. K Dalziel 1969 *Biochem. J.* 114:547
54. B Ren, W Huang, B Akesson. R Ladenstein 1997 *J Mol Biol* 268:869
55. L Flohé, KD Aumann, R Brigelius-Flohé, D Schomburg, W Straßburger, F Ursini 1993 *Active Oxygen, Lipid Peroxides, and Antioxidants* K Yagi (Ed) CRC Press, Boca Raton, p 299
56. F Ursini, M Maiorino, R Brigelius-Flohé, KD Aumann, A Roveri, D Schomburg. L Flohé 1995 *Methods Enzymol* 252:38
57. A Grossmann, A Wendel 1983 *Eur J Biochem* 135:549

58. A Sevanian, SF Muakkassah-Kelly, S Montestruque 1983 *Arch Biochem Biophys* 223:441
59. Y Yamamoto, K Takahashi 1993 *Arch Biochem Biophys* 305:541
60. L Flohé, W Günzler, G Jung, E Schaich, F Schneider 1971 *Hoppe Seylers Z Physiol Chem* 352:159
61. KD Aumann, N Bedorf, R Brigelius-Flohé, D Schomburg, L Flohé 1997 *Biomed Environ Sci* 10:136
62. M Björnstedt, J Xue, W Huang, B Akesson, A Holmgren 1994 *J Biol Chem* 269:29382
63. A Roveri, M Maiorino, C Nisii, F Ursini 1994 *Biochim Biophys Acta* 1208:211
64. C Godeas, F Tramer, F Micali, A Roveri, M Maiorino, C Nisii, G Sandri, E Panfili 1996 *Biochem Mol Med* 59:118
65. F Ursini, S Heim, M Kiess, M Maiorino, A Roveri, J Wissing. L Flohé 1999 *Science* 285:1393
66. M Maiorino, L Flohé, A Roveri, P Steinert, JB Wissing, F Ursini 1999 *BioFactors* 10:251
67. R Brigelius-Flohé 1999 *Free Radic Biol Med* 27:951
68. L Flohé, W Schlegel 1971 *Hoppe Seylers Z Physiol Chem* 352:1401
69. FF Chu, RS Esworthy, YS Ho, M Bermeister, K Swiderek, RW Elliott 1997 *Biomed Environ Sci* 10:156
70. L Flohé, I Brand 1970 *Klin Chem Klin Biochem* 8:156
71. RE Pinto, W Bartley 1969 *Biochem J* 115:449
72. L Flohé, E Wingender, R Brigelius-Flohé 1997 *Oxidative stress and signal transduction* HJ Forman, E Cadenas (Eds) Chapman & Hall, New York, p 415
73. DB Cowan, RD Weisel, WG Williams, DAG Mickle 1993 *J Biol Chem* 268:26904
74. C Jimenez, NB Ghyselinck, A Depeiges, JP Dufaure 1990 *Biol Cell* 68:171
75. F-F Chu, RS Esworthy 1995 *Arch Biochem Biophys* 323:288
76. RS Esworthy, KM Swiderek, YS Ho FF Chu 1998 *Biochim Biophys Acta* 1381:213
77. FF Chu, RS Esworthy, L Lee, S Wilczynski 1999 *J Nutr* 129:1846
78. S Florian, K Wingler, K Schmehl, G Jacobasch, OJ Kreuzer, W Meyerhof, R Brigelius-Flohé 2000 submitted
79. A Roveri, M Maiorino, F Ursini 1994 *Methods Enzymol* 233:202
80. C Godeas, G Sandri, E Panfili 1994 *Biochim Biophys Acta* 1191:147
81. A Roveri, M Maiorino, C Nisii, F Ursini 1994 *Biochim Biophys Acta* 1208:211
82. TR Pushpa-Rekha, AL Burdsall, LM Oleksa, GM Chisolm, DM Driscoll 1995 *J Biol Chem* 270:26993
83. M Arai, H Imai, D Sumi, T Imanaka, T Takano, N Chiba, Y Nakagawa 1996 *Biochem Biophys Res Commun* 227:433
84. T Nakane, K Asayama, K Kodera, H Hayashibe, N Uchida, S Nakazawa 1998 *Free Radic Biol Med* 25:504
85. F Weitzel, F Ursini, A Wendel 1990 *Biochim Biophys Acta* 1036:88
86. K Wingler, M Böcher, L Flohé, H Kollmus, R Brigelius-Flohé 1999 *Eur J Biochem* 259:149
87. K Wingler, R Brigelius-Flohé 1999 *BioFactors* 10:245
88. YS Ho, JL Magnenat, RT Bronson, J Cao, M Gargano, M Sugawara, CD Funk 1997 *J Biol Chem* 272:16644
89. JB de Haan, C Bladier, P Griffiths, M Kelner, RD O'Shea, NS Cheung, RT Bronson, MJ Silvestro, S Wild, SS Zheng, PM Beart, PJ Hertzog, I Kola 1998 *J Biol Chem* 273:22528
90. RS Esworthy, JR Mann, M Sam, FF Chu 2000 *Am J Physiol Gastrointest Liver Physiol* 279:G426
91. X Sun, PM Moriarty, LE Maquat 2000 *EMBO J* 19:4734
92. MJ Christensen, KW Burgener 1992 *J Nutr* 122:1620
93. G Bermano, JR Arthur, JE Hesketh 1996 *FEBS Lett* 387:157
94. RA Sunde, JA Dyer, TV Moran, JK Evenson, M Sugimoto 1993 *Biochem Biophys Res Commun* 193:905

95. XG Lei, JK Evenson, KM Thompson, RA Sunde 1995 *J Nutr* 125:1438
96. K Forchhammer, K Boesmiller, A Böck 1991 *Biochimie* 73:1481
97a. RM Tujebajeva, PR Copeland, X-M Xu, BA Carlson, JW Harney, DM Driscoll, DL Hatfield, MJ Berry 2000 *EMBO Reports* 1:158
97b. D Fagegaltier, N Hubert, K Yamada, T Mizutani, P Carbon, A. Krol 2000 *EMBO J* 19:4796
98. D Fagegaltier, N Hubert, K Yamada, T Mizutani, P Carbon, A Krol *EMBO J* 19:4796
99. PR Copeland, JE Fletcher, BA Carlson, DL Hatfield, DM Driscoll 2000 *EMBO J* 19:306
100. E Beutler 1983 *Biomed Biochim Acta* 42:S234
101. WH Cheng, YS Ho, BA Valentine, DA Ross, GF Combs, Jr., XG Lei 1998 *J Nutr* 128:1070
102. Y Fu, WH Cheng, JM Porres, DA Ross, XG Lei 1999 *Free Radic Biol Med* 27:605
103. H Jaeschke, YS Ho, MA Fisher, JA Lawson, A Farhood *Hepatology* 29:443
104. [Editorial] 1979 Selenium in the heart of China *Lancet* 2:889
105. MA Beck, RS Esworthy, YS Ho, FF Chu 1998 *FASEB J* 12:1143
106. MA Beck, PC Kolbeck, LH Rohr, Q Shi, VC Morris, OA Levander 1994 *J Med Virol* 43:166
107. MA Beck, Q Shi, VC Morris, OA Levander 1995 *Nat Med* 1:433
108. H Guanqing 1979 *Chin Med J (Engl)* 92:416
109. Y Xiu communicated at the *7th International Symposium on Selenium in Biology and Medicine*, October 1-5, 2000, Venice, Italy
110. C Kretz-Remy, P Mehlen, ME Mirault, AP Arrigo 1996 *J Cell Biol* 133:1083
111. V Makropoulos, T Bruning, K Schulze-Osthoff 1996 *Arch Toxicol* 70:277
112. K Hori, D Hatfield, F Maldarelli, BJ Lee, KA Clouse 1997 *AIDS Res Hum Retroviruses* 13:1325
113. PA Sandström, J Murray, TM Folks, AM Diamond 1998 *Free Radic Biol Med* 24:1485
114. Y Yamamoto, K Takahashi 1993 *Arch Biochem Biophys* 305:541
115. Y Saito, T Hayashi, A Tanaka, Y Watanabe, M Suzuki, E Saito, K Takahashi 1999 *J Biol Chem* 274:2866
116. D Steinberg 1997 *J Biol Chem* 272:20963
117. H Sies, A Wahlländer, C Waydhas, S Soboll, D Häberle 1980 *Adv Enzyme Regul* 18:303
118. RJ Kulmacz, WE Lands 1983 *Prostaglandins* 25:531
119. M Haurand, L Flohé 1988 *Biol Chem Hoppe Seyler* 369:133
120. DA Parks, GB Bulkley, DN Granger 1983 *Surgery* 94:428
121. RS Esworthy, J Mann, M Sam, J Doroshow, FF Chu Abstract submitted to the *7th International Symposium on Selenium in Biology and Medicine*, October 1-5, 2000, Venice, Italy
122. K Nomura, H Imai, T Koumura, M Arai, Y Nakagawa 1999 *J Biol Chem* 274:29294
123. H Imai, K Narashima, M Arai, H Sakamoto, N Chiba, Y Nakagawa 1998 *J Biol Chem* 273:1990
124. R Brigelius-Flohé, S Maurer, K Lötzer, G Böl, H Kallionpää, P Lehtolainen, H Viita, S Ylä-Herttuala 2000 *Atherosclerosis* 152:307
125. M Maiorino, M Coassin, A Roveri, F Ursini 1989 *Lipids* 24:721
126. F Weitzel, A Wendel 1993 *J Biol Chem* 268:6288
127. K Schnurr, J Belkner, F Ursini, T Schewe, H Kühn 1996 *J Biol Chem* 271:4653
128. R Brigelius-Flohé, B Friedrichs, S Maurer, M Schultz, R Streicher *Biochem J* 328:199
129. PA Baeuerle, T Henkel 1994 *Annu Rev Immunol* 12:141
130. L Flohé, R Brigelius-Flohé, C Saliou, MG Traber, L Packer 1997 *Free Radic Biol Med* 22:1115
131. M Conrad, U Heinzmann, W Wurst, GW Bornkamm, M Brielmeier Abstract submitted to the *7th International Symposium on Selenium in Biology and Medicine*, October 1-5, 2000, Venice, Italy
132. FH Hawker, PM Stewart, PJ Snitch 1990 *Crit Care Med* 18:442

133. R Gärtner, MW Angstwurm, J Schottdorf 1997 *Med Klin* 92 Suppl 3:12
134. R Gärtner, MW Angstwurm, J Schottdorf Abstract submitted to the *7th International Symposium on Selenium in Biology and Medicine*, October 1-5, 2000, Venice, Italy
135. P Knekt, M Heliövaara, A Reunanen, A Aromaa Abstract submitted to the *7th International Symposium on Selenium in Biology and Medicine*, October 1-5, 2000, Venice, Italy
136. RJ Shamberger, DV Frost 1969 *Can Med Assoc J* 100:682
137. GN Schrauzer 1976 *Bioinorg Chem* 5:275
138. CC Clayton, CA Baumann 1949 *Cancer Res* 9:575
139. HE Ganther 1999 *Carcinogenesis* 20:1657

# Chapter 15. Selenoproteins of the thioredoxin system

Arne Holmgren

*Medical Nobel Institute for Biochemistry, Department of Medical Biochemistry and Biophysics, Karolinska Institute, SE-171 77 Stockholm, Sweden*

**Summary:** Human and mammalian thioredoxin reductases are selenoproteins containing an essential catalytically active selenocysteine (Sec) residue. In contrast to the enzymes from bacteria, yeast and plants, the mammalian enzymes are larger and entirely different in structure and mechanism. They are homologous to glutathione reductase, but with a C-terminal elongation of 16 residues containing the conserved C-terminal active site sequence -Gly-Cys-Sec-Gly. The active site is a selenenylsulfide formed from the conserved cysteine-selenocysteine sequence, which is reduced to a selenolthiol by electrons from the redox active disulfide of the other subunit. The essential role of selenium in thioredoxin reductase explains the very broad substrate specificity including reduction of thioredoxin, selenite, dehydroascorbic acid and ascorbyl free radical, hydrogen peroxide or lipid hydroperoxides. The essential role of selenium in human thioredoxin reductases explains roles of selenium in cell growth via pleiotropic effects in reduction of thioredoxin which has multiple roles in electron transport to essential biosynthetic enzymes, thiol redox control of transcription factors, and in defense against oxidative stress. Clinically used inhibitors of cell growth or inflammation like gold thioglucose are targeted to the selenocysteine residue of the enzyme.

**Introduction**

The thioredoxin system comprised of NADPH, thioredoxin (Trx) and the flavoprotein thioredoxin reductase (TR) is ubiquitously present from Archea to man [1,2]. Thioredoxin with a redox-active dithiol/disulfide is an electron donor for essential enzymes such as ribonucleotide reductase and a general protein disulfide reductase with numerous functions in control of intracellular redox potential, defense against oxidative stress and signal transduction by thiol redox control [2]. Thioredoxin reductases from mammalian cells and higher eukaryotes are selenoenzymes [3,4] and very different from the smaller selenium-independent enzymes of archea, bacteria, yeast and plants [5]. This chapter will discuss reactions between selenium compounds and the thioredoxin system and some of the structure-function relationships of mammalian thioredoxin reductases.

## General properties of thioredoxin systems

All thioredoxin reductases reduce oxidized thioredoxin (Trx-S$_2$) at the expense of NADPH [1,2] (Reaction 1) and reduced thioredoxin [Trx-(SH)$_2$] is reoxidized by disulfides in proteins generating thiols (Reaction 2):

$$\text{Trx-S}_2 + \text{NADPH} + \text{H}^+ \xrightarrow{\text{TR}} \text{Trx-(SH)}_2 + \text{NADP}^+ \quad (1)$$

$$\text{Trx-(SH)}_2 + \text{Protein-S}_2 \xrightarrow{\text{spontaneous}} \text{Trx-S}_2 + \text{Protein-(SH)}_2 \quad (2)$$

Generally, the $K_m$-value for NADPH is low or in the range 10 µM and that of Trx-S$_2$ is typically from 1 to 3 µM.

Isolation and characterization of mammalian thioredoxin and thioredoxin reductase started about 30 years ago [6-8]. As shown in Table 1, there are some major differences between the thioredoxin systems of prokaryotes like E. coli and that of mammalian organisms.

**Table 1.** Properties of thioredoxin systems

|  | E. coli | Human |
|---|---|---|
| Thioredoxin | $M_r$= 12,000<br>108 aa<br>-CGPC-active site<br>Trx-S$_2$ stable upon storage | $M_r$= 12,000<br>104 aa<br>-CGPC-active site<br>+3 structural SH-groups, Trx-S activity reversibly lost by additional disulfide formation upon aerobic storage |
| Thioredoxin reductase | $M_r$= 70,000<br>2 subunits<br>High specificity<br>Stable | $M_r$= 114,000 or larger<br>2 subunits<br>Broad specificity, selenoenzymes<br>Labile to oxidation - reduction cycles |

E. coli and mammalian cytosolic thioredoxins are homologous proteins with a conserved -Cys-Gly-Pro-Cys-active site. However, mammalian thioredoxin must be purified in the fully reduced form since they contain structural SH-groups which form additional disulfides upon oxidation. This may have autoregulatory function of thioredoxin activity in resting cells or upon oxidative stress yet incompletely known in vivo. Thioredoxin reductases from mammalian cells have very different properties when compared with the enzymes from *E. coli*, yeast or plants [see 5 for review]. The cytosolic enzyme has subunits with 55 kDa or larger instead of the 35 kDa in the *E. coli* enzyme with known three-dimensional structure [5]. As will be described below, the mammalian enzyme also has a very broad substrate specificity entirely different from the generally species-specific enzymes only reducing Trx-$S_2$ present in prokaryotes, yeast and plant cytosol.

**Selenium reduction by the thioredoxin system**
The fact that administration of selenium compounds like selenite ($SeO_3^{2-}$) cause the inhibition of tumor cell proliferation in vivo and the knowledge that thioredoxin reductase appeared to be more highly expressed in malignant cells prompted us to start investigations on the reactions to selenium compounds with the mammalian thioredoxin system. Contrary to expectations, we discovered that selenite is a direct substrate for thioredoxin reductase as well as an efficient oxidant of thioredoxin [9,10].

With 200 µM NADPH and 50 nM calf thymus thioredoxin reductase, addition of 10 µM selenite caused oxidation of 40 µM NADPH in 12 min and 100 µM NADPH after 30 min demonstrating a direct reduction of selenite with redox cycling by oxygen [9,10]. This was demonstrated by incubation under anaerobic conditions where only 3 mol of NADPH was oxidized per mol of selenite according to Reaction [3]:

$$SeO_3^{2-} + 3\ NADPH + 3H^+ \xrightarrow{TR} Se^{2-} + 3\ NADP^+ + 3\ H_2O \quad (3)$$

Addition of thioredoxin stimulated the reaction further since selenite rapidly reacts with Trx-$(SH)_2$ to oxidize it to Trx-$S_2$ [11-13]. Since glutathione reductase will not react with selenite [13], Reaction 3 should provide cells with selenide, a required precursor for selenophosphate and selenocysteine synthesis [14]. Selenite and glutathione react to form selenodiglutathione (GS-Se-SG) which has been suggested to be a major metabolite of inorganic selenium salts in mammalian tissues [15]. Reaction of selenodiglutathione by NADPH and glutathione reductase was demonstrated by Ganther [16] and it has been proposed to be a source of selenide in cells as well as an inhibitor of neoplastic growth [17]. We synthesized GS-Se-SG [11,18] and discovered that is a direct efficient substrate for mammalian thioredoxin reductase and a highly efficient oxidant of reduced thioredoxin. Since GSSG is not a substrate for mammalian thioredoxin reductase [7,8] the insertion of the selenium atom in the GSSG molecule to form GS-Se-SG makes this molecule highly reactive with the enzyme.

Reduction of GS-Se-SG to yield selenide by glutathione reductase requires two mol of NADPH. We found only the first stoichiometric reduction to be fast with GS-Se$^-$ as a product [11]. The second reaction was slow and relatively inefficient. These results strongly suggest that the major selenide generation in cells is via thioredoxin reductase and thioredoxin. Thus, in mammalian cells the selenoenzyme thioredoxin reductase is also responsible for the synthesis of selenide required for its own synthesis.

An oxygen dependent non-stoichiometric consumption of NADPH is given by the thioredoxin system in the presence of selenite, selenodiglutathione and selenocystine [9-11,18]. The latter is an efficient substrate for mammalian thioredoxin reductase with a $K_m$ of 6 µM [18]. The mechanism may be that the XSe$^-$ reacts with a dithiol (or selenolthiol) to catalyze oxidation according to Reaction 4:

$$XSe^- + R-(SH)_2 + (O) \longrightarrow XSe^- + R-S_2 + H_2O \qquad (4)$$

The effect will be $O_2$-dependent consumption of NADPH and provides an explanation for the lack of an autooxidizable free pool of selenocysteine as well as the acute toxic effects of selenium compounds on cells, for example, leading to apoptosis.

## Substrate specificity of thioredoxin reductase

Mammalian thioredoxin reductases display a surprisingly very wide substrate specificity as first observed during purification [7,8]. This is in contrast to the smaller prokaryotic enzymes, which do not react with mammalian thioredoxins despite their identical active sites and closely related three-dimensional structures [19]. As summarized in Table 2, a truly wide range of direct reductions are catalyzed by the mammalian cytosolic thioredoxin reductases. Thioredoxin from *E. coli* is a substrate with a similar $K_{cat}$, but with a 15-fold higher $K_m$ value (35 µM) compared with the rat liver protein [8]. The mammalian cytosolic thioredoxins generally show full crossreactivity with the enzymes from different sources and *vice versa*. In many instances free selenocyst(e)ine will stimulate reduction of substrates [24,13,23,29].

## Structure and mechanism of mammalian thioredoxin reductase

Recent biochemical studies, sequencing and cloning of mammalian thioredoxin reductases has revealed that the enzymes are selenoproteins and entirely different from the corresponding enzymes in bacteria, yeast and plants [see 5 for review]. Stadtman and coworkers serendipitously discovered that human tumor cell thioredoxin reductase is a selenoprotein using labeling of selenoproteins with radioactive selenite [3]. This also explained [30] why a previously putative clone of the human enzyme [31], where the TGA codon for selenocysteine (Sec) was interpreted as the stop codon (Figure 1) gave no enzyme activity. The TGA acts as a stop codon in *E. coli* due to the fact that the species-specific machinery for synthesis of selenoproteins is different in bacteria and mammalian cells [14].

Table 2. Reactions catalyzed by cytosolic mammalian thioredoxin reductases (rat, bovine and human)

| Reaction | Ref |
|---|---|
| 5,5-dithiobis-(2-nitrobenzoic acid) reduction | [7] |
| Thioredoxin-$S_2$ reduction | [8] |
| Protein disulfide isomerase (PDI) | [20] |
| Selenite ($SeO_3^{2-}$) and selenocysteine reduction | [10,13] |
| Selenodiglutathione reduction | [11] |
| Nitrosoglutathione (GSNO) reduction | [21] |
| Electron donor to plasma glutathione peroxidase | [22] |
| $H_2O_2$ and lipid hydroperoxide reductase | [23,24] |
| Reduction of alloxan and vitamin K | [8, 25] |
| NK-lysin disulfide reduction and inactivation of cytotoxic activity | [26] |
| Lipoic acid and lipoamide reduction | [27] |
| Reduction of dehydroascorbic acid | [28] |
| Reduction of ascorbyl free radical | [29] |

By sequencing large parts of the cytosolic bovine enzyme, we also directly identified the C-terminal peptide as containing selenocysteine. The bovine peptides were used to identify a rat cDNA clone which was sequenced [4]. The results showed a polypeptide chain with a high homology to glutathione reductase [4,23], including an identical active site disulfide (CVNVGC) (Figure 1), but with a 16-residue elongation containing the conserved C-terminal sequence -Gly-Cys-Sec-Gly. A selenocysteine insertion sequence (SECIS) was identified in the 3'untranslated region [4]. Furthermore, digestion of thioredoxin reductase by carboxypeptidase after reduction by NADPH released selenocysteine with loss of activity; the oxidized form of the enzyme was resistant to carboxypeptidase digestion [4]. Redox titrations with dithionite and NADPH demonstrated that the mechanism of the human placenta enzyme is similar to that of lipoamide dehydrogenase and glutathione reductase and distinct from the mechanism of thioredoxin reductase from *E. coli* [32].

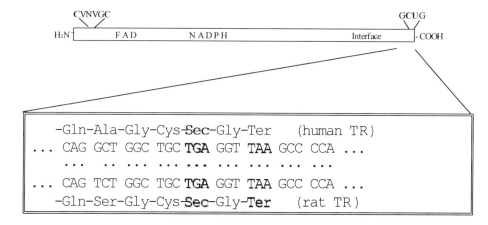

**Figure 1.** Schematic subunit structure of the human and rat thioredoxin reductases [4,30,31]. The N-terminal glutathione reductase-like active site disulfide (CVNVGC) is shown in the upper portion of the figure as well as the FAD, NADPH and Interface domains. The active site is shown in the C-terminus with GCUG denoting Gly-Cys-Sec-Gly. Below, that region of the human and rat cytosolic genes with the TGA codon encoding Sec is shown.

The results also demonstrated that the Sec residue of human thioredoxin reductase is redox active and communicates with the redox active disulfide, since more than 4 electrons per subunit are required to completely reduce the FAD of the oxidized enzyme. Furthermore, the Sec residue is alkylated with loss of activity only after reduction by NADPH [4,33,34]. The Sec residue is also the target of the irreversible inhibitor 1-chloro-2,4-dinitrobenzene only after reduction by NADPH [35] as shown by peptide analysis [34].

The essential role of selenium in the catalytic activities of mammalian thioredoxin reductase was revealed by characterization of recombinant enzymes with selenocysteine mutations [23]. This was done by removing the selenocysteine insertion sequence in the rat gene and changing the $Sec_{498}$ encoded by TGA to Cys or Ser codons by mutagenesis. The truncated protein having the C-terminal dipeptide deleted, a condition expected to mimic that which occurs during extreme selenium deficiency, was also engineered. All three mutants were successfully overexpressed in *E. coli* and purified to homogeneity with 1 mol of FAD per monomeric subunit. All three mutant proteins rapidly generated the $A_{540}$ absorbance resulting from the thiolate-flavin charge transfer complex characteristic of mammalian TR. Only the $Sec_{498}$ Cys enzyme showed catalytic activity in reduction of thioredoxin, with a 100-fold lower $K_{cat}$ and a 10-fold lower $K_m$ compared to the wild type rat enzyme. The pH-optimum of the Sec-containing wild type enzyme was 7 whereas the $Sec_{498}$ Cys enzyme showed a pH optimum of 9. This strongly suggested the involvement of the low pKa Sec selenol in the enzyme mechanism. Also selenium was required for hydrogen peroxide reductase activity [23]. Thus, selenium is required for the catalytic activities of thioredoxin reductase explaining the essential role of this trace element in cell growth.

Based on the homology to glutathione reductase, we have recently proposed a model of mammalian thioredoxin reductase (Figure 2).

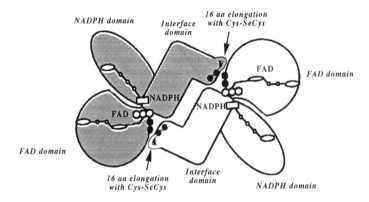

*Thioredoxin Reductase*

**Figure 2.** Structural model of mammalian thioredoxin reductase based on the homology to glutathione reductase. The 16-residue C-terminal extension with the active site is shown as well as the head to tail arrangement of the subunits in the dimer. Taken from [36] the FAD, NADPH and Interface domains are shown (see also Figure 1).

The enzyme is a head to tail dimer with the 16-residue elongation in principle taking the place of GSSG in glutathione reductase. The active site of the enzyme is a selenolthiol in its reduced form and a selenenylsulfide formed from the conserved cysteine-selenocysteine sequence in the oxidized form [36]. The selenenylsulfide was isolated by peptide sequencing and also confirmed by mass spectrometry [36]. Mechanisms of the enzyme have also been postulated involving a reductive half-reaction similar to that of glutathione reductase leading to reduction of the active site disulfide (Figures 1 and 2.). Electrons are thereafter transferred from the redox-active dithiols to the selenenyl-sulfide of the other subunit generating the selenolthiol. Characterization of the Cys mutant enzyme revealed that the selenium atom with its larger radius is critical for the formation of the unique selenenylsulfide [36], since the C-terminal dithiol remains reduced in the Cys mutant [36]. Similar results comfirming these results have also been obtained by others [37] confirming the model. The structure of the enzyme is presently solved by x-ray crystallography, since the Cys mutant enzyme has been crystallized [38], also confirming the model [38, unpublished results].

**Isoenzymes of thioredoxin reductase**
Apart from the cytosolic thioredoxin reductase, two genes encoding additional forms of human and mouse selenoprotein thioredoxin reductases have recently been identified [39]. One is a mitochondrial enzyme [40,41] and the other is preferentially expressed in testis. All these enzymes have extensions in the N-terminal region, but share the C-terminal active site sequence. Additional complexity is given by the identification of enzymes with mRNA variants differening in the 5'-untranslated region [42] and by 5'-exon splicing [43]. The nematode *C. elegans* contains two homologues related to mammalian thioredoxin reductase, one with Cys and the other with Sec. The Sec containing enzyme with 74 kDa subunits is the major selenoprotein in *C. elegans*.

**Medical aspects of selenium in thioredoxin reductase**
Human thioredoxin reductase is a general reducing enzyme with a wide substrate specificity contributing to cellular redox homeostasis and is a major pathophysiological factor and drug target. Together with thioredoxin it is involved in prevention, intervention and repair of damage caused by hydrogen peroxide-based oxidative stress. As a selenite reducing enzyme with a selenol containing active site, human thioredoxin reductase plays a central role in selenium physiology. As recently covered in an extensive review [45], a range of human diseases and conditions are now known or suspected to be related to the activity and function of thioredoxin reductase. This involves diseases like reumatoid arthritis, Sjögren´s syndrome, AIDS and malignancies. The close homology between human thioredoxin reductase and glutathione reductase has lead to the realization that several clinically used drugs like nitrosurea derivatives are targeted to thioredoxin reductase [45]. Furthermore, studies on the regulation of thioredoxin reductase mRNA [46] and the development of specific inhibitors for use in antitumor therapy [47-49] make the enzyme a major target for drug development. In this context it will be important in future studies to also establish the role of the glutathione-dependent glutaredoxin system [50,51], which is an alternative non-selenium pathway of transferring electrons to essential biosynthetic reactions like ribonucleotide reductase. Thus, determining if a malignant cell is dependent on the thioredoxin system or the glutaredoxin system should be essential in drug selection in tumor therapy.

The fact that the thioredoxin system is ubiquitous and present in quite highly variant forms in pathogenic bacteria makes the enzyme a particularly attractive drug target. There is a surprising diversity in the structure and mechanism of action of the enzyme in several important pathogenic bacteria as reviewed in [45]. This may lead to the development of specific inhibitors of bacterial infections like in Lepra, parasitic diseases or malaria. Treatment of an inflammatory disease like reumatoid arthritis with drugs such as gold thioglucose and auranofin, which are strong inhibitors of thioredoxin reductase, might also be considered due to the probable binding of these drugs to the reduced Sec residue in the enzyme.

Since thioredoxin reductase is involved in central biosynthetic reactions and serves as a defense against oxidative stress via thioredoxin, this enzyme should have a high priority in synthesis of selenoproteins. Of particular interest is whether the truncated enzyme expected in selenium deficiency is present in cells. This may be of great importance for understanding the effects of selenium supplementation as an anticancer agent. Obviously, much research remains to understand the pathophysiological implications of the selenium-containing mammalian thioredoxin reductases.

**Acknowledgement:** The research support from the Swedish Medical Research Council (3529), the Swedish Cancer Society and the KA Wallenberg foundation is greatfully acknowledged.

**References**

1. A Holmgren 1985 *Annu Rev Biochem* 54:237
2. ESJ Arnér, A Holmgren 2000 *Eur J Biochem* 267:6102
3. T Tamura, T C Stadtman 1996 *Proc Natl Acad Sci USA* 93:1006
4. L Zhong, ESJ Arnér, J Ljung, F Åslund, A Holmgren 1998 *J Biol Chem* 273:8581
5. CH Williams Jr, LD Arscott, S Müller, BW Lennon, ML Ludwig, P-F Wang, DM Veine, K Becker, RH Schirmer 2000 *Eur J Biochem* 267:6110
6. NE Engström, A Holmgren, A Larsson, S Söderhäll 1974 *J Biol Chem* 249:205
7. A Holmgren 1977 *J Biol Chem* 252:4600
8. M Luthman, A Holmgren 1982 *Biochemistry* 21:6628
9. A Holmgren, S Kumar 1989 *Selenium in Biology and Medicine* A Wendel (Ed) Springer-Verlag, Berlin, 47
10. S Kumar, M Björnstedt, A Holmgren 1992 *Eur J Biochem* 207:435
11. M Björnstedt, S Kumar, A Holmgren 1992 *J Biol Chem* 267:8030
12. X Ren, M Björnstedt, B Shen, M Ericson, A Holmgren 1993 *Biochemistry* 32:9701
13. M Björnstedt, S Kumar, L Björkhem, G Spyrou, A Holmgren 1997 *Biomed Environ Sci* 10:271
14. TC Stadtman 1996 *Annu Rev Biochem* 65:83
15. HS Hsieh, HE Ganther 1975 *Biochemistry* 14:1632
16. HE Ganther 1971 *Biochemistry* 10:4089
17. RJ Shamberger 1985 *Mutat Res* 154:29
18. M Björnstedt, S Kumar, A Holmgren 1995 *Methods Enzymol* 252:219
19. A Holmgren 1995 *Structure* 3:239
20. J Lundström, A Holmgren 1990 *J Biol Chem* 265:9114
21. D Nikitovic, A Holmgren 1996 *J Biol Chem* 271:19180

22. M Björnstedt, J Xue, W Huang, B Åkesson, A Holmgren 1994 *J Biol Chem* 269:29382
23. L Zhong, A Holmgren 2000 *J Biol Chem* 275:18121
24. M Björnstedt, M Hamberg, S Kumar, J Xue, A Holmgren 1995 *J Biol Chem* 270:11761
25. A Holmgren,C Lyckeborg 1980 *Proc Natl Acad Sci USA* 77:5149
26. M Andersson, A Holmgren, G Spyrou 1996 *J Biol Chem* 271:10116
27. ESJ Arnér, J Nordberg, A Holmgren 1996 *Biochem Biophys Res Commun* 225:268
28. JM May, S Mendiratta, KE Hill, RF Burk 1997 *J Biol Chem* 272:22607
29. JM May, CE Cobb, S Mendiratta, KE Hill, RF Burk 1998 *J Biol Chem* 273:23039
30. VN Gladyshev, K-T Jeng, TC Stadtman 1996 *Proc Natl Acad Sci USA* 93:6146
31. PY Gasdaska, JR Gasdaska, S Cochran, G Powis 1995 *FEBS Letters* 373:5
32. LD Arscott, S Gromer, RH Schirmer, K Becker, CH Williams Jr 1997 *Proc Natl Acad Sci USA* 94:9621
33. SN Gorlatov, TC Stadtman 1998 *Proc Natl Acad Sci USA* 95:8520
34. J Nordberg, L Zhong, A Holmgren, ESJ Arnér 1998 *J Biol Chem* 273:10835
35. ESJ Arnér, M Björnstedt, A Holmgren 1995 *J Biol Chem* 270:3479
36. L Zhong, ESJ Arnér, A Holmgren 2000 *Proc Natl Acad Sci USA* 97:5854
37. SR Lee, S Bar-Noy, J Kwon, RL Levine, TC Stadtman, SG Rhee 2000 *Proc Natl Acad Sci USA* 97:2521
38. L Zhong, K Persson, T Sandalova, G Schneider, A Holmgren 2000 *Acta Cryst D* 56:1191
39. Q-A Sun, W Yalin, F Zappacosta, K-T Jeang, BJ Lee, DL Hatfield, VN Gladyshev 1999 *J Biol Chem* 274:24522
40. SR Lee, JR Kim, KS Kwon, HW Yoon, RL Levine, A Ginsburg, SG Rhee 1999 *J Biol Chem* 274:4722
41. A Miranda-Vizuete, AE Damdimopoulos, JR Pedrajas, J-Å Gustafsson, G Spyrou 1999 *Eur J Biochem* 261:405
42. A-K Rundlöf, M Carlsten, MMJ Giacobini, ESJ Arnér 2000 *Biochem J* 347:661
43. QA Sun, F Zappacosta, VM Factor, PJ Wirth, DL Hatfield, VN Gladyshev 2000 *J Biol Chem* in press.
44. VN Gladyshev, M Krause, X-M Xu, KV Korotkov, GV Kryukov, Q-A Sun, BJ Lee, JC Wootton, DL Hatfield 1999 *Biochem Biophys Res Commun* 259:244
45. K Becker, S Gromer, RH Schirmer, S Müller 2000 *Eur J Biochem* 267:6118
46. DL Kirkpatrick, S Watson, M Kunkel, S Fletcher, S Ulhag, G Powis 1999 *Anticancer Drug Res* 5:421
47. JR Gasdaska, JW Harney, PY Gasdaska, G Powis, MJ Berry 1999 *J Biol Chem* 274:25379
48. MM Berggren, JF Mangin, JR Gasdaska, G Powis 1999 *Biochem Pharmacol* 57:187
49. G Powis, DL Kirkpatrick, M Angulo, A Baker 1998 *Chem Biol Interact* 111-112:23
50. A Holmgren 1999 *Redox Regulation of Cell Signaling and its Clinical Application* L Packer, J Yodoi (Eds) Marcel Dekker, New York, 279
51. A Holmgren 1989 *J Biol Chem* 264:13963

# Chapter 16. Selenium, deiodinases and endocrine function

Donald L. St. Germain

*Departments of Medicine and of Physiology, Dartmouth Medical School, Lebanon, NH 03756, USA*

**Summary:** Selenium status influences a number of endocrine processes, most notably those involved in thyroid hormone synthesis and metabolism. Thyroid follicular cells maintain a highly oxidative environment as required for thyroid hormone synthesis. Glutathione peroxidase expression in the thyroid gland is thus important to prevent oxidation-induced cellular toxicity. In addition, the iodothyronine deiodinases, which catalyze the principal reactions of thyroid hormone metabolism, are selenoproteins. Selenium may also impact carbohydrate metabolism and female reproduction, though these effects are less well characterized.

## Introduction

The trace element selenium has been implicated as playing an important role in several endocrine processes. For example, the anti-oxidative actions of the glutathione peroxidase selenoproteins likely serve a protective function in the thyroid gland, an organ that generates large amounts of $H_2O_2$ for use during thyroid hormone synthesis [1]. In addition, the deiodinase family of selenoenzymes catalyzes the metabolism of thyroid hormones both in the thyroid gland and in extrathyroidal tissues [2]. Finally, selenium has been implicated in oxidative processes that may play a role in the pathogenesis of complications associated with diabetes mellitus [3]. Selenium deficiency thus has the potential to disrupt thyroid hormone synthesis and metabolism and exacerbate the complications of diabetes. At the opposite end of the spectrum, selenium toxicity may also affect endocrine processes, as it has been associated with impaired growth [4] and with reproductive failure [5].

This chapter will examine these and other aspects of selenium's role in endocrine function. Particular emphasis will be placed on selenium's involvement in thyroid hormone metabolism, an area where there has been considerable progress in our understanding during the last decade [6].

## Selenium in endocrine tissues

Indirect evidence supporting an important role of selenium in endocrine function has come from studies examining the levels of selenium, and its rate of retention, in various tissues [7]. Thus, the selenium content of the thyroid gland in adult humans is as high, or higher, than that found in other organs, and is roughly twice that observed in the liver [8,9]. These findings hold even in populations where selenium intake is relatively restricted [10].

Additional information concerning the tissue distribution of selenium comes from animal experiments. Typical of such data is a recent report by Bates et al. [11]. Using a neutron activation technique, the selenium levels in several tissues of rats of various ages and during pregnancy were determined (Figure 1). The highest selenium content in neonatal animals was observed in the pituitary gland, with the thyroid gland, cerebral cortex, and liver having the next most abundant levels. Selenium levels in most organs during pregnancy and in the adult were higher than in younger animals, with the pituitary and liver exhibiting the highest values. Relatively high selenium content was also noted in other adult endocrine organs such as the ovary and testes, and in the placenta and the pregnant uterus at the site of blastocyst implantation.

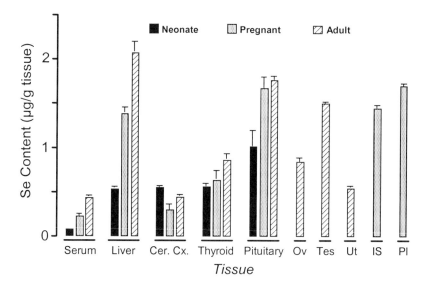

**Figure 1.** Selenium content of serum and selected tissues in neonatal (12 day old), pregnant, and non-pregnant adult rats. Cer. Cx., cerebral cortex; Ov, ovary; Tes, testes; Ut, uterus; IS, uterine implantation site; and Pl, placenta. (Adapted from Bates et al. [11].)

The relatively high selenium content of endocrine organs may be due in part to their high rate of retention of this element, especially in the face of limited selenium availability. This was demonstrated by Behne et al. [7] by injecting $^{75}$Se (as selenite) into selenium-deficient and selenium-replete animals. The degree of selenium retention (calculated as the percentage dose of $^{75}$Se retained per gram of tissue) at six weeks post injection was far greater in endocrine organs (testes, pituitary, thyroid, adrenals, ovaries) and the brain, than in the liver, muscle, heart, or erythrocytes. Using polyacrylamide gel electrophoresis, these investigators identified twelve selenium-containing proteins, in addition to glutathione peroxidase, in the various rat tissues. In selenium repletion studies, the element appeared to be preferentially diverted to the synthesis of proteins other than glutathione peroxidase. This suggests a hierarchy of selenium utilization both in terms of tissue distribution and selenoprotein synthesis (Figure 2). This concept is further supported by the results of both in vivo [12,13] and cell culture [14-16] experiments showing the preferential synthesis of the type 1 iodothyronine deiodinase (D1) selenoenzyme over glutathione peroxidase under conditions of limited selenium supply.

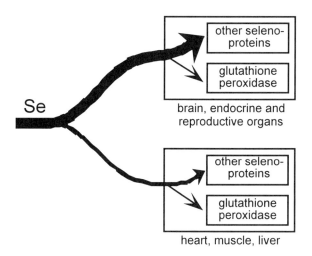

**Figure 2.** Schematic diagram showing the hierarchical supply of selenium to different tissues for use in the synthesis of selenoproteins. (Taken from Behne et al. [7].)

## Selenium and the thyroid gland

Hydrogen peroxide is an essential cofactor for the synthesis of thyroid hormones [17]. It is utilized by thyroid peroxidase for the oxidation of iodide, the first step in the synthetic pathway. The amount of $H_2O_2$ generated by thyrocytes during this process is large and similar to that produced in activated leukocytes [1]. Among the eleven selenoproteins expressed in thyroid follicular cells are cytosolic glutathione peroxidase (cGSHPx), phospholipid hydroperoxide glutathione peroxidase (phGSHPx), and extracellular glutathione peroxidase (eGSHPx) [18]. These proteins likely play a critical role in preventing $H_2O_2$-induced cellular toxicity.

Although the selenium content of the thyroid is high relative to other tissues, nutritional selenium deficiency in the rat does result in an approximate 30-50% decrease in the selenium level in this organ [11,19]. This is associated with significant decreases in the expression of cGSHPx and phGSHPx in both the basal state and under conditions of iodine deficiency [20] where $H_2O_2$ production is increased due to thyrotropin (TSH) stimulation [17]. It has thus been suggested that combined selenium and iodine deficiency may result in irreversible damage to the thyroid [1]. This mechanism has been invoked to explain the finding of small, fibrotic thyroid glands in myxedematous cretins living in central Africa. In this region, as well as in areas of China, Southeast Asia, Russia, and Egypt, severe dietary deficiencies of selenium and iodine coexist [21].

In support of a causal relationship between selenium deficiency and thyroidal damage, Contempre et al. [22] studied the effects of selenium deprivation with and without concurrent iodine deficiency on thyroid gland histology in 8 week-old Wistar rats. They observed that selenium deficiency resulted in damage to the thyroid gland. Furthermore, in the setting of a combined selenium and iodine deficiency, an acute toxic load of iodine resulted in greater cellular necrosis and fibrosis than that observed in iodine deficient animals that were selenium replete. These results, however, were not confirmed by Colzani et al. [23]. In their studies, no evidence of inflammatory changes, necrosis or fibrosis was observed in rats with combined deficiency to which iodine was acutely administered. In the latter studies, however, pregnant and neonatal rats of a different strain (Sprague-Dawley) were used which makes a direct comparison of the data problematic.

Of potential clinical importance, Zimmermann et al. [24] recently reported the results of iodine supplementation, in the form of an oral dose of iodized oil, given to goitrous children with normal thyroid function living in an area of combined selenium and iodine scarcity. During the one-year follow-up period, TSH levels decreased significantly and thyroid volume was reduced an average of 50%. No adverse effects were noted with this regimen suggesting that iodine repletion is well tolerated in individuals with moderately severe deficiencies of both these trace elements.

**Iodothyronine deiodinases**
In addition to glutathione peroxidase, a 27.8 kDa selenoprotein, designated by Behne et al. [7] as protein #7, is highly expressed in the rodent thyroid gland, and to a lesser extent in the liver and kidney. Subsequent studies by these investigators and others identified this protein as the D1, an oxidoreductase capable of removing iodine from thyroxine and other thyroid

hormones [25,26].

Two other iodothyronine deiodinases, types 2 and 3 (D2 and D3, respectively), have also been identified and demonstrated to be selenoproteins [27-30]. Of note, the D2 contains two in-frame UGA codons, although it is unclear if the second codon, located near the extreme 3' end of the coding region, actually directs incorporation of selenocysteine, or serves as a termination signal [30]. There has been some controversy as to whether the D2 selenoenzyme identified by Croteau et al. [30] is responsible for D2 activity in mammalian tissues [31-33]. It has recently been reported, however, that mice with a targeted disruption of the selenoenzyme *DIO2* gene are completely devoid of D2 activity in all tissues [34]. This finding verifies that the D2 activity found in all mammalian tissues is a selenoenzyme.

The deiodinases catalyze the removal of iodine from iodothyronine substrates [2]. As shown in Figure 3, deiodination can occur at either the 5', or the chemically equivalent 3' position, on the phenolic (outer) ring, or at the 5 or 3 positions on the tyrosyl (inner) ring. The former reaction is, in essence, an activating one in that thyroxine (T4) is converted to the metabolically more active thyroid hormone 3,5,3'-triiodothyronine (T3). In contrast, inner ring deiodination is an inactivating process that converts T4 and T3 to the metabolically inactive compounds, 3,3',5'-triiodothyronine (reverse T3, rT3) and 3,3'-diiodothyronine.

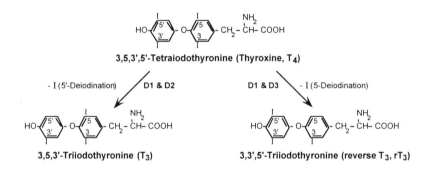

**Figure 3.** Deiodination processes as catalyzed by the D1, D2, and D3 selenoenzymes.

The deiodinases also differ in other respects, most notably in their tissue patterns of expression, susceptibility to various chemical inhibitors, and the changes that occur in activity levels with alterations in thyroid hormone status (Table 1.) This "autoregulation" of deiodinase activities in response to thyroid hormone levels results from both pre- and post-translational mechanisms [30,37-41], and appears to have adaptive value. For example, in response to hypothyroidism, 5'-deiodination, as catalyzed by D1 activity in the thyroid and D2 activity in extrathyroidal tissues, increases, whereas the expression of the D3 inactivating deiodinase declines [6]. Such alterations in deiodinase activities assist in maintaining relatively normal T3 levels in the serum [42]. In addition to alterations in thyroid status, a number of other hormonal and nutritional factors can influence deiodinase expression [2].

Table 1. Characteristics of the iodothyronine deiodinases

| Characteristic | D1 | D2 | D3 |
|---|---|---|---|
| Reaction catalyzed | 5' or 5 | 5' | 5 |
| Substrate preference | 5': rT3 > T2S > T4<br>5: T4S > T3S | T4 > rT3 | T3 > T4 |
| Location (rat) | liver, kidney, thyroid, pituitary | pituitary, brain, brown fat * | brain, skin, uterus, placenta, fetal tissues |
| Inhibitors<br>  propylthiouracil<br>  aurothioglucose<br>  iopanoic acid | <br>++++<br>++++<br>+++ | <br>+<br>++<br>++++ | <br>+/-<br>++<br>+++ |
| Activity in hypothyroidism | ↑ thyroid,<br>↓ liver, kidney | ↑ all tissues | ↓ brain |

* In humans, D2 appears to be expressed more broadly than in rats with mRNA and/or activity observed in the thyroid gland, skeletal muscle, and heart. Original references are cited in review articles by Leonard and Visser [43] and St. Germain and Galton [2].

D1, D2, and D3 have molecular masses, as predicted from their amino acid sequences, of 29, 30, and 32 kDa, respectively, and have several structural features in common [28,30,44]. The most important of these is the presence of a selenocysteine residue at the mid-portion of each molecule which is surrounded by a highly homologous region of 15 amino acids that represents the active catalytic site [30]. Several investigators, using site-directed mutagenesis, have demonstrated the importance of the selenocysteine residue to deiodinase function [28,30,44]. Thus, substitution of cysteine for selenocysteine in D1 reduces catalytic activity by approximately two orders of magnitude, whereas the substitution of other amino acids renders the enzyme inactive [44,45]. Similar findings have been noted for the D3 [27]. In the case of the D2, cysteine substitution markedly increases the Km value for iodothyronine substrates, thereby rendering it less efficient in catalyzing the deiodination reaction [46]. These results demonstrate that the potent nucleophilic properties of selenium are essential for the efficient catalysis of deiodination. This fact is emphasized by the observation that all deiodinases isolated to date, from vertebrate species as divergent as amphibians and mammals, are selenoproteins [2].

Several other structural features shared by the deiodinases have been defined and their significance investigated. These include two conserved histidine residues in the carboxy-terminal half of the molecules that are essential for catalytic activity [47] and hydrophobic regions near the amino-termini that appear to represent transmembrane domains used for anchoring the proteins in the endoplasmic reticulum [48]. Recent studies have indicated that the D1 and D2 are oriented such that the amino-terminus is in the lumen of the endoplasmic reticulum, and the active catalytic site and carboxy-terminus are in the cytoplasm [48,49]. Other residues in the amino-terminal half of the D1 appear to be important for substrate specificity [50], whereas a cysteine residue located viscinal to the selenocysteine residue may be important for thiol cofactor utilization [51].

The determination of deiodinase activity in tissue homogenates requires the addition of thiol cofactors [52], and usually dithiothreitol at concentrations of 1 to 50 mM are used for this purpose. Glutathione and a reconstituted thioredoxin system can also be used to support D1, but not D2 or D3, activity in vitro [53-55], though there is evidence that these may not be endogenous cofactors for this enzyme [51]. The cofactor(s) that supports deiodination in intact cells remains undefined.

The *Dio1*, *Dio2*, and *Dio3* genes have been localized to human chromosomes 1p32-33, 14q24.3, and 14q32, respectively, and their genomic structures have been characterized [56-60]. The human, as well as the mouse genes are relatively simple and contain 4 (D1), 2 (D2) or a single (D3) exon. The predominant mRNA species transcribed from these genes are approximately 2.0 kb, 6.5 kb, and 2.2 kb in length, respectively [28,30,61]. Each has been demonstrated to contain within its 3' untranslated region a functional selenocysteine insertion sequence (SECIS) [62-65]. In the case of the D2, the SECIS element is located a remarkable 5.4 kb from the first selenocysteine-encoding UGA codon in both the human and the mouse mRNAs [64,65].

The physiological roles of the deiodinases have been inferred largely from their catalytic properties and expression patterns. As noted above, changes in deiodinase activities induced by alterations in thyroid hormone levels appear to assist in maintaining serum levels of the active hormone T3 within relatively narrow bounds [2]. In addition, the combination of deiodinase types expressed in different tissues may allow for tissue specific patterns of thyroid hormone metabolism, which may result in organ specific levels of thyroid hormone action. Dramatic examples of such effects have been demonstrated in the eye and other tissues during metamorphosis of the amphibian tadpole [66,67], and in the developing mammalian cochlea [68].

**Selenium and thyroid hormone metabolism**
The effects of selenium status on thyroid hormone metabolism have been investigated primarily by inducing selenium deficiency in rats by nutritional deprivation. Changes in serum thyroid hormone levels in this model system are surprisingly mild given that all the enzymes responsible for catalyzing deiodination are selenoproteins. Feeding rats a low selenium diet (typically containing 3–10 µg selenium/kg chow) for several weeks results in a 40-60% increase in serum T4 levels, T3 levels which are unchanged or slightly decreased, unchanged rT3 levels, and TSH concentrations that are unchanged or increased as much as 2-fold [69-71]. Compared to adult male rats, these changes tend to be less pronounced in adult females and in the fetus [11,71].

The impact of selenium deficiency on deiodinase activities is dependent on both the tissue examined and the degree of selenium deprivation [11,12,19,72]. For example, D1 activity is markedly decreased in the liver and kidney of selenium deficient animals [25,69], primarily because of impaired translational efficiency [73]. One consequence of this is an increase in the serum levels of the sulfated conjugates of T4 and T3 due to both their enhanced synthesis and impaired clearance [74].

In tissues that have a high rate of selenium retention such as the thyroid, pituitary, and gonads, D1 expression is preserved during selenium deficiency [11,12]. Indeed, in some selenium deprivation studies, an increase in D1 activity in the thyroid has been observed, most likely secondary to an increase in TSH stimulation [13,19]. This effect may serve to maintain thyroidal T3 secretion and thus compensate for any decrease in T3 production due to impaired 5'-deiodination processes in other organs.

D2 activities in brain and brown adipose tissue have been noted to decrease by as much as 40–80% in selenium deficient adult rats [69,72,75], though the effects in the fetus and neonatal rodent are less pronounced [11]. In brown fat there is also impaired D2 induction in response to cold exposure [76] and iodine deficiency [75]. These decreases in D2 activity likely reflect both impaired synthesis of this rapidly turning-over protein in the face of selenium deficiency, as well as an enhanced rate of proteasome-mediated enzyme degradation induced by the elevated T4 levels [77,78]. However, in the brain, D2 activity does increase normally in response to the hypothyroidism of iodine deficiency [75].

Selenium deficiency appears to have less effect on D3 than on D1 and D2 expression. In the brain, D3 activity has been reported to be unchanged [11] or decreased by 20% [72]. D3 activity is also unchanged in the uterine implantation site and the placenta [11,71].

The physiological consequences of these changes in deiodinase activity remain uncertain, though some data suggest they might influence thyroid hormone action. Thus, Campos-Barros et al. [79] have observed modest, but significant decreases in the T3 content of various brain regions of selenium deficient rats, and Beckett et al. [80] have noted similar findings in the liver. Furthermore, Mitchell et al. [81] have described a significant decrease in brain-derived neurotrophic factor in selenium-deficient neonatal rats. Whether this results directly from alterations in brain thyroid hormone metabolism remains uncertain. A summary of the various changes in thyroid hormone economy noted in rodent model systems is shown in Table 2.

**Table 2.** Changes in thyroid hormone economy accompanying selenium deficiency in rats.

| Increased | Largely Unchanged | Decreased |
| --- | --- | --- |
| Serum T4 | Serum T3 (may be decreased) | D1 (liver, kidney) |
| Serum free T4 | Serum rT3 | D2 (brain, pituitary, BAT) |
| Serum T4 Sulfate | Serum TSH (may be increased) | D3 (brain, skin, uterus) |
| Serum T3 Sulfate | D1 (thyroid, pituitary, gonads) | Tissue T3 levels (brain, liver) |
|  | D3 (placenta, implantation site) | cGSHPx (liver, thyroid) |
|  |  | phGSHPx (liver, thyroid) |

BAT, brown adipose tissue; cGSHPx, cytosolic glutathione peroxidase activity; phGSHPx, phospholipid hydroperoxide glutathione peroxidase activity. Data from references 11, 13, 15, 19, 20, 69, 70, 74-77, 79, 80, 82.

## Selenium and thyroid hormone economy in humans

Clinical evidence suggests that the above observations in rodents are relevant to humans. For example, studies have demonstrated decreased serum T3/T4 ratios in individuals with modest selenium deficiency associated with cystic fibrosis [83], phenylketonuria [84], or aging [85]. Selenium supplementation typically results in a modest lowering of serum T4 levels and an increase in the T3/T4 ratio to control levels [83,84,86]. Similar effects have been seen with selenium supplementation in a population with combined selenium and iodine deficiency in Zaire [82]. Presumably, these changes reflect correction by selenium of impairments in D1 and/or D2 activities.

However, the response to selenium is more complicated in individuals with iodine-deficient cretinism. In this setting of impaired thyroid function, selenium supplementation worsens the hypothyroid state as evidenced by decreases in serum T4 and T3 levels and increases in TSH [87]. Although the reason for this response is not certain, it has been suggested that selenium-induced increases in thyroidal glutathione peroxidase activity lower $H_2O_2$ levels and thus compromise thyroid hormone synthesis [1]. The finding that the D1 is a mixed-function enzyme capable of 5-, as well as 5'-deiodination, suggests that a second factor may be involved. Thus, a selenium-induced increase in extrathyroidal D1 activity might serve to degrade sulfated conjugates of T4 and T3 thereby effectively enhancing the clearance rate of these hormones. These observations have prompted recommendations that iodine supplementation should take priority over selenium supplementation when deficiencies of both trace elements coexist.

**Selenium, diabetes mellitus and female reproduction**
An early indication that selenium status might have an impact on the course of diabetes mellitus was reported by Souness et al. in 1983 [88]. These investigators observed that insulin-stimulated glucose oxidation was decreased in adipocytes isolated from selenium-deficient rats. Several additional studies have demonstrated an insulin-like effect of selenium in cultured adipocytes [89], or when selenium is administered orally [90,91] or by injection to rats rendered diabetic with streptozotocin [92]. The recent results of Battell et al. [93], where the daily intraperitoneal injection of sodium selenate essentially normalized plasma glucose levels after two weeks in a significant fraction of diabetic rats, are typical of these reports. These investigators also noted that selenium treatment prevented the development of a cardiomyopathy that typically occurs in these animals. The mechanism of the glucose lowering effect of selenium is not known, but nearly toxic doses of the element are required to induce the effect.

Other studies have addressed the issue of whether selenium supplementation might have a positive effect on the complications of diabetes mellitus. For example, Douillett et al. observed that selenium supplementation in diabetic rats corrected renal hyperperfusion and diminished the number and severity of renal glomerular lesions [3].

Several studies have examined plasma selenium levels in patients with diabetes mellitus and the results have been inconsistent with decreased, normal, or elevated levels observed (reviewed in Navarro-Alarcon et al. [94]). To date, clinical studies examining the effects of selenium supplementation on the control and complications of patients with diabetes mellitus have not been reported.

Selenium deficiency has been associated with miscarriage in pigs and cattle [95]. However, as induced by dietary manipulation in rodents, selenium deprivation has no apparent effect on female reproduction, even when animals are carried through six generations [11]. This contrasts sharply with observations that selenium is essential for male fertility [96].

In humans, a possible association between selenium deficiency and miscarriage has been reported in a single study from the United Kingdom [97]. The investigators observed that women undergoing a first trimester spontaneous abortion had lower serum selenium concentrations than either non-pregnant women, or those with a viable pregnancy of the same gestational age. Whether this association is causal is unknown, but Nicoll et al. [98] recently examined plasma selenium concentrations in non-pregnant women with a history of recurrent miscarriage and found the levels to be no different from a control group. They concluded that selenium status was not a factor in the pathogenesis of recurrent miscarriage.

## Summary

Studies in several species have demonstrated that selenium status may impact on a number of aspects of endocrine function. This has been demonstrated most convincingly with regard to effects of selenium deficiency on thyroid gland function and thyroid hormone metabolism. These changes appear to be mediated through alterations in the expression of the deiodinase and glutathione families of selenoproteins, which are expressed in both the thyroid gland and in extrathyroidal tissues. The clinical significance in humans of such changes, however, remains uncertain and may be confined to populations living in regions of severe combined selenium and iodine deprivation, or to individuals with disease states that predispose to selenium deficiency. An understanding of the importance of selenium in other endocrine systems requires additional study.

## References

1. B Corvilain, B Contempré, AO Longombé, P Goyens, C Gervry-Decoster, F Lamy, JB Vanderpas, JE Dumont 1993 *Am J Clin Nutr Suppl* 57:244S
2. DL St. Germain, VA Galton 1997 *Thyroid* 7:655
3. C Douillet, A Tabib, M Bost, M Accominotti, F Borson-Chazot, M Ciavatti 1996 *Proc Soc Exp Biol Med* 211:323
4. O Thorlacius-Ussing, A Flyvbjerg, J Esmann 1987 *Endocrinology* 120:659
5. RK Parshad 1999 *Indian J Exp Biol* 37:615
6. DL St. Germain 2001 *Thyroid hormone metabolism* LJ DeGroot, JL Jaimeson (Ed) 4th Ed W. B. Saunders Co. Philadelphia p1320
7. D Behne, H Hilmert, S Scheid, H Gessner, W Elger 1988 *Biochem Biophys Acta* 966:12
8. RC Dickson, RH Tomlinson 1967 *Clin Chim Acta* 16:311
9. J Aaseth, H Frey, E Glattre, G Norheim, J Ringstad, Y Thomassen 1990 *Biol Trace Elem Res* 24:147
10. B Tiran, E Karpf, A Tiran 1995 *Arch Environ Health* 50:242
11. JM Bates, VL Spate, JS Morris, DL St Germain, VA Galton 2000 *Endocrinology* 141:2490
12. S Vadhanavikit, HE Ganther 1993 *J Nutr* 123:1124
13. G Bermano, F Nicol, JA Dyer, RA Sunde, GJ Beckett, JR Arthur, JE Hesketh 1995 *Biochem J* 311:425
14. M Gross, M Oertel, J Köhrle 1995 *Biochem J* 306:851
15. SG Beech, SW Walker, GJ Beckett, JR Arthur, F Nicol, D Lee 1995 *Analyst* 120:827
16. S Villette, G Bermano, JR Arthur, JE Hesketh 1998 *FEBS Letters* 438:81
17. JT Dunn 2001 *Biosynthesis and secretion of thyroid hormones* LJ DeGroot, JL Jaimeson (Ed) 4th Ed W. B. Saunders Co. Philadelphia p1290
18. J Köhrle 1999 *Biochimie* 81:527

19. J Chanoine, LE Braverman, AP Farwell, M Safran, S Alex, S Dubord, JL Leonard 1993 *J Clin Invest* 91:2709
20. JH Mitchell, F Nicol, GJ Beckett, JR Arthur 1996 *J Mol Endocrin* 16:259
21. R Utiger 1998 *N Engl J Med* 339:1156
22. B Contempré, JE Dumont, JF Denef, MC Many 1995 *Eur J Endocrinol* 133:99
23. RM Colzani, S Alex, SL Fang, S Stone, LE Braverman 1999 *Biochimie* 81:485
24. MB Zimmermann, P Adou, T Torresani, C Zeder, RF Hurrell 2000 *Eur J Clin Nutr* 54:209
25. D Behne, A Kyriakopoulos, H Meinhold, J Köhrle 1990 *Biochem Biophys Res Commun* 173:1143
26. JR Arthur, F Nicol, GJ Beckett 1990 *Biochem J* 272:537
27. DL St. Germain, R Schwartzman, W Croteau, A Kanamori, Z Wang, DD Brown, VA Galton 1994 *Proc Natl Acad Sci USA* 91:7767 Correction 1994 *Proc Natl Acad Sci USA* 91:11282
28. W Croteau, SL Whittemore, MJ Schneider, DL St. Germain 1995 *J Biol Chem* 270:16569
29. JC Davey, KB Becker, MJ Schneider, DL St. Germain, VA Galton 1995 *J Biol Chem* 270:26786
30. W Croteau, JC Davey, VA Galton, DL St. Germain 1996 *J Clin Invest* 98:405
31. M Safran, AP Farwell, JL Leonard 1991 *J Biol Chem* 266:13477
32. S Pallud, A-M Lennon, M Ramauge, J-M Gavaret, W Croteau, M Pierre, F Courtin, DL St. Germain 1997 *J Biol Chem* 272:18104
33. JL Leonard, DM Leonard, M Safran, R Wu, ML Zapp, AP Farwell 1999 *Endocrinology* 140:2206
34. M Schneider, S Fiering, DL St. Germain, VA Galton 2000 *Endocrine J* 47 (Suppl):184 (abstract)
35. M Moreno, MJ Berry, C Horst, R Thoma, F Goglia, JW Harney, PR Larsen, TJ Visser 1994 *FEBS Let* 344:143
36. F Santini, RE Hurd, IJ Chopra 1992 *Endocrinology* 131:1689
37. JL Leonard, JE Silva, MM Kaplan, SA Mellen, TJ Visser, PR Larsen 1984 *Endocrinology* 114:998
38. DL St. Germain, C W 1989 *Endocrinology* 125:2735
39. BA O'Mara, W Dittrich, TJ Lauterio, DL St. Germain 1993 *Endocrinology* 133:1715
40. LA Burmeister, J Pachucki, DL St. Germain 1997 *Endocrinology* 138:5231
41. J Steinsapir, AC Bianco, C Buettner, J Harney, PR Larsen 2000 *Endocrinology* 141:1127
42. SM Lum, JT Nicoloff, CA Spencer, EM Kaptein 1984 *J Clin Invest* 73:570
43. JL Leonard, TJ Visser 1986 *Biochemistry of deiodination* G Hennemann (Ed) Marcel Dekker New York p189
44. MJ Berry, L Banu, PR Larsen 1991 *Nature* 349:438
45. MJ Berry, AL Maia, JD Kieffer, JW Harney, PR Larsen 1992 *Endocrinology* 131:1848
46. C Buettner, J Harney, PR Larsen 2000 *Endocrinology* 141:4606
47. M Berry 1992 *J Biol Chem* 267:18055
48. N Toyoda, MJ Berry, JW Harney, PR Larsen 1995 *J Biol Chem* 270:12310
49. MMA Baqui, B Gerebon, JW Harney, PR Larsen, AC Bianco 2000 *Endocrinology* 141:4309
50. N Toyoda, JW Harney, MJ Berry, PR Larsen 1994 *J Biol Chem* 269:20329
51. W Croteau, JE Bodwell, JM Richardson, DL St. Germain 1998 *J Biol Chem* 237:25230
52. I Chopra 1977 *Endocrinology* 101:453
53. A Goswami, IN Rosenberg 1987 *Endocrinology* 121:1937
54. A Goswami, IN Rosenberg 1988 *Endocrinology* 123:192
55. J Sharifi, DL St. Germain 1992 *J Biol Chem* 267:12539
56. AL Maia, MJ Berry, R Sabbag, J Harney, PR Larsen 1995 *Mol Endocrinol* 9:969

*Selenium, deiodinases and endocrine function* 201

57. TC Jakobs, MR Koehler, C Schmutzler, F Glaser, M Schmid, J Köhrle 1997 *Genomics* 42:361
58. FS Celi, G Canettieri, DP Yarnell, DK Burns, M Andreoli, AR Shuldiner, M Centanni 1998 *Mol Cell Endocrinol* 141:49
59. A Hernández, J Park, GJ Lyon, TK Mohandas, DL St. Germain 1998 *Genomics* 53:119
60. A Hernández, GJ Lyon, MJ Schneider, DL St. Germain 1999 *Endocrinology* 140:124
61. DL St. Germain, W Dittrich, CM Morganelli, V Cryns 1990 *J Biol Chem* 265:20087
62. MJ Berry, L Banu, Y Chen, SJ Mandel, JD Kieffer, JW Harney, PR Larsen 1991 *Nature* 353:273
63. D Salvatore, SC Low, M Berry, AL Maia, JW Harney, W Croteau, DL St. Germain, PR Larsen 1995 *J Clin Invest* 96:2421
64. C Buettner, JW Harney, PR Larsen 1998 *J Biol Chem* 273:33374
65. JC Davey, MJ Schneider, KB Becker, VA Galton 1999 *Endocrinology* 140:1022
66. KB Becker, KC Stephens, JC Davey, MJ Schneider, VA Galton 1997 *Endocrinology* 138:2989
67. N Marsh-Armstrong, H Huang, BF Remo, TT Liu, DD Brown 1999 *Neuron* 24:871
68. A Campos-Barros, LL Amma, JS Faris, R Shailam, MW Kelley, D Forrest 2000 *Proc Natl Acad Sci USA* 97:1287
69. GJ Beckett, DA MacDougal, F Nicol, JR Arthur 1989 *Biochem J* 259:887
70. J Chanoine, M Safran, AP Farwell, S Dubord, S Alex, S Stone, JR Arthur, LE Braverman, JL Leonard 1992 *Endocrinology* 131:1787
71. J Chanoine, S Alex, S Stone, SL Fang, I Veronikis, JL Leonard, LE Braverman 1993 *Pediatric Res* 34:288
72. H Meinhold, A Campos-Barros, B Walzog, R Köhler, F Müller, D Behne 1993 *Exp Clin Endocrinol* 101:87
73. D DePalo, WB Kinlaw, C Zhao, H Engelberg-Kulka, DL St. Germain 1994 *J Biol Chem* 269:16223
74. SY Wu, WS Huang, IJ Chopra, M Jordan, D Alvarez, F Santini 1995 *Am J Physiol* 268:E572
75. JH Mitchell, F Nicol, GJ Beckett, JR Arthur 1997 *J Endocrinol* 155:255
76. JR Arthur, F Nicol, GJ Beckett, P Trayhurn 1991 *Can J Physiol Pharmacol* 69:782
77. JP Chanoine, M Safran, AP Farwell, P Tranter, DM Ekenbarger, S Dubord, S Alex, JR Arthur, GJ Becker, LE Braverman, JL Leonard 1992 *Endocrinology* 130:479
78. J Steinsapir, J Harney, PR Larsen 1998 *J Clin Invest* 102:1895
79. A Campos-Barros, H Meinhold, B Walzog, D Behne 1997 *Eur J Endocrinol* 136:316
80. GJ Beckett, A Russell, F Nicol, P Sahu, CR Wolf, JR Arthur 1992 *Biochem J* 282:483
81. JH Mitchell, F Nicol, GJ Beckett, JR Arthur 1998 *J Mol Encrinol* 20:203
82. B Contempré, NL Duale, JE Dumont, B Ngo, AT Diplock, J Vanderpass 1992 *Clin Endocrinol* 36:579
83. E Kauf, H Dawczynski, G Jahreis, E Janitzky, K Winnefeld 1994 *Biol Trace Elem Res* 40:247
84. MR Calomme, JB Vanderpas, B Francois, M Van Caille-Bertrand, A Herchuelz, N Vanovervelt, C Van Hoorebeke, DA Vanden Berghe 1995 *Experientia* 51:1208
85. O Olivieri, D Girelli, AM Stanzial, L Rossi, A Bassi, R Corrocher 1996 *Biol Trace Elem Res* 51:31
86. O Olivieri, D Girelli, M Azzini, AM Stanzial, C Russo, M Ferroni, R Corrocher 1995 *Clin Sci* 89:637
87. B Contempré, JE Dumont, B Ngo, CH Thilly, AT Diplock, J Vanderpas 1991 *J Clin Endocrinol Metab* 73:213
88. JE Souness, JE Stouffer, V Chagoya de Sanchez 1983 *Biochem J* 214:471
89. O Ezaki 1990 *J Biol Chem* 265:1124
90. R Ghosh, B Mukherjee, M Chatterjee 1994 *Diabetes Res* 25:165

91. DJ Becker, B Reul, AT Ozcelikay, JP Buchet, JC Henquin, SM Brichard  1996 *Diabetalogia* 39:3
92. JH McNeill, HL Delgatty, ML Battell  1991 *Diabetes*  40:1675
93. ML Battell, HL Delgatty, JH McNeill  1998 *Mol Cell Biochem*  179:27
94. M Navarro-Alarcon, H Lopez-G de la Serrana, V Perez-Valero, C Lopez-Martinez  1999 *Sci Total Environ*  228:79
95. LD Stuart, FW Oehme  1982 *Vet Hum Toxicol*  24:435
96. RS Bedwal, A Bahuguna  1994 *Experientia*  50:626
97. JW Barrington, P Lindsay, D James, S Smith, A Roberts  1996 *Brit J Obst Gynecol*  103:130
98. AE Nicoll, J Norman, A Macpherson, U Acharya  1999 *Br J Obst Gyn*  106:1188

# — Part III —

Selenium and human health

# Chapter 17. Selenium as a cancer preventive agent

Gerald F. Combs, Jr.

*Division of Nutritional Sciences, Cornell University, Ithaca, NY 14853, USA*

Junxuan Lü

*AMC Cancer Center, 1600 Pierce St., Denver, CO 80214, USA*

**Summary:** A confluence of different types of evidence indicates that the essential nutrient selenium can affect cancer risk. Most epidemiological studies have shown inverse associations of selenium status and cancer risk; almost all experimental animal studies have shown that supranutritional exposures of selenium can reduce tumor yield; and each of the limited number of clinical intervention trials conducted to date has found selenium treatment to be associated with reductions in cancer risks. The known metabolic functions of selenium, which appear to be discharged by a fairly small and likely incomplete number of selenoproteins may not fully explain these effects. Certainly, selenoproteins may play cancer preventive roles, namely, those involved in antioxidant protection (the glutathione peroxidases), redox regulation (the thioredoxin reductases) and hormonal regulation of metabolism (iodothyronine 5'-deiodinases). However, only minimal evidence is currently available to address the question of whether selenoenzymes are, in fact, important in cancer prevention. More abundant empirical evidence has shown anti-carcinogenic effects of selenium in individuals with apparently full selenoenzyme expression. In fact, it appears that a number of selenium-metabolites (particularly, hydrogen selenide, methylselenol and, under some circumstances, selenodiglutathione) can be directly anti-carcinogenic. Therefore, while the hypothesis remains plausible that selenium-deprivation may increase cancer risk, there is strong support for the hypothesis that supranutritional exposures to selenium can reduce cancer risk. These hypotheses are not mutually exclusive, and it is likely that selenium can function as a cancer preventive agent through both nutritional and supranutritional mechanisms.

**Emergence of a selenium-cancer link**
The nutritional essentiality of selenium was recognized in the late 1950's when the element was found to be the active principle in liver that could replace vitamin E in the diets of rats and chicks for the prevention of

vascular, muscular and/or hepatic lesions [1]. The first suggestion that selenium may be anti-carcinogenic came a decade later and was based on an empirical observation of an inverse relationship of cancer mortality rates and forage crop selenium contents in the United States [2,3]. The corpus of scientific evidence that has subsequently been developed indicates that, indeed, selenium can play a role in cancer prevention.

The epidemiological literature on selenium and cancer has been reviewed [4,5]. Most, but not all, such literature has found selenium status to be inversely associated with cancer risk. Prospective cohort studies in several countries have all shown cancer cases to have significantly lower mean pre-diagnostic serum selenium levels than controls [6-13], and negative associations have been found for various parameters of selenium status and risks to cancers or pre-cancerous lesions of the bladder [6], brain [14], esophagus [15], lung [16-18], head and neck [19], ovary [7], pancreas [20], thyroid [21], stomach [22,23], melanoma [24], prostate [25] and colon [26].

Studies with animal tumor models have shown that selenium treatment can reduce tumor yields. Some years ago, Combs [27] estimated that, of what was then more than 100 studies in which tumor production and/or preneoplastic endpoints had been measured, two-thirds showed that supranutritional selenium doses reduced the incidences of such outcomes, with half showing reductions of 50% or more. Further studies have demonstrated similar reductions in tumor yields [see 28] or experimental metastases [29,30]. Four studies have found selenite-selenium treatment to enhance tumorigenesis; but the interpretation of these is not straightforward, as three [31-33] found increases in tumors at one site to be accompanied by reductions at another site, and one [34] found such enhancement only when the carcinogen was administered in a certain way.

The strongest evidence of anti-cancer efficacy of selenium in humans comes from a limited number of clinical trials, the most informative of which has been the Nutritional Prevention of Cancer (NPC) Trial [35,36]. This randomized, placebo-controlled trial showed that selenium-supplementation (200 mcg selenium/day as selenium-yeast) in skin cancer patients did not reduce their risks to recurrent basal/squamous cell carcinomas. At the same time, the results suggested that selenium-supplementation did reduce risks to other cancers: 37% fewer total non-skin cancers, 45% fewer total carcinomas, 50% fewer total cancers, 63% fewer cancers of the prostate, 58% fewer cancers of the colon-rectum, and 46% fewer cancers of the lung. Those findings are consistent with the reports of Yu et al. [37] that the four-year use of selenium-yeast prevented new cases of primary liver cancer among hepatitis B surface antigen carriers, and that the eight-year use of selenite-fortified table (15 ppm mg selenium) in an entire township was associated with a 75% decrease in the rate of primary liver cancer from the pre-trial rate. A more robust study conducted in an area of China with a high prevalence of esophageal cancer found a treatment containing selenium (50 mcg selenium/day as selenium-yeast, plus vitamin E and ß-carotene) to produce modest reductions in risks of total cancer mortality (13%), stomach cancer mortality (21%) and total mortality (9%) among subjects in the general population [38]. The same group, however, found no protective effects among subjects with esophageal dysplasia given a treatment containing multiple vitamins, minerals and selenite [39].

**Mechanisms of anti-carcinogenicity**
It is highly relevant to public health considerations to determine whether the apparent relationship of selenium and cancer risk involves increased risks due to selenium-deficiency, reduced risks associated with adequate to luxus selenium status, or whether both types of relationship may occur. The possibility that selenium deficiency may increase cancer risk might be expected on the basis of limited expression of selenoenzymes involved in antioxidant protection (glutathione peroxidases [GPxs]), and redox regulation (thioredoxin reductases [TRs]). Thus, the metabolic defense against carcinogenic free radicals would be expected to be reduced. That this may occur is supported by experimental findings. For example, the induction of skin tumors by either ultraviolet irradiation [40-42] or phorbol esters [34] varied inversely with skin GPx activity in animals, and protection by selenite against (2-oxoproyl)amine-induced intrahepatic cholangio-carcinomas in Syrian golden hamsters correlated with the restoration of hepatic GPx activity [43].

Alternatively, anti-carcinogenic effects of selenium have been observed under conditions of maximal selenoprotein expression and these observations suggest other anti-carinogenic mechanisms that most likely involving selenium-metabolites. On this point, a large body of evidence makes it abundantly clear that selenium intake in *excess* of the nutritional requirement can inhibit and/or retard tumorigenesis in experimental animals. Anti-tumorigenically effective selenium-exposures in animal models (e.g., at least 1 mg/kg diet) have typically been an order of magnitude greater than those required to prevent clinical signs of selenium deficiency or to support the maximal expression of known selenoproteins (less than 0.2 mg/kg diet). Accordingly, it is significant that selenium-supplements were effective in reducing cancer risks in the NPC Trial [35] where few, if any, subjects had nutritionally limiting selenium intakes as judged by their baseline plasma selenium levels[1]. In fact, subjects entering the NPC Trial with plasma selenium levels below ca. 120 ng/ml showed the higher risks of subsequent

---

[1] The cohort level was 114±23 ng/ml; very few subjects had levels below 80 ng/ml, the level Nève [45] found to be the upper limit for GPx responses to supplemental selenium in healthy adults. These levels suggest an average daily intake of at least 85 mcg selenium/day, or at least 155% of the RDA [46].

cancer as well as the strongest apparent protective effects of selenium-yeast supplementation[2] [36,44].

Available evidence addresses the cancer-impacts of supranutritional selenium exposures far more completely than it does those of selenium-deficiency. Thus, while it remains plausible that selenium-deprivation may enhance tumorigenesis, it is well documented that at least some forms of selenium can, in supranutritional doses, reduce cancer risk. Accordingly, while anti-carcinogenic functions of selenoproteins cannot be excluded, it is probable that that one or more selenium-compounds/metabolites can function in directly anti-carcinogenic ways. Individually and collectively, these two general mechanisms would appear to underlie the anti-carcinogenic effects described for selenium in cellular antioxidant protection, carcinogen metabolism, gene expression, immune surveillance, cell cycle/death regulation and neo-angiogensis [see 4,5 for review].

## Metabolic bases for selenium anti-carinogenesis

### Roles of selenoenzymes

Because the etiologies of at least some cancers are believed to involve mutagenic oxidative stress, the antioxidant GPxs and the redox-regulatory TRs are expected to have anti-carcinogenic impact by way of genetic and epi-genetic pathways. For example, these pathways may operate by removing DNA-damaging $H_2O_2$ and lipid hydroperoxides, by blocking the production of reactive oxygen species and malonyldialdehyde or by regulating the redox signaling system that is critical to the growth of many cancers [45-47]. Such mechanisms may underlie the protection by selenium against both the carcinogenic and cytotoxic effects of UV-irradiation, which are thought to be due to the oxidative stress of $H_2O_2$ generated photochemically, which has been observed at doses within the range (0.1-0.5 mg/kg diet) of maximization of selenoprotein expression [40-42,49-51].

As an essential component of iodothyronine 5-deiodinases (DIs), selenium may also be important in the regulation of thyroid hormone metabolism, thereby affecting cancer cell growth. Several findings point to this possibility. For example, thyroid hormones were shown to oppose the proliferative action of estrogen on breast cancer [54], breast cancer incidence was higher in areas of endemic iodine deficiency than in non-endemic areas [55], and breast cancer patients had lower plasma thyroid hormone ($T_3$) levels as well as a tendency toward lower toenail selenium concentrations than controls [56]. Nevertheless, the epidemiological evidence relating thyroid hormone status and breast cancer risk is not consistent [56-61].

---

[2]These findings suggest that plasma selenium level is appropriate as an eligibility upper limit for future cancer prevention trials.

Correction of nutritional selenium-deficiency might, therefore, be expected to have anti-carcinogenic effects by increasing the expression of the antioxidant GPxs, the redox-regulatory TRs and, perhaps, the hormone-regulating DIs (see Figure 1). That possibility not withstanding, the documented efficacy of selenium-supplementation in reducing cancer risk in non-deficient individuals (i.e., at supranutritional intakes) would suggest the involvement of other anti-carcinogenic mechanisms that function in addition to or in lieu of mechanisms involving selenoproteins.

**Roles of selenium-metabolites.**
There is evidence of anti-carcinogenic activities for several intermediary metabolites of selenium. These include: 1) selenodiglutathione (GSSeSG), the reductive metabolite of the oxidized inorganic salts (selenite, selenate); 2) hydrogen selenide ($H_2Se$), the common intermediate of that reductive pathway and the catabolism of selenoamino acids; and 3) the methylated metabolites of selenide ($[CH_3]_xSe$) that have hitherto been thought of only as excretory forms of the element. The anti-carcinogenic activities attributable to each of these metabolites are summarized in Figure 1.

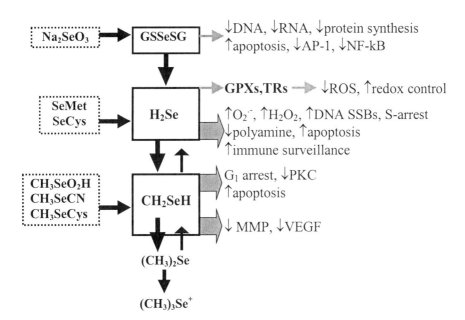

**Figure 1.** Selenium-metabolites apparently active in cancer prevention. See text for discussion.

The product of the thiol-dependent reduction of selenite, GSSeSG, would appear to be relevant only under conditions of exposure to selenite and/or selenate. GSSeSG has been shown to block protein synthesis by inhibiting eukaryotic initiation factor 2 [62], to inhibit ribonucleotide reductase [63], to serve as an efficient oxidant of thioredoxin [64], and to suppress the mRNAs for several GPx isoforms [65]. Support that these effects can be anti-carcinogenic is suggested by findings that GSSeSG can inhibit DNA-binding of the transcription factor AP-1 [66], inhibit cell proliferation [62,65,67-69], enhance apoptosis [68], and be more effective than either selenite or selenomethionine in inhibiting the growth of Ehrlich ascites tumors in mice [70]. Both selenite and GSSeSG have been shown to induce apoptosis in vitro; however, unlike selenite, GSSeSG did not induce widespread tyrosine phosphorylation characteristic of oxidative stress [71]. Since GSSeSG is unstable under physiological conditions, it is unlikely to accumulate in cells and breaks down to glutathione selenol (GSSeH) and $H_2Se$.

Hydrogen selenide appears to be an important player in selenium-anti-carcinogenesis by way of its further metabolism. Its oxidative metabolism produces superoxide anion ($O_2^-$) and $H_2O_2$, the formation of which appear to stimulate apoptosis. That a mechanism mediated by reactive oxygen species (ROS) may be involved in selenite-induced apoptosis is suggested by the findings that the genotoxic and pro-apoptotic effects of selenite on leukemia or mammary cancer cells could be blocked by a superoxide dismutase mimetic, but not by an hydroxy free radical scavenger [72]. Further, catalase added to the cell culture medium blocked the induction of DNA single strand breaks by selenite [73]. In addition, $H_2Se$ can be methylated to produce a string of metabolites that, although being readily excreted, include some that are anti-carcinogenic.

Ip, Ganther and coworkers [74-79] have produced strong experimental evidence that the anti-tumorigenic effects of selenium, at least in the 7,12-dimthylbenzanthracene (DMBA)-induced rat mammary tumor model, are mediated by methylselenol ($CH_3SeH$) or its derivatives (see Figure 1). They found that the $CH_3SeH$-precursors selenobetaine ($CH_3SeO_2H$) and methyl-selenocysteine ($CH_3SeCys$) are anti-carcinogenic, each being somewhat more efficacious in that regard than selenite. In contrast, dimethyl selenoxide, which is metabolized to dimethylselenide ($[CH_3]_2Se$) and very rapidly excreted in the breath, was very poorly chemopreventive, and the rapidly excreted urinary metabolite trimethylselenonium ($[CH_3]_3Se^+$) was completely ineffective. Further work has shown that the $CH_3SeH$-precursors methyl-selenocyanate ($CH_3SeCN$) and $CH_3SeCys$ can each inhibit mammary cell growth, arresting cells in the $G_1$ or early S phase and inducing apoptosis [80-82]. In addition, $CH_3SeCys$ has been shown to inhibit the cell cycle regulatory enzymes $CDK_2$ and protein kinase C (PKC) [83,84].

Because methylated selenium-compounds can be demethylated ultimately to feed the $H_2Se$-exchangeable metabolic pool (see Figure 1), both $CH_3SeCys$ and dimethylselenoxide can support GPx expression [78]. Despite that phenomenon, evidence indicates that $CH_3SeH$ and its precursors have anti-carcinogenic actions independent of those associated with the $H_2Se$ pool. Ip et al. [74-79] found that arsenic, which competitively inhibits both the methylation of $H_2Se$ and the demethylation of $CH_3SeH$ (and the analogous di- and tri-methylated species), greatly reduced the anti-tumorigenic effects of selenite while enhancing those of selenobetaine or methylselenocysteine ($CH_3SeCys$) which yields $CH_3SeH$ metabolically. Specifically, $CH_3SeH$-precursors were shown to lack the genotoxic (i.e., production of DNA single-strand breaks [80,81,85]) or DNA-oxidative damaging [86] effects of selenite or selenide. Unlike the proximal $H_2Se$-precursors, $CH_3SeH$-precursors were potent inhibitors of the expression of matrix matalloproteinases in vascular endothelial cells and vascular endothelial growth factor in cancer cells [87].

There is no evidence that the common forms of selenium in foods and feedstuffs, such as the selenoamino acids selenomethionine and selenocysteine, are anti-carinogenic. However, both these amino acids can be metabolized first to $H_2Se$ and, then, to $CH_3SeH$ (see Figure 1). That conversion occurs directly for selenocysteine, which cannot be used directly in general protein synthesis, is catablolized by a lyase to yield $H_2Se$. Degradation of selenomethionine is not direct, however, and it can enter the general protein pool by replacing methionine. In fact, the degradation of selenomethionine from either dietary or protein-turnover sources necessarily involves its first being converted to selenocysteine by the methionine-transsulfuration pathway. For this reason, most studies have found selenomethionine to be generally less anti-carcinogenically efficacious than selenocysteine or selenite [88-92], as would be expected in short-term studies and, particularly, in the absence of luxus amounts of methionine. However, under steady-state conditions effected by long-term use, and particularly with high-methionine diets, one would expect the anti-tumorigenic efficacy of selenomethionine to approach that of selenocysteine and selenite.

A number of synthetic selenium-compounds have also been found to be anti-carcinogenic. Ip et al. [92] tested a series of alkylselenocyanates ($H[CH_2]_xSeCN$) using the DMBA-induced murine mammary carcinogenesis model and found that anti-carcinogenic efficacy varied directly with increasing chain length up to five carbons. The same group [93] also showed that allyl-selenocysteine, which is expected to yield allylselenol, a fairly hydrophobic metabolite, is more anti-carcinogenic than the corresponding alkylseleno-cysteine. Several aryl selenocyanates have also been found to be anti-tumorigenic. The more effective of these are benzylselenocyanate [94-96], *p*-methoxybenzyl-selenocyate [95], and *p*-phenylselenocyanate [95-98]. These compounds are thought to undergo initial metabolism through arylselenol, which may explain their similar responses to the alkylselenocyanates and other $CH_3SeH$-precursors. Each induces apoptosis of cancer cells in vitro without inducing DNA single strand breaks. When compared to selenite on a molar basis, these forms are not only less effective in supporting GPx expression but also less toxic; yet, they offer comparable anti-tumorigenic efficacy [99,100]. It would appear that the anti-carcinogenic efficacies of theses synthetic selenium-compounds is related to their relative lipophilicities and, thus, to uptake/retention by transformed cells. Accordingly, their anti-tumorigenic efficacies would appear to be affected by dietary fat intakes, being enhanced by the use of low-fat diets [101]. That anti-carcinogenicity need not involve selenoprotein expression is again evidenced, this time by triphenylselenonium chloride (TPSe) [102]. This compound is anti-tumorigenic at fairly high levels of exposure (dietary $EC_{50}$=15 ppm for preventing DMBA-induced mammary cancer [101]). The selenium in TPSe is tightly bonded to three unsubstituted benzene rings rendering it unavailable to metabolism [102,103], ineffective in supporting GPx expression in the selenium-deficient rats, and without adverse effects on rat growth at dietary levels as high as 200 ppm [102].

The anti-carcinogenic activities of the methylated selenium-metabolites and synthetic selenium-compounds are likely related to reactions with critical proteins as well as to redox cycling, which effects may selectively impact the transformed phenotype. Ganther [103] described four general reactions through which selenium-compounds may affect cellular proteins (see also Chapter 23): (i) formation of selenotrisulfide bonds (-S-Se-S-); (ii) formation of selenylsulfide bonds (-S-Se-); (iii) catalysis of disulfide bond formation or its reversal; and (iv) formation of diselenide bonds (-Se-Se-). The first three reactions would affect the activities of many enzymes with critical sulfhydryl groups, while the last would specifically affect activities of the selenoproteins which have selenocysteine residues at these active centers. Selenium-induced inhibition, presumably due to one or more of these reactions, has been demonstrated for a variety of relevant enzymes: ribonuclease [104], Na,K-ATPase [105], and PKC [84,106,107]. Inhibition of PKC would be particularly important, as that enzyme system is known both to activate nuclear transcriptional factors and to serve as a receptor for phorbol ester-type tumor promoters. The inhibition of PKC by a selenium-metabolite such as $CH_3SeH$ would be expected to trigger a number of downstream effects including cell cycle arrest, apoptosis and angiogenic switch regulation. Evidence for at least some of these effects has been reported in response to the $CH_3SeH$-precursors: decreased cdk2 kinase activity [83]; decreased DNA synthesis and elevated *gadd* gene expression [82]; inhibition of vascular endothelial MMPs and VEGF expression [87]. Thus, it appears that selenium-doses large enough to support high, steady-state concentrations of $CH_3SeH$ can effect anti-carcinogenesis by inhibiting critical redox-sensitive factors including PKC and, probably, NF-κB and AP-1, thus, impairing tumor cell metabolism and transformation. These effects would appear to be fairly targeted to certain factors, rather than involving wider perturbations in cellular redox control. After all, selenium-metabolites are typically present in tissues in much lower (nano- to micro-molar) concentrations than those (millimolar) of thiols. In fact, susceptibility to redox modification by selenium-attack seems to be limited to structures containing clustered cysteinyl residues [106,107].

Many of the effects of selenium-compounds on cell proliferation may result from their abilities to form catalytically active, redox-cylcing intermediates. Selenite, diselenides and the oxidation product of $H_2Se$, selenium dioxide, for example, can each react with GSH to produce the selenolate ion (RSe$^-$) [108-110]. In the presence of GSH and molecular oxygen, RSe$^-$ can cycle continuously to generate $O_2^-$ and $H_2O_2$. This redox cycling is thought to be the basis of selenium-toxicity, and it is possible that it may also contribute to anti-carcinogenesis. Spallholz et al (2000) found that dimethyldiselenide ($[CH_3Se]_2$) was the most catalytically active of 19 selenium-compounds[3] in its ability to generate $O_2^-$ in the presence of GSH and $O_2$. They attributed this activity to $CH_3SeH$ produced by the reduction of ($[CH_3]Se)_2$ presumably generating the radical anion $CH_3Se^-$. Thus, it is possible that catalytically active species may be generated intracellularly as the result of the metabolism of proximal (e.g., $CH_3SeCys$, $CH_3SeO_2H$) and/or upstream (e.g., selenomethionine, selenocysteine) precursors. Accordingly, it may be possible to target selenium-anticarcinogenesis to cells with such metabolic capabilities.

## Conclusion

It is clear that selenium-compounds can reduce the risks for at least some kinds of cancer. These anti-carcinogenic effects may involve the protective, nutritional function of selenium as an essential constituent of a number of metabolically important selenoenzymes. In addition to such effects, selenium appears to have a significant supranutritional function as a source of selenium-metabolites that can, themselves, inhibit carcinogenesis. While the former activities are not nearly as well studied as the latter, a prudent perspective would hold that while deprivation of selenium may enhance cancer risk, even greater cancer-preventive benefits are to be obtained through the use selenium-compounds at supranutritional levels. According to this view, limiting selenium-supplementation only to those levels required to

---

[3]In addition to ($[CH_3]Se)_2$, these included nine other catalytically active selenium-compounds: selenite, selenium dioxide, selenocystine, selenocystamine, diselenopropionic acid, diphenyldiselenide, dibenzyldiselenide, $p$XSC and 6-propylselenouracil; and nine selenium-compounds that were not catalytically active: elemental selenium, selenate, SeMet, $CH_3SeCys$, selenobetaine, dimethylselenoxide, selenopyridine, TPSe and potassium selenocyante.

optimize selenoprotein expression would be to limit the prospects for realizing the potential of selenium in cancer prevention.

**References**

1. K Schwarz, CM Foltz 1957 *J Am Chem Soc* 79:3292
2. RJ Shamberger, DV Frost 1969 *Can Med Assn J* 104:82
3. J Shamberger, CE Willis 1971 Clin Lab Sci 2:211
4. GF Combs Jr, WP Gray 1998 *Pharmacol Ther* 79:179
5. GF Combs Jr, LC Clark 1999 *Nutritional Oncology* D Heber, GL Blackburn, VLW Go (Eds) Academic Press, Inc New York, 215
6. KJ Helzlsouer, AJ Alberg, EP Norkus, JS Morris, SC Hoffman, GW Comstock 1996 *J Nat Cancer Inst* 88:32
7. KJ Helzlsouer, GW Comstock, JS Morris 1989 *Cancer Res* 49:6144
8. FJ Kok, AM De Bruijn, A Hofman, R Vermeeren, HA Valkenburg 1987 *Am J Epidemiol* 125:12
9. JT Salonen, R Salonen, R Lappeteläinen, PH Mäenpää, G Alfthan, P Puska 1985 *Br Med J* 290:417
10. J Nayini, K El-Bayoumy, S Sugie, LA Cohen, BS Reddy 1989 *Carcinogen* 10:509
11. W Willett, B Polk, S Morris, M Stampfer, S Preissel, B Rosner, J Taylor, K Schneider, C Hames 1983 *Lancet* 2:130
12. Van Den Brandt, R Goldbohm, P Van't Veer, P Bode, E Dorant, R Hermus, F Sturmans 1993 *J Nat Cancer Inst* 85:224
13. AMY Nomura, J Lee, GN Stemmernann, GF Combs Jr 2000 *Cancer Epidemiol Biomarkers Prev* 9:883
14. P Philipov, K Tzatchev 1988 *Zentrabl Neurochir* 49:344
15. K Jaskiewicz, WFO Marasas, JW Rossouw, FE Van Niekerk, EWP Heinetech 1988 *Cancer* 62:263
16. L. Geardsson, D Brune, IGF Nordberg, PO Wester 1985 *Br J Industrial Med* 42:617.
17. H Miyamoto, Y. Araya, M Ito, H Isobe, H Dosaka, T Shimizu, F Kishi, I Yamamoto., H Honma, Y Kawakami 1987 *Cancer* 60:1159
18. P Knekt, A Aromaa, J Maatela, G Alfthan, R Aaran, M Hakama, T Hakulinen, R Peto, L Teppo 1990 *J Nat Cancer Inst* 82:864
19. T Westin, E Ahlbom, E Johansson, B Sandstrom, I Karlberg, S. Edstrom 1989 *Arch Otolaryngol Head Neck Surg* 115:1079
20. PGJ Burney, GW Comstock, JS Morris 1989 *J Clin Nutr* 49:895
21. E. Glattre, Y Thomassen, SO Thoresen, T Haldorsen, PG Lund-Larsen, L Theodorsen, J. Aaseth 1989 *Int J Epidemiol* 18:45
22. CP Caygill, K Lavery, PA Judd, MJ Hill, AT Diplock 1989 *Food Addit Contam* 6:359.
23. P Knekt, A Aromaa, J Maatela ,G Alfthan, RK Aaran, L Teppo, M Hakama 1988 *Int J Cancer* 42:846
24. U Reinhold, H Blitz, W Bayer, KH Schmidt 1989 *Acta Derm Venerol* 69:132
25. K Yoshizawa, WC Willett, SJ Morris, MJ Stampfer, D Spiegelman, EB Rimm, E Gioiovannucci 1998 *J Nat Cancer Inst* 90:1219
26. LC Clark, LJ Hixson, GF Combs Jr, ME Reid, BW Turnbull, RE Sampliner 1993 *Cancer Epidemiol Biomarkers Prev* 2:41
27. GF Combs Jr 1989 *Nutrition and Cancer Prevention* T Moon, M Micozzi (Eds) Marcel Dekker New York, 389
28. C Ip 1998 *J Nutr* 128:1845.
29. L Yan, JA Yee, MH McGuire, GL Graef 1997 *Nutr Cancer* 28:165
30. L Yan, JA Yee, D Li, MH McGuire, GL Graef 1999 *Anticancer Res* 19:1337
31. J Ankerst, H Sjogren 1982 *Int J Cancer* 29:707

32. R.D. Dorado, EA Porta, TM Aquino 1985 *Hepatol* 5:1201
33. H Nakadaira, T Ishizu, M Yamamoto 1996 *Cancer Lett* 106:279
34. JP Perchellet, NL Abney, RM Thomas, YL Guislan, EM Perchellet 1987 *Cancer Res* 47:477
35. LC Clark, GF Combs Jr, BW Turnbull, E Slate, D Alberts, D Abele, R Allison, J Bradshaw, D Chalker, J Chow, D Curtis, J Dalen, L Davis, R Deal, M Dellasega, R Glover, G Graham, E Gross, J Hendrix, J Herlong, F Knight, A Krongrad, J Lesher, J Moore, K Park, J Rice, A Rogers, B Sanders, B Schurman, C Smith, E Smith, J Taylor, J Woodward 1996 *J Am Med Assoc* 276:1957
36. LC Clark, B Dalkin, A Krongrad, GF Combs Jr, BW Turnbull, EH Slate, R Witherington, JH Herlong, E Janosko, D Carpenter, C Borosso, S Falk, J Rounder 1998 *Brit J Urol* 81:730
37. SY Yu, YJ Zhu, WG Li 1997 *Biol Trace Elem Res* 56:117
38. JY Li, PR Taylor, B Li, WJ Blot, W Guo, S Dawsey, GQ Wang, CS Yang, SF Zheng, M Gail, GY Li, BQ Liu, J Tangrea, YH Sun, F Liu, F Fraumeni Jr, YH Zhang 1993 *J Nat Cancer Inst* 85:1492
39. WJ Blot, JY Li, PR Taylor, W Guo, S Dawsey, GQ Wang, CS Yang, SF Zheng, M Gail, GY Li, BQ Liu, J Tangrea, YH Sun, F Liu, F Fraumeni Jr, YH Zhang, B Li 1993 *Nat Cancer Inst* 85:1483
40. KE Burke, GF Combs Jr, EG Gross, KC Bhuyan, H Abu-Libdeh 1992 *Nutr Cancer* 17:123
41. BC Pence, E Pelier, DM Dunn 1994 *J Invest Dermatol* 102:759
42. AM Diamond, P Dale, JL Murray, DJ Grdina 1996 *Mutat Res* 356:147
43. Y Kise, M Yamamura, M Kogata, M Nakagawa, S Uetsuji, H Takada, K Hioki, M Yamamoto 1991 *Nutr Cancer* 16:153
44. GF Combs Jr, LC Clark, BW Turnbull 2001 *Proc 7$^{th}$ Internat Symp Selenium Biol Med* (In press)
45. J Nève 1995 *J. Trace Elements Med. Biol.* 9:65-73
46. Panel on Dietary Antioxidants and Related Compounds 2000 *Dietary Reference Intakes for Vitamin C, Vitamin E, Selenium and Beta-Carotene and other Carotenoids* National Academy Press Washington, D.C.
47. M Berggren, A Gallegos, JR Gasdaska, PY Gasdaskya, J Warneke, G Powis 1996 *Anticancer Res* 16:3459
48. G Powis, DL Kirkpatrick, M Angulo, JR Gasdaska, MB Powell, SE Salmon, WR Montfort 1996 *Anticancer Drugs* 7 (suppl 3):121
49. QA Sun, Y Wu, F Zappacosta, KT Jeang, BJ Lee, DL Hatfield, VN Gladyshev 1999 *J Biol Chem* 35:24522
50. MT Leccia, MJ Richard, JC Beani, H Faure, AM Monjo, J Cadet, P Amblard, A Favier 1993 *Photochem Photobiol* 58:548
51. A Moysan, P Morliere, I Marquis, A Richard, L Dubertret 1995 *Skin Pharmacol* 8:139.
52. CJ Bertling, F Lin, AW Girotti 1996 *Photochem Photobiol* 64:13
53. N Emonet, MT Leccia, A Favier, JC Beani, MJ Richard 1997 *J Photochem Photobiol B* 40:84
54. H Vorherr 1987 *Eur J Cancer Clin Oncol* 23:255
55. P Mustacchi, FS Greenspan 1986 *Wesner's The Thyroid* SH Ingbar, LE Braverman (Eds) Lippincott Philadelphia, 1453
56. JJ Strain, E Bokje, P Van't Veer, J Coulter, C Stewart, H Logan, W Odling-Smee, RAJ Spence, K Steele 1977 *Nutr Cancer* 27:48
57. DP Rose, TE Davis 1978 *Cancer* 41:666
58. DP Rose, TE Davis 1979 *Cancer* 43:1434
59. M Lemaire, L Baugnet-Mahieu 1986 *Eur J Cancer Clin Oncol* 22:301
60. B Rasmussen, U Freedt-Rasmussen, L Hededüs, H Perrild, K Bech 1987 *Eur J Cancer Clin Oncol* 23:553

61. O Takatani, T Okumoto, H Kosana, M Nishidi, M Hiruide 1989 *Cancer Res* 49:3109
62. LN Vernie 1987 *Proc 3rd Internat Symp Selenium Biol Med* GF Combs Jr, JE Spallholz, OA Levander, JE Oldfield (Eds) AVI Publ Co Westport, CT 1074
63. G Spyrou, M Bjornstedt, S Skog, A Holmgren 1996 *Cancer Res* 56:4407
64. M Bjornstedt, S Kumar, A Holmgren 1992 *J Biol Chem* 267:8030
65. L Wu, J Lanfear, PR Harrison 1995 *Carcinogenesis* 16:1579
66. M Bjornstedt, S Kumar, L Bjorkhem, G Spyrou, A Holmgren 1997 *Biomed Environ Sci* 10:271
67. LN Vernie, CJ Hamburg, WS Bont 1981 *Cancer Lett* 14:303
68. J Lanfear, JJ Flemming, L Wu, G Webster, PR Harrison 1994 *Carcinogenesis* 15:1387
69. BC Pence, M Stewart, L Walsh, G Cameron 1996 *Proc 6th Internat Symp Selenium Biol Med* ILSI, Beijing 82
70. KA Poirier, JA Milner 1983 *J Nutr* 113:2147
71. PR Harrison, J Lanfear, L Wu, J Fleming, L McGarry, L Blower 1997 *Biomed Environ Sci* 10:235
72. J Lü, M Kaeck, C Jiang, AC Wilson, HJ Thompson 1994 *Biochem Pharmacol* 47:1531
73. Z Zhu, M Kimura, Y Itokawa, T Aoki, JA Takahasi, S Nakatsu,Y Oda, H Kikuchi 1996 *Biol Trace Elem Res* 54:123
74. C Ip, H Ganther 1990 *Cancer Res* 50:1206
75. C Ip, H Ganther 1991*Carcinogenesis* 12:365
76. C Ip, HE Ganther 1992 *Carcinogenesis* 13:1167
77. C Ip, HE Ganther 1992 *J Inorgan Biochem* 46:215
78. C Ip, C Hayes, RM Budnick, HE Ganther 1991 *Cancer Res* 51:595
79. S Vanhanavikit, C Ip, HE Ganther 1993 *Xenobiotica* 23:731
80. J Lü, C Jiang, M Kaeck, H Ganther, S Vadhanavikit, C Ip, HJ Thompson 1995 *Biochem Pharmacol* 252:739
81. J Lü, Pei, C Ip, D Lisk, H Ganther, HJ Thompson 1996 *Carcinogenesis* 17:1903
82. M Kaeck, J Lü, R Strange, C Ip, HE Ganther, HJ Thompson 1997 *Biochem Pharmacol* 53:921
83. R Sinha, D Medina 1997 *Carcinogenesis* 18:1541
84. R Sinha, SC Kiley, JX Lü, R Moraes, HJ Thompson, S Jaken, D Medina 1999 *Cancer Lett* 146:135
85. AC Wilson, HJ Thompson, PJ Schedin, NW Gibson, HE Ganther 1992 *Biochem Pharmacol* 43:1137
86. MS Stewart, RL Davis, LP Walsh, BC Pence 1997 *Cancer Lett* 117:35
87. C Jiang, W Jiang, C Ip, H Ganther, J. Lü 1999 *Mol Carcinogenesis* 26(4):213
88. C Ip, G White 1987 *Carcinogenesis* 8:1763
89. C Ip, C Hayes 1989 *Carcinogenesis* 10:921
90. HJ Thompson, C Ip 1991 *Biol Trace Elem Res* 30:163
91. B Siwek, E Bahbouth, MA Serra, E Sabbioni, MC De Pauw-Gillet, R Bassleer 1994 *Arch Toxicol* 68:246
92. C Ip, S Vadhanavkit, H Ganther 1995 *Carcinogenesis* 16:35
93. C Ip, Z Zhu, HJ Thompson, D Lisk, HE Ganther 2000 *Anticancer Res* 19 (In press)
94. BS Reddy, S Sugie, H Maruyama, K El-Bayoumy, P Marra 1987 *Cancer Res* 47:5901
95. Z Ronai, JK Tillotson, F Traganos, Z Darzynkiewice, CC Conaway, P Upadhyaya, K El-Bayoumy 1995 *Int N Cancer* 63:428
96. V Adler, MR Pincus, S Posner, P Upadhyay, K El-Bayoumy, Z. Ronai 1996 *Carcinogenesis* 17:1849
97. B Prokopczyk, JE Cox, P Upadhyaya, S Amin, D Desai, D Hoffmann, K El-Bayoumy 1996 *Carcinogenesis* 17:749
98. C Ip, H Ganther 1993 *Selenium in Biology and Human Health* RF Burk (Ed) Springer-Verlag New York 170

99. C Ip, K El-Bayoumy, P Upadhyaya, H Ganther, S Vadanavikit, H Thompson 1994 *Carcinogenesis* 15:187
100. BS Reddy, A Rivenson, K El-Bayoumy, P Upadhyaya, B Pittman, CV Rao 1997 *J Nat Cancer Inst* 89:506
101. C Ip, H Thompson, HE Ganther 1994 *Carcinogenesis* 15:2879
102. C Ip, HJ Thompson, HE Ganther 1998 *Anticancer Res* 18:9
103. HE Ganther 1999 *Carcinogenesis* 20:1657
104. HE Ganther, C Corcoran 1969 *Biochem* 8:2557
105. PL Bergad, WB Rathbun 1986 *Curr Eye Res* 5:919
106. R Gopalakrishna, U Gundimeda, ZH Chen 1997 *Arch Biochem Biophys* 348:25
107. R Gopalakrishna, ZH Chen, AU Gundimed 1997 *Arch Biochem Biophys* 348:37
108. J Chaudiere, O Courtin, J Leclaire 1992 *Arch Biochem Biophys* 296:328
109. JE Spallholz 1997 *Biomed Environ Sci* 10:260
110. MS Stewart, JE Spallholz, KH Neldner, BC Pence 2000 *Free Radic Biol Med* 26: 42

# Chapter 18. Selenium deficiency and human disease

Ruth J. Coppinger and Alan M. Diamond

*Department of Human Nutrition, University of Illinois at Chicago, Chicago, IL 60612, USA*

**Summary:** The beneficial role of selenium became apparent in the 1950's, when it was shown to prevent a variety of diseases in animals, often when these animals were already exhibiting vitamin E deficiency. In humans, there are several examples of disease that can be attributed to selenium deficiency, and for some, the ultimate causal relationship has been established by the demonstration that symptoms can be alleviated with selenium supplementation of the at risk population. This is true both in cases where diseases associated with selenium deficiency are endemic in areas of low selenium consumption, such as Keshan disease, Keshin-Beck disease and myxedematous cretinism, as well as in circumstances where selenium deficiency results from other pre-existing health conditions. The consequences of too little dietary selenium can be considered in the context of the decline of specific selenium-containing proteins, many of which have only recently been identified.

## Introduction

Evidence of the toxic consequences of selenium over-consumption has been demonstrated throughout the history of the element. The mechanisms of selenium toxicity appear to be due to general effects as a consequence of either the mis-incorporation of selenium in sulfur-containing amino acids or a pro-oxidant effect of selenium-containing compounds [1,2]. While incidences of selenium toxicity have been documented in animals including humans in current times, this chapter will focus on the opposite situation, the consequences of selenium deficiency that occur due to low selenium intake or to other predisposing health conditions. Among the diseases associated with regional dietary deficiency, the most convincing data for an etiological role of selenium deficiency exists for Keshan and Kashin-Beck diseases. A form of cretinism associated with hypothyroidism appears also be related to selenium deficiency. These diseases will be discussed in detail in this as well as other chapters. Less clear is whether an endemic nephropathy that occurs in the Balkans is also due to selenium deficiency [3].

## The geography of endemic selenium deficiency

The selenium content of food is largely dependent on the selenium content of the soil in which it was grown. Volcanic soils that undergo significant runoff are particularly susceptible to mineral leaching. Although selenium content of soils varies greatly around the world, only two regions are so selenium-poor that overt deficiency syndromes are endemic. A belt of land, stretching from northeast China, northern North Korea and far eastern Siberia to south central China, Nepal and Tibet is very iodine deficient and contains several regions of selenium deficiency [4]. Central Africa, particularly the Democratic Republic of Congo (formerly Zaire) has similar overlapping deficiencies [5 and references therein]. Due to the patchy distribution of selenium poor soil, residents experience deficiency-related diseases only when the diet is derived almost entirely from local foods, with little or no importation of foods from outside areas. All three deficiency syndromes discussed below are associated with additional probable etiological risk factors, including compounded iodine deficiency, viral pathogens, and environmental toxins.

## Myxedematous cretinism

Myxedematous cretinism is a disorder of skeletal growth, neurological development, hypothyroidism and thyroid atrophy that occurs in regions with overlapping deficiencies of iodine and selenium. This contrasts with neurological cretinism, which is characterized by severe mental retardation, abnormalities in speech and hearing, neuromuscular disorders such as ataxia and spasticity, thyroid hormone levels similar to individuals without cretinism, and frequent goiter. Neurological cretinism occurs in many areas with very severe iodine deficiency with adequate or deficient selenium [6]. People with cretinism may exhibit symptoms of one classification, or the symptoms may be mixed.

Endemic myxedematous cretinism occurs in areas with widespread goiter where mean serum selenium in children is below 500 nmol/L (compared to normal values >2500) [7]. The greatest concentration occurs in central Africa, while neurological cretinism predominates in Latin America, Asia, and New Guinea [8]. In 1990, goiter was present in 65-85% of the population of northern Democratic Republic of Congo, and cretinism, mostly of the myxedematous type, in 2-6% of the population [7]. Symptoms include dry, scaly skin (from the excessive mucin that the syndrome derives its name), infantile facial features, distortions of long bone growth leading to short stature, lack of fusion of the cranial plates, irregular tooth implantation, delayed sexual maturation, and hypothyroidism associated with thyroid atrophy (involution) [9]. Skeletal abnormalities have been described as "stippled epiphysis" because mineralization takes place from multiple scattered locations in the epiphyseal cartilage [10]. Like other symptoms of

*Selenium deficiency and human disease* 221

both forms of cretinism, this can be attributed to insufficient thyroid hormone at the time needed to promote transcription of genes for development of the relevant organ system [11]. Differences between the two forms suggests that thyroid hormones are insufficient during different periods of development; the neurological form results from inadequate thyroid hormone during fetal neural development and myxedematous cretinism results from insufficiency later in gestation and childhood.

**Etiology**
Thyroid hormone is synthesized in the thyroid gland as the prohormone 3,5,3',5'-tetraiodothyronine (T4, thyroxine) and is converted in the periphery to the active 3,5,3'-triiodothyronine (T3) [for a complete review of thyroid metabolism, see ref. 6 and Chapter 16]. T4 synthesis begins with the iodination of the amino acid tyrosine; this step requires the production of $H_2O_2$ by thyroperoxidase. 3,5-Diiodotyrosine is stored in the colloid core of the thyroid follicle, and when the gland is stimulated, it is dimerized and modified to form T4 that is released into the circulatory system. T4 is converted to T3 by 5' thyronine deiodinase types I and II in tissues and plasma. 5 thyronine deiodinase type III converts T4 to the inactive isomer 3,3',5'-triiodothyronine (rT3) and also inactivates T3 by further deiodination. All three deiodinase enzymes are selenoproteins, but deiodinase II is less affected by selenium deficiency, and was not identified as a selenoenzyme until it was cloned in 1996 [12, for a review of the deiodinases, see ref 13 and Chapter 16]. During selenium deficiency, deiodinase activity is maintained in the brain and other critical organs, probably by the predominance of deiodinase II, and the possible sequestration of selenium [14,15].

Neurological development in myxedematous cretinism appears to be preserved by the interactions of the maternal iodine and selenium deficiencies (see figure 1). The fetal thyroid gland is not fully functional until about 20 weeks of gestation, but fetal tissue deiodinases become active much sooner [11]. Prior to activation of T4 synthesis, the fetus depends on placental transport of maternal T4 [11 and references therein]. In an iodine and selenium deficient mother, circulating T4 concentrations are elevated due to low deiodinase activities, allowing greater placental transport of T4 to the fetus (see figure 1). In pregnancies with only iodine deficiency or with more severe iodine deficiency combined with selenium deficiency, circulating maternal T4 levels are too low to support normal neurological development in the fetus [8].

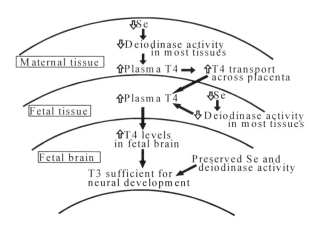

Figure 1. The role of selenium deficiency in the preservation fetal neurological development in iodine deficiency. Diminished maternal selenium levels lead to reduced deiodinase activity and elevated plasma T4 available for placental transport. Fetal brain deiodinase activity is preserved in selenium deficiency, and with the increased fetal plasma T4 levels, provides sufficient T3 for neurological development.

Thyroid hormone synthesis and thyroid tissue growth are stimulated by the presence of iodine and by the pituitary thyroid stimulating hormone (TSH), which is negatively regulated by circulating T3 and T4 [6]. Subjects with myxedematous cretinism have very low plasma T3 and T4 levels and TSH concentrations over 20 times that of other residents of the community without cretinism, suggesting that the thyroid gland is incapable of response to TSH stimulation [7]. Noncretinous iodine and selenium deficiencies, on the other hand, lead to goiter and elevated T4 levels with more typical TSH [7]. Indeed, the thyroid gland of people with myxedematous cretinism loses responsiveness to iodine supplementation during childhood and becomes involuted, highly fibrotic and necrotic and incapable of supporting growth and sexual maturation [8].

Thyroid involution occurs in rats deficient in both iodine and selenium after the administration of a physiological thyroid-stimulating dose of iodine (see figure 2) [16]. Under these conditions, the colloid core of thyroid follicle collapses, rapid necrosis occurs, macrophages infiltrate the tissue, and fibrosis ensues [17]. Inhibition of thyroperoxidase eliminates this effect, suggesting the involvement of $H_2O_2$ [17]. This implies that thyroid involution may be caused by lowered activity of the selenoprotein glutathione peroxidase during selenium deficiency, allowing greater oxidative damage and production of toxic aldehydes [17]. It is thought that in myxedematous cretinism, a burst of TSH that is secreted soon after birth contributes to the involution of the thyroid gland by similarly stimulating the activity of thyroperoxidase [11,16].

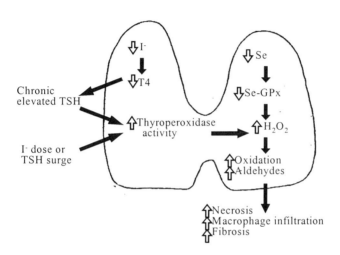

**Figure 2.** The proposed roles of selenium and iodine deficiencies in involution of the thyroid gland in myxedematous cretinism. Iodine deficiency leads to chronic elevation of thyroid stimulating hormone (TSH) and thyroperoxidase activity, which may be exacerbated by acute intake of iodine or the TSH surge in early infancy. The resulting increase of $H_2O_2$ is not neutralized by the low levels of the selenoprotein glutathione peroxidase during selenium deficiency. The excess $H_2O_2$ leads to oxidative damage of biomolecules and increased concentration of aldehyde compounds, resulting in thyrocyte necrosis, macrophage response and fibrosis proliferation.

## Therapy

Iodine supplementation will restore the thyroid function in many patients with hypothyroid cretinism under the age of five [8]. Selenium supplementation of individuals deficient in both iodine and selenium can be hazardous, however. Selenium deficiency with iodine deficiency contributes to increased plasma T4 through inhibition of deiodinases [4]. When selenium is replenished, T4 levels plummet due to conversion to T3 and the involuted thyroid is unable to increase T4 output in response to the increased T4 metabolism [8]. Thus, supplementation of selenium must be accompanied by iodine supplementation in populations deficient in both minerals.

Unfortunately, due to ongoing political instability and poverty of the nations in which endemic myxedematous cretinism is present, some populations with endemic cretinism have received supplementation, but such programs have not become widespread [18].

**Keshan disease**
Keshan disease was identified as a syndrome following an epidemic outbreak in 1935 in Keshan county, China. It occurs in the selenium poor areas of the iodine deficient band of land running across China and neighboring nations and only in remote, hilly areas where concentrations of selenium in the soils, food supply and human plasma are all low [19]. The disease is an endemic cardiomyopathy that afflicts the peasant population, appearing in women of childbearing age and preschool children.

Seasonal and annual variability in disease incidence suggested pathogenic involvement in the development of this disease. Endemic communities in northern areas of China experience higher rates of the disease in winter months while in southern China occurrence of the disease is concentrated in the summer [19]. Annual prevalence rates can range from fewer than ten to several hundred cases per 100,000 residents [19]. A myocarditic Coxsackievirus has been associated with the pathogenesis of Keshan disease [reviewed in 20]. The novel interaction of selenium deficiency with Coxsackievirus virulence is addressed in detail in Chapter 19 of this book.

Keshan disease occurs with symptoms of congestive heart failure or, less frequently, as sudden death or stroke from diffuse cardiac thromboses [22]. Symptoms include ECG changes such as right branch block, and cardiogenic shock [23]. Autopsy typically reveals cellular edema, mitochondrial swelling, and overlapping striations of fibrotic tissue, indicating multiple bouts of localized necrosis [22].

The average intake of selenium in Keshan disease endemic areas has been estimated at 10 µg/day, approximately half that found in low selenium areas without the cardiomyopathy [24]. Soil selenium concentrations in Keshan endemic areas are typically about 3-4% of selenium levels in areas without Keshan disease [25]. Hair selenium concentration in endemic communities has been reported to be below 0.12 ppm while in nearby nonendemic communities it was between 0.12 and 0.2 ppm; concentrations from areas of China outside the band of iodine and selenium deficiency were 0.25-0.6 ppm [26]. Dietary analyses indicate that less diversified diets are more conducive to the disease development [19]. Localities of endemic Keshan incidence have diets comprised mostly of home grown corn, while peasant diets in Keshan disease-free communities contained greater variety of grains and more soybeans [19]. Supplementation of the diets of over one million peasants with selenium-enriched salt has reduced incidence of Keshan disease, thereby validating the causal relationship between the mineral deficiency and the disease [21].

**Kashin-Beck disease**
Kashin-Beck disease is an osteoarthropathy that is endemic in a selenium and iodine deficient area of China. Like Keshan disease, this disease is limited to poor, isolated communities with very low concentrations of both iodine and selenium in the soil, agricultural products and human tissues. Urinary iodine in Kashin-Beck patients is 1.2 μg/dL, compared to 1.8 μg/dL in neighboring, disease-free villages [27]; average serum selenium among residents of the Kashin-Beck endemic areas averages less than 11 ng/ml, while among residents of areas of China without mineral deficiency, concentrations are 60-105 ng/ml [27]. Selenium concentrations in staple foods of Kashin-Beck endemic communities have been reported to be about one-eighth of that found in areas outside the band of iodine and selenium deficiency [4]. That this disease only occurs in areas of combined mineral deficiency implicates both iodine and selenium as causes. However, the complete absence of this disease in other selenium and iodine deficient areas, such as central Africa, suggests multiple etiological factors [28]. Though the complete etiology of this syndrome has not been fully established, environmental factors such as mycotoxins and soil decomposition products have been investigated at length and are thought to interact with mineral deficiencies to contribute to the disease development.

Most communities with endemic Kashin-Beck disease lie in cold mountainous regions where the soil may be poor in minerals and nitrogen, and where residents consume diets consisting largely of homegrown grains [26]. The symptoms of the disease make the manual farmwork painful and difficult, and have therefore drawn the attention of international researchers and interventions by humanitarian organizations, the World Health Organization and the Chinese government [29]. The prevalence of Kashin-Beck disease had dropped considerably in the recent decades in which it has been studied. In 1992, the number of cases of the disease was estimated to be 2.2-3 million in China with some endemic villages having prevalence rates of 80% [30], but with 20-40% being more common [4]. Chinese prevention programs, such as the selenium enrichment of table salt for five million persons have led to a drastic reduction in new cases [29 and references therein], with the rate of incidence dropping to nearly zero in many formerly endemic communities [4].

The disease is characterized by a range of bone and joint malformations that appear during childhood or the pubertal growth spurt and progress until growth ceases. Kashin-Beck disease is characterized by varying degrees of joint deformation and immobility. Mild cases display enlargement and cracking (crepitus) of small joints; in serious cases, large joints are enlarged and may be distorted and growth of the long bones may end prematurely, resulting in short stature [29]. Symptoms tend to be bilaterally symetrical [31]. The hands or feet are usually affected, and bones of the wrists and ankles may be reduced in size, or some may be absent entirely [31].

Normal growth at the end of long bones takes place when chondrocytes on the underside of the epiphysial growth plate proliferate toward the end of the growing bone [see 32 for a review of bone physiology]. The chondrocytes secrete procollagen which matures and forms extracellular collagen matrix. The chondrocytes then disintegrate and the epiphyseal collagen matrix is mineralized by osteoblasts on the surface of the bone.

Radiographic examination of individuals afflicted with Kashin-Beck disease reveals deformation and fragmentation of the epiphyseal plate, uneven mineralization of the extracellular matrix and irregulatity of the bone surface [31]. More advanced symptoms include pockets of unmineralized collagen surrounded by bone [33]. A diagnostic feature of the disease is the fragmented, blocky or conical shape of the growth apparatus as a result of necrosis in one area of chondrocytes [31].

Histological observations indicate that the disease-associated joint malformations result from necrosis of chondrocytes during bone growth, and from the secondary repair and remodeling that follows clearance of necrotic tissue by the circulatory system [31]. Broadspread necrosis throughout the growth area may lead to premature closing of the growth plate and cessation of growth [31]. Impaired maturation of collagen has also been observed and has been suggested to result from fulvic acid in well water (see below) [33].

**Etiology and therapy**
Familial trends are not discernable in Kashin-Beck disease and immigrants into endemic areas develop symptoms in rates inversely proportionate to age at immigration [27]. This argues against a genetic component to the disease. Surveys of environmental conditions and animal research have identified four etiological factors for the disease: iodine deficiency, selenium deficiency, well water contamination by organic decomposition products (fulvic acids), and fungal grain contamination. Removal of the causative conditions in the early stages of the disease development through selenium supplementation, improved grain storage and drinking water purification may eliminate symptoms by allowing the return of normal bone growth [28]. Once bone growth has ended, few therapeutic options are available. Surgical reshaping of the bones of the ankle and the knee has been employed by Chinese physicians to improve the alignment of the weight bearing structures [34]. This procedure has improved mobility and relieved pain in many cases. In 1992, the international physicians' group Doctors Without Borders (Médecins Sans Frontières) began training local physicians in physiotherapy techniques to reduce the pain and immobility of the disease symptoms [30].

Many research efforts have focused on prevention and, as mentioned above, interventions have been successful in reducing new case incidence in many communities.

### Iodine deficiency

Affected and unaffected residents of Kashin-Beck endemic villages have much lower serum thyroxine and triiodothyronine and higher rates of goiter than residents of non-Kashin-Beck endemic villages [27]. Within communities with endemic Kashin-Beck, elevated thyrotropin (TSH) and lower urinary iodine are predictors of the disease occurrence [27]. The mechanism of iodine involvement in the disease may be related to a dependence of growth plate chondrocytes on locally produced triiodothyronine (T3). Because of diminished thyroxine (T4) availability and reduced deiodinase activity from low selenium concentrations, the availability of active thyroid hormone may be insufficient to maintain normal growthplate function or increase susceptibility to free radical damage from other environmental contributors [5].

### Selenium deficiency

Kashin Beck disease occurs only in regions with severe soil deficiency of both iodine and selenium. Both serum selenium concentrations and the activity of serum glutathione peroxidase are lower in residents of villages where the disease is endemic [27]. However, disease-free residents of endemic villages have lower plasma selenium concentrations and glutathione peroxidase activities than sufferers of the disease [27]. Whatever the mechanism, selenium status is not a predictor of the disease within communities where the disease is endemic, but low selenium status is predictive of communities having Kashin-Beck disease present. Selenium deficiency alone does not induce joint malformities in animal models [33]. In conjunction with fulvic acid contaminants found in water sources in endemic villages, selenium deficiency does induce joint morphology in rats similar to that found in the disease [35]. Supplementation of selenium as a preventative is controversial, not only because of its disputed efficacy alone without additional measures, but also because selenium supplementation can exacerbate hypothyroidism in iodine deficient subjects [8].

### Fulvic acid

Fulvic acid is the water soluble fraction of the products of microbial and chemical degradation of plant and animal matter. It is found throughout the environment in soil, peat, water, coal, and sediment and is present in higher concentrations in well water of Kashin-Beck endemic communities than in disease free areas [28, 33]. Fulvic acids are a heterogeneous group of complex polymers of highly oxygen-substituted benzene rings [36]. They contain many highly reactive functional groups including carboxyls, hydroxyls, carbonyls, phenols and quinones [36]. These contribute to a high concentration of stable free radicals, estimated at $10^{18}$ per gram of fulvic acid [36].

Fulvic acid has been found to localize to the bone and cartilage, tissue with low selenium concentrations, especially in deficient state [37]. When extracts of soils of Kashin-Beck endemic areas are administered to selenium deficient mice in concentrations similar to those found in well water, the knees develop skeletal irregularities [35]. In addition, collagen maturation in these joints is inhibited due to over-hydroxylation of lysine residues, an abnormality that has been observed in Kashin-Beck patients [35].

**Grain contamination mycotoxins**
Less research has been devoted to mycotoxin involvement in the development of Kashin-Beck disease than to the other diet-related contributors. The stored grain of families with Kashin-Beck disease has been found to have significantly higher rates of fungal contamination than that of healthy families [28, 38]. Contamination of barley with any of four fungal taxa: *Alternaria*, *Cladosporium*, *Dreschlera*, and *Trichothecium roseum* is associated with elevated risk that increases with the presence of each additional variety [38]. Families with all four fungi present in barley stores are more than 1000 times as likely to have children with Kashin-Beck disease [38].

The corn contaminant *Fusarium oxysporum* and the wheat fungus *Alternaria alternata*, are found more frequently in endemic areas than in disease-free areas [28]. These two contaminants greatly increase the semiquinone radical concentration of contaminated grain [28]. Based on chemical and spectral similarities between the semiquinone species in contaminated grain and fulvic acid, it has been proposed that grain contaminants may play a role in the oxidative damage that contribute to the development of Kashin-Beck disease [28]. More work is needed to identify the effect of mycotoxins on the metabolic processes in cartilage.

Additional research would be needed before the process of Kashin-Beck osteoarthropathy could be fully understood. However, due to intervention programs and increased importation of food into endemic areas, incidence of this disease has dropped dramatically in recent decades, and it may become obsolete before it is fully comprehended [29 and references therein].

**Selenium deficiency as a consequence of disease**
In addition to the role of selenium deficiency in specific diseases associated with the endemic regions discussed above, selenium deficiency may also result from circumstances where inadequate selenium consumption is a consequence of altered nutritional status due to illness. This is particularly

*Selenium deficiency and human disease* 229

evident for patients receiving either enteral nutrition (EN, tube feeding) or total parenteral nutrition (TPN, intravenous feeding), many of which receive their nutrition by this route for extended periods of time. Due to either low levels of selenium in the feeding solutions or poor absorption, selenium deficiency in such individuals have been associated with muscle weakness and tenderness, nail bed changes, and cardiomyopathy with associated declines in cardiac function [39-44]. The relationship between these symptoms and selenium deficiency has been supported by data indicating that selenium supplementation alleviates many of these symptoms [39-42,44]. Selenium deficiency has also been noted in individuals required to reduce protein intake due to phenylketanuria [45-47], gastrointestinal disorders including Crohn's disease [41,48], and renal failure [49]. Similarly, selenium deficiency associated with chronic pancreatitis [50] and patients with sepsis may contribute to increased risk and morbidity [51,52].

**Other possible consequences of selenium deficiency**
Numerous other human health concerns have been suggested to result from selenium deficiency. These include intractable seizures, rheumatoid disease, arteriosclerosis, poor respiratory outcome in premature infants, miscarriage, neurological disorders and depression and cancer risk [53-55, and an excellent review in 56]. In addition, selenium deficiency may increase the risk of myocardial damage that occurs during ischemia and subsequent reperfusion [57]. In each of these examples, it remains unclear whether selenium deficiency is the primary risk factor or one of several parameters that increase susceptibility to disease initiated by other etiological agents.

For patients infected with human immunodeficiency virus, selenium deficiency has been shown to develop with AIDS progression and selenium status has been shown to be a strong indicator of disease outcome [58]. The role of selenium deficiency in AIDS progression will be discussed in detail in Chapter 20.

**Selenium deficiency and selenoprotein biosynthesis**
As the details of selenoprotein biosynthesis and the identity of resulting proteins are topics covered in other chapters in this book, this subject will receive only an abbreviated review here. All mammalian selenoproteins contain selenium in the form of the amino acid selenocysteine, which is encoded by the UGA triplet in selenoprotein mRNAs. As would be expected, limitations of available selenium for the synthesis of selenocysteine would be expected to result in the reduction of the levels of those proteins containing that amino acid. As described in detail in Chapter 3, there are two major forms of tRNA$^{[Ser]Sec}$, which are essential for the synthesis of all selenoproteins. This is because tRNA$^{[Ser]Sec}$ isoforms are both the site of selenocysteine synthesis and the adaptor molecules which recognizes

appropriate UGA codons in selenoprotein mRNAs. These tRNA species differ by methylation of the ribose portion of the wobble nucleotide of the anticodon. The amount of the unmethylated precursor tRNA relative to the methylated isoacceptor is typically reduced during selenium deficiency, and this phenomenon has been observed both in mammalian cells grown in culture and in most organs in the rat model where this was investigated (see Chapter 3). Since the methylation of tRNA[Ser]Sec has been argued to be relevant for the translational efficiency of glutathione peroxidase and perhaps other selenoproteins as well [59], this process represents one means by which selenoprotein levels can be reduced in response to selenium deficiency. Selenium deficiency can result also in a dramatic reduction in selenoprotein mRNA levels in animal models. For example, severe selenium deficiency in a rat model has been shown to almost entirely suppress the transcription of the cytosolic glutathione peroxidase in the liver and heart, while levels of the mRNA for thyronine deiodinase I decreased by 50% [60]. The reduction in selenoprotein mRNA levels under conditions of reduced selenium availability differs in various organs and for distinct selenoprotein genes [61]. Moriarty et al. have shown that the reduction in glutathione peroxidase mRNA levels during selenium deficiency is likely to be due to an enhancement of the decay of mRNAs containing nonsense codons [62]. The reduction in specific selenoproteins as a result of selenium deficiency may therefore contribute to the causes of symptoms associated with the diseases described above.

Approximately half of the characterized selenoproteins have been implicated to have anti-oxidant function [63, for a review], suggesting that increased risks of human diseases associated with selenium deficiency may be attributed to increased oxidative stress and alterations in redox signaling. In 1996, Kayanoki et al. demonstrated that selenium deficient bovine renal epithelial cells were more susceptible to hydrogen peroxide-induced apoptosis than selenium supplemented cells [64]. Since that observation, several investigators have reported that specific selenoproteins with anti-oxidant activities could suppress apoptosis induced by a variety of pro-apoptotic signals. Over-expression of the selenium-containing glutathione peroxidase from the poxvirus *Molluscum contagiosum* inhibited UV-induced cell death in human keratinocytes [65]. Glutathione peroxidase over-expression has also been shown to prevent apoptosis in T cells following interleukin withdrawal [66]. The ability of glutathione peroxidase over-expression to stimulate HIV replication during an acutely spreading infection has been attributed to the inhibition of a cellular apoptotic response that otherwise would limit viral spread [67]. Several signaling pathways have been shown to be influenced by reactive oxygen species [68]. Selenium has been reported to have direct effects on signaling molecules, such as NFκB and caspase 3, which are important to cellular processes including apoptosis, transformation, and immune function, [69-72]. Direct effects of selenium on signaling involving the mitogen-activated protein (MAP) kinase have also been reported [73-76]. These studies indicate that selenium supplementation and increased selenoprotein levels can influence apoptosis and alter redox-sensitive signaling. They further suggest that the reduction in selenoprotein activity as a consequence of selenium deficiency is a plausible mechanism by which too little selenium intake/utilization can contribute to a wide variety of human diseases.

**References**

1. DG Barceloux 1999 *Clin Toxicol* 37:145
2. JE Spallholz 1997 *Biomed Environ Sci* 10:260
3. ZJ Maksimovic, I Djujic 1997 *Biomed Environ Sci* 10:300
2. K Ge, G Yang 1993 *Am J Clin Nutr Supp* 57:259S
3. RD Utiger 1998 *N Eng J Med* 339:1156
4. JR Arthur, GJ Beckett 1999 *Brit Med Bull* 55:658
5. JB Vanderpas, B Contempre, NL Duale, W Goossens, N Bebe, R Thorpe, K Ntambue, Ch Thilly, AT Diplock 1990 *Am Journal Clin Nutr* 52:1087
6. JB Vanderpas, B Contempre, NL Duale, H Deckx, N Bebe, AO Longombe, CH Thilly, AT Diplock, JE Dumont 1993 *Am J Clin Nutr Supp*; 57:271S
7. JE Dumont, AM Ermans, PA Bastenie 1963 *J Clin Edocrinol Metab* 23:325
8. HJ Andersen 1961 *Acta Pediatrica* 50:Supp 125.
9. JE Dumont, B Corvilain, B Contempre 1994 *Mol Cell Endocrinol* 100:163
10. W Croteau, JC Davey, VA Galton, DL St Germain 1996 *Journal of Clin Invest* 98:405
11. D St. Germain, VA Galton 1997 *Thyroid* 7:655
12. JM Bates, VL Spate, JS Morris, DL St. Germain, VA Galton, 2000 *Endocrinology* 141:2490
13. TD Buckman, MS Sutphin, CD Eckhert 1993 *Biochem et Biophys Acta* 1163:176
14. B Contempre, JF Denef, JE Dumont, MC Many 1993 *Endocrinology* 132:1866
15. B Contempre, JE Dumont, JF Denef, MC Many 1995 *Eur J Endocrinol* 133:99
16. B Corvilain, B Contempré, AO Longombé, P Goyens, C Gervy-Decoster, F Lamy, JB Vanderpas, JE Dumont 1993 *Am J Clin Nutr Suppl* 57:244S
17. WH Yu 1982 *Jap Circ J* 46:1201
18. OA Levander, MA Beck 1997 *Biol Trace Elem Res* 56:5.
19. YY Cheng, PC Qian 1990 *Biomed Environ Sci* 3:422
20. VJ Ferrans 1989 *Am J Cardio* 64:9C
21. Y Li , T Peng, Y Yang, C Niu, LC Archard, H Zhang 2000 *Heart*: 83:696
22. A Aro, J Kumpulainen, G Alfthan, AV Voshchenko, VN Ivanov 1994 *Biol Trace Element Res* 40: 277
23. C Chai, J Tian, Q Qian, P Zhang, Q Xu, D Mao 1994 *Biol Trace Element Res*; 43-45:177
24. Keshan Disease Research Group of the Chinese Academy of Medical Sciences 1979 *Chinese Med J* 92:477
25. R Moreno-Reyes, C Suetens, F Mathieu, F Begaux, D Zhu, M Rivera, M Boelaert, J Nève, N Perlmutter, J Vanderpas 1998 *N Engl J Med* 339:1112
26. A Peng, C Yang, H Rui H Li 1992 *J Toxicol Environ Health* 35:79
27. E Allander 1994 *Scand J Rheum* 23 (Supp 99):1
28. R Tomlinson, 1999 *BMJ* 318:485
29. Y Wang, Zyang, L Gilula, C Zhu 1996 *Radiology* 201:265
30. J Vaughan 1981 *The Physiology of Bone, Third Edition* Clarendon Press, Oxford

31. C Yang, C Niu, M Bodo, E Gabriel, H Notbohm, E Wolf, P Müller 1993 *Biochem J* 289:829
32. FD Liu, ZL Wang, M Hinsenkamp 1998 *I Orthopaedics* 22:87
33. C Yang, E Wolf, K Röser, G Delling, PK Müller 1993 *Virchows Archiv A Pathol Anat Histopath* 423:483
34. R Hartenstein 1981 *Science* 212:743
35. A Peng, WH Wang, CX Wang, ZJ Wang, HF Rui, WZ Wang, ZW Yang 1999 *Envirn Health Perspect* 107:293
36. C Chasseur, C Suetens, N Norland, F Bagaux, E Haubruge 1997 *Lancet* 350:1074
39. RW Marcus 1993 *MD Med J* 42:669
40. JB Levy, HW Jones, AC Gordon 1994 *Prostgrad Med J* 70:235
41. CK Abrams, SM Siram, C Galsim, H Johnson-Hamilton, FL Munford, H Mezghebe 1992 *Nutr Clin Pract* 7:175
42. M Yagi, T Tani, T Hashimoto, K Shimuzu, T Hagakawa, K. Miwa, I Miyazaki 1996 *Nutrition* 12:40
43. AL Buchman, A. Moukarzal, ME Ament 1994 *JPEN J Parenter Enteral Nutr* 18:231
44. Y Saito, T Hashimoto, M Sasaki, S Hanaoka, K Sugai 1998 *Dev Med Child Neurol* 40:743
45. G Darling, P Mathias, M O'regan, E Naughten 1992 *J Inherit Metab Dis* 15:769
46. TH Best, DN Franz, DL Gilbert, DP Nelson, MR Epstein 2000 *Neurology* 54:2328
47. MM van Bakel, G Printzen, B Wermuth, UN Wiesmann 2000 *Am J Clin Nutr* 72:976
48. T Rannem, K Ladefoged, E Hylander, J Hegnoj, S. Jarnum 1992 *Am J Clin Nutr* 56:933
49. T Zima, VT Tesar, O Mestek, K Nemecek 1999 *Blood Purification* 17:187
50. DJ Bowrey, GJ Morris-Stiff, MC Puntis 1999 *HPB Surg* 11:207
51. X Forceville, D. Vitoux, R Gauzit, A Combes, P Lahilaire, P Chappuis 1998 *Crit Care Med* 26:1536
52. R Gartner, M Angstwurm 1999 *Med Klin* 94 Suppl 3:54
53. VT Ramaekers, M Calomme, D Vanden Berghe, W Makropoulos 1994 *Neuropediatrics* 25:217
54. BA Darlow, TE Inder, PJ Graham, KB Sluis, TJ Malpas, BJ Taylor, Winterbourn 1995 *Pediatrics* 96:314
55. JW Barrington, P Lindsay, D James, S Smith, A Roberts 1996 *Br J Obstet Gynaecol* 103:130
56. MP Rayman 2000 *Lancet* 356:233
57. M-C Toufektsian, F Boucher, S Pucheu, S Tanguy, C Ribout, D Sanou, Tresallet, J de Leiris 2000 *Toxicol* 148:125
58. MK Baum, G Shor-Posner, L Shenghan, Z Guoyan, L Hong, MA Fletcher, H Sauberlich, JB Page 1997 *J Acquir Immune Defic Syndrom Human Retrovirol* 15:370
59. ME Moustafa, MA El-Saadani, KM Kandeel, DB Mansur, BJ Lee, DL Hatfield, AM Diamond 1998 *RNA* 4:1
60. MS Saedi, CG Smith, J Frampton, I Chambers, PR Harrison, RA Sunde 1988 *Biochem Biophys Res Commun* 153:855
61. G Bermano, F Nicol, JA Dyer, RA Sunde, GJ Beckett, JR Arthur, JE Hesketh 1995 *Biochem J* 311 (pt 2):425
62. PM Moriarty, CC Reddy, LE Maquat 1998 *Mol Cell Biol* 18:2932
63. VN Gladyshev, DL Hatfield 1999 *J Biomed Sci* 6:151
64. Y Kayanoki, J Fuji, KN Islam, K Suzuki, S Kawata, Y Matsuzawa, N Taniguchi 1996 *J Biochem (Tokyo)* 119:817
65. JL Shisler, TG Senkevich, MJ Berry, B Moss 1998 *Science* 279:102
66. DM Hockenbery, ZN Oltavi, X-M Yin, CL Millman, SJ Korsmeyer 1993 *Cell* 75:241
67. PA Sandstrom, J Murray, TM Folks, AM Diamond 1998 *Free Rad Biol Med* 24:1485
68. V Adler, Z Yin, KD Tew, Z Ronai 1999 *Oncogene* 18:6104
69. K Nomura, H Imai, T Koumura, M Arai, Y Nakagawa 1999 *J Biol Chem* 274:29294

70. H-S Park, S-H Huh, Y Kim, J Shim, S-H Lee, I-S Park, Y-K Jung, IY Kim, E-J Choi 2000 *J Biol Chem* 275:8487
71. R Brigelius-Flohe, R Maurer, K Lotzer, G Bol, H Kallionpaa, P Lehtolainen, H Viita, S Yla-Herttuala 2000 *Atherosclerosis* 152:307
72. C Kretz-Remy, P Mehlen, ME Mirault, AP Arrigo 1996 *J Cell Biol* 133:1083
73. V Adler, MR Pincus, S Posner, P Upadhayaya, K El-Bayoumy, Z Ronai 1996 *Carcin* 17:1849
74. SR Stapelton, GL Garlock, L Foellmi-Adams, RF Kletzien 1997 *Biochem Biophys Acta* 1355:259
75. YJ Hei, S Farahbakhshian, X Chen, ML Battell, JH McNeil 1998 *Mol Cell Biochem* 178:367
76. SM Shieke, K Briviba, LO Klotz, H Sies 1999 *FEBS Lett* 448:301

# Chapter 19. Selenium as an antiviral agent

Melinda A. Beck

*Departments of Pediatrics and Nutrition, University of North Carolina at Chapel Hill, Chapel Hill, NC 27599, USA*

**Summary:** Recent work with selenium has demonstrated that a deficiency in this trace mineral will lead to increased viral pathogenesis. Selenium-deficient animals infected with a viral pathogen demonstrate immune dysfunction, including altered chemokine and cytokine expression patterns. A benign coxsackievirus infection of selenium-deficient mice leads to the development of myocarditis and further experiments demonstrated that the change in virulence was due to point mutations in the viral genome. Thus, replication in a selenium-deficient host led to a normally benign virus acquiring virulence due to viral mutations. A deficiency in selenium is also associated with disease progression in HIV-infected individuals and with hepatitis C virus-induced liver cancers. It appears that adequate levels of selenium help to protect the host against viral infection.

**Keshan disease**
In the early 1930's, a cardiomyopathy was first recognized in China that affected mainly children and women of child-bearing age. Called Keshan disease after the county where it was first described, epidemiological studies discovered that a deficiency in selenium was the main cause of the disease [1]. Specific geographic regions in China were found to have soils deficient in selenium and grains grown in this soil were selenium-deficient.

Subsequent supplementation of the selenium-deficient population with sodium selenite eradicated the disease, although a deficiency in selenium was not the entire explanation. Several aspects of the disease suggested that a co-factor(s) was required along with the deficiency in selenium. It was noted that Keshan disease occurred during specific times of the year, rather than occurring throughout the year. In addition, not everyone with a low selenium status developed the disease. These traits of the disease suggested that an infectious agent, along with a selenium deficiency, might be associated with Keshan disease development.

Several viruses are known to infect heart muscle. In particular, the entero-enteroviruses are most often associated with myocarditis, or inflammatory heart disease. Scientists in China were able to isolate enteroviruses from the blood and tissue of some Keshan disease victims [2]. Recently, using the powerful technique of the polymerase chain reaction (PCR), scientists have been able to identify enterovirus RNA fragments from archival tissue of Keshan disease victims [3]. Coxsackieviruses, particularly coxsackievirus B3, have been frequently identified in tissue from Keshan disease victims.

**Animal model for Keshan disease**
Mouse models for coxsackievirus-induced myocarditis have been studied for years, and indeed, the development of disease in the mouse closely mimics the human condition. Scientists in China used the mouse as a model for Keshan disease by feeding mice grains grown in Keshan endemic and non-endemic areas. Mice were then infected with a strain of coxsackievirus B4 that was originally isolated from the heart of a Keshan disease victim. Hearts from infected mice fed the grains from Keshan endemic areas developed severe myocarditis, whereas mice fed the grains from non-endemic areas (and hence contained selenium) did not develop severe disease [4]. These results suggested that infection with a coxsackievirus coupled with a lack of selenium led to the development of Keshan disease.

To further understand the role of selenium in the development of Keshan disease, we fed mice a diet deficient or adequate in selenium for four weeks prior to infection with coxsackievirus B3/20, a myocarditic strain of the virus. Histopathologic analysis of heart tissue from infected mice revealed that hearts from mice fed a diet deficient in selenium had much more damage than hearts from mice fed a diet adequate in selenium [5]. Thus, a deficiency in selenium led to increased pathogenicity of a myocarditic strain of coxsackievirus B3.

A second experiment was then designed in which a normally amyocarditic strain of coxsackievirus B3 (CVB3/0) was used to infect selenium-adequate and selenium-deficient mice. As expected, hearts from the selenium-adequate mice showed little or no damage following infection. However, hearts from selenium-deficient mice demonstrated a moderate degree of inflammation [7]. Thus, a host deficiency in selenium led to a normally benign coxsackievirus becoming virulent.

**Effect of selenium-deficiency on the immune response**
Because a host deficiency of selenium led to an increase in pathogenicity of a myocarditic viral strain and a normally avirulent strain became virulent, the immune system of the animals was analyzed. It may be that a deficiency in selenium leads to impaired immunity, thus explaining the enhanced pathogenicity of the virus.

Neutralizing antibodies are an important defense against reinfection with coxsackievirus, as well as providing a means to deal with replicating virus in a primary infection. However, a deficiency in selenium was not found to affect the ability of the mouse to secret neutralizing antibodies. In contrast, spleen cells isolated from selenium-deficient animals were greatly impaired in their ability to proliferate in response to either a mitogen or specific viral antigen. Natural killer cell activity, also important in clearing virus from target tissue, was not affected by a deficiency in selenium [6].

Myocarditis induced by infection with coxsackievirus is characterized by an influx of inflammatory cells to the heart [7]. In the mouse, the development of inflammation occurs within six days of infection. When the virus replicates in the heart tissue following infection, a brisk immune response is induced, characterized by the appearance of macrophages and NK cells early post infection, then later replaced with a T cell infiltrate. Early experiments demonstrated the importance of the immune infiltrate to the development of pathology and that T cells were the primary cell phenotype involved in the development of CVB3-induced myocarditis [8]. Nude mice are resistant to development of myocarditis when infected with coxsackievirus B3 [9]. These results highlight the idea that the inflammatory response to CVB3 induces the myocarditis, rather than virus replication in and of itself [10,11]. Indeed, the amyocarditic strain of CVB3, CVB3/0, does not induce an inflammatory response, even though the virus replicates in and can be recovered from the heart tissue of infected animals.

There are many signals which direct the trafficking of immune cells to the site of infection. For example, chemoattractant molecules secreted by immune cells as well as other types of cells [12] provide a chemical gradient by which immune cells migrate to the site of infection. The absolute requirement for the involvement of chemokines in the inflammatory response to infection with coxsackievirus was demonstrated with macrophage inflammatory protein-1$\alpha$ (MIP-1$\alpha$) knockout mice. MIP-1$\alpha$ is a chemokine that is chemotactic for activated CD8+ T cells and NK cells. The knockout mice infected with a myocarditic strain of coxsackievirus did not develop any myocarditis, whereas the wildtype mice developed myocarditis [13]. Thus, an absence of MIP-1$\alpha$ completely protected the mice from coxsackievirus-induced cardiac inflammation. Although virus was recoverable from the hearts of MIP-1$\alpha$ knockout mice, it failed to induce an immune response. These results demonstrate that the production of MIP-1$\alpha$ is an essential component of the immune response to infection with CVB3.

What is the role for chemokines in selenium-deficient and selenium-adequate mice infected with the amyocarditic strain of CVB3, CVB3/0? To answer this question, mice at weaning were fed selenium-adequate or selenium-deficient diets for four weeks prior to infection with CVB3/0. At various times post infection, mice were killed and the hearts removed for study. We measured cardiac mRNA levels for various chemokines using an RNAse protection assay.

We found no differences in the expression of RANTES, eotaxin, MIP-1α and MIP-β between mice fed selenium-deficient or selenium-adequate diets. The expression of mRNA for RANTES, MIP-1β and eotaxin peaked between days seven and ten post infection, whereas expression of mRNA for MIP-1α was highest on day ten post infection, the last day of measurement [14].

Significant differences between selenium-adequate and selenium-deficient mice were found with respect to mRNA expression of MCP-1. As shown in Figure 1, mRNA levels for MCP-1 peaked at day seven for the selenium-adequate mice, then began to decline by day ten post infection. In contrast, the MCP-1 mRNA level for selenium-deficient mice was low at day seven, but showed a dramatic increase at day ten post infection.

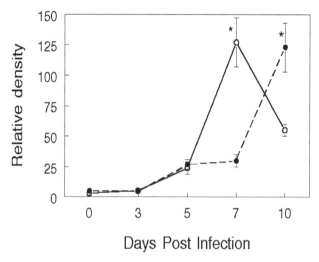

**Figure 1.** Heart mRNA levels for MCP-1 from selenium-adequate (open circles) and selenium-deficient (closed circles). Each point represents the mean ± S.D. of five mice. The * indicates statistical difference between selenium-adequate and selenium-deficient group (P< 0.05).

These results suggest that MCP-1 may play a role in the development of myocarditis in selenium-deficient mice infected with CVB3/0. Recall that the selenium-adequate mice do not develop myocarditis when infected with this benign strain of virus. Therefore, altered expression of the chemokine MCP-1 may play a role in the susceptibility of the selenium-deficient mice to developing myocarditis.

In addition to chemokines, cytokines also play a role in the development of inflammation. We found that cardiac mRNA expression for a number of pro-inflammatory cytokines including IL (interleukin)-1, IL-2, IL-4, IL-5, IL-6 and IL-15 were similar between selenium-adequate and selenium-deficient mice. However, there were significant differences in the expression of γ-IFN. The mRNA levels for γ-IFN, a potent anti-viral and macrophage activator, were greatly reduced when compared with selenium-adequate mice [14]. Peak expression occurred on day seven post infection for the selenium-adequate mice, and reduced expression on day ten. However, the selenium-deficient mice expressed very little mRNA for γ-IFN at day seven, although some expression occurred on day ten post infection.

## Changes in the CVB3 genome

The increase in virulence of the amyocarditic virus in the selenium-deficient animals may have been due to alterations in the host immune response, as described in the previous section. However, an alternative explanation is that the virus itself was influenced by the host deficiency in selenium. In order to address this possibility, virus obtained from selenium-deficient mice were passed back into selenium-adequate mice. If the change in virulence was due to immune dysfunction in the selenium-deficient mice, then the selenium-adequate mice infected with virus from the selenium-deficient mice should not develop myocarditis. However, the infected selenium-adequate mice did develop myocarditis. Passing the virus from selenium-adequate to selenium-adequate mice did not induce a change in the viral phenotype, demonstrating that passage alone could not cause a change in viral virulence.

This result clearly demonstrated that changes in the viral genome had occurred as a consequence of the virus replicating in a selenium-deficient host. In order to confirm that the phenotype change was due to a change in the viral genome, virus isolated from both selenium-deficient and selenium-adequate mice were sequenced. No changes were found in the virus isolated from selenium-adequate mice when compared with the sequence of the virus used to inoculate the mice. However, six nucleotide changes, all point mutations, were found in the viruses isolated from selenium-deficient mice [15]. All of these mutations were also common to CVB3 strains which are virulent in mice. Thus, a deficiency in selenium allowed a normally benign virus to become virulent by alterations in the viral genome.

## Glutathione peroxidase and viral mutations

How did a host deficiency in selenium lead to viral mutations? One possibility is the role of selenium in the function of selenium-dependent glutathione peroxidase (GPx). There are four isozymes of GPx, and GPx-1, or classical GPx, is found in almost all cells of the body. GPx-1 functions as an antioxidant by catalyzing the reduction of peroxides. A deficiency of selenium in the diet will lead to a decrease in the activity of glutathione peroxidase. Indeed, GPx levels were markedly decreased in mice fed the selenium-deficient diets for four weeks. To determine if the change in viral virulence could be associated with a decrease in GPx activity, GPx-1 knockout mice were utilized.

GPx-1 knockout and wildtype control mice were infected with the amyocarditic strain of CVB3, CVB3/0. At various times post infection, mice were killed and their tissues removed for study [16]. A little over 50% of the GPx-1 knockout mice infected with CVB3/0 developed myocarditis, whereas none of the wildtype mice developed any cardiac inflammation. Cardiac viral titers were similar between the groups. This experiment suggested that the increase in virulence of the normally amyocarditic virus in the selenium-deficient mice was likely due to a deficiency in GPx-1 activity.

As shown in Table 1, sequencing of isolates obtained from GPx-1 knockout and wildtype mice revealed seven point mutations in the virus isolated from GPx-1 knockout mice when compared with the stock virus. No changes were found in virus isolated from the wildtype mice. Six of the seven changes were found to be identical to the changes found in the selenium-deficient animals. Interestingly, viral mutations were found only in viral genomes recovered from GPx-1 knockout mice which demonstrated myocarditis. This result demonstrated that a lack of GPx-1 activity may be the mechanism by which a deficiency in selenium alters the viral genome.

**Influenza virus and selenium-deficiency**
Each year, worldwide infection with influenza virus results in a great deal of morbidity and mortality. In the United States alone, over 20,000 deaths occur each year as a consequence of infection with influenza virus and its complications [17]. Influenza virus is able to cause disease each year due to its ability to change its surface antigens recognized by the immune system. Both small changes (antigenic drift) and large changes (antigenic shift) in the viral proteins allow the virus to avoid neutralization by the host's antibody response which may have been primed to the virus before the viral changes occurred [18]. Each year, scientists make an estimate of the surface proteins that may be circulating during the fall-winter influenza season. These estimates must be made 6-8 months before influenza epidemics begin in order to provide enough time for vaccine development. Thus, depending on how well scientists were able to predict the influenza strain that will circulate in any given year will influence the efficacy of the vaccine.

The precise mechanism(s) by which influenza virus mutates each year is not well understood. Changes in two influenza virus surface proteins, the hemagglutinin and the neuraminidase, which are involved in the ability of the virus to bind to and be released from host cells, are primarily responsible for the yearly epidemics. These two proteins are also involved in virulence of the virus. Because the HA and the NA are external proteins, their exposure to the immune system is thought to be a driving force for inducing viral mutations.

The influenza virus matrix protein is also involved in virulence. However, because this protein is internal and therefore not exposed to the host antibody response, this protein is more stable and exhibits few nucleotide changes over time [19].

Table 1. Point mutations in the CVB3/0 genome as a consequence of replicating in selenium-deficient or GPx-1 knockout mice.

| Nt Number[b] | Stock[c] | Host Condition[a] | | | | |
|---|---|---|---|---|---|---|
| | | Se+[d] | Se-[e] | KO-P[f] | KO-NP[g] | WT[h] |
| 234 | C | C | T | T | C | C |
| 788 | G | G | A | A | G | G |
| 2271 | A | A | T | T | A | A |
| 2438 | G | G | C | C | G | G |
| 2690 | G | G | G | A | G | G |
| 3324 | C | C | T | T | C | C |
| 7334 | C | C | T | T | C | C |

[a]Virus was isolated and sequenced from a minimum of five mice.
[b]Nucleotide number read in a 5' to 3' direction.
[c]Stock virus used to inoculate the animals.
[d]Three-week old mice were fed a selenium-adequate diet for four weeks prior to inoculation with stock virus. These mice did not develop myocarditis.
[e]Three-week old mice were fed a selenium-deficient diet for four weeks prior to inoculation with stock virus. These mice developed myocarditis.
[f]GPx-1 knockout mice which developed myocarditis post inoculation with stock virus.
[g]GPx-1 knockout mice which did not develop myocarditis post inoculation with stock virus
[h]Wildtype control mice for the GPx-1 knockout mice. These mice did not develop myocarditis.

Because of the earlier work with coxsackievirus, it was reasoned that a deficiency in selenium may also have an effect on the influenza virus genome. To test this hypothesis, mice were fed diets either adequate or deficient in selenium for four weeks prior to infection with a mild strain of influenza, influenza A/Bangkok/1/79. Mice were killed at various times post infection.

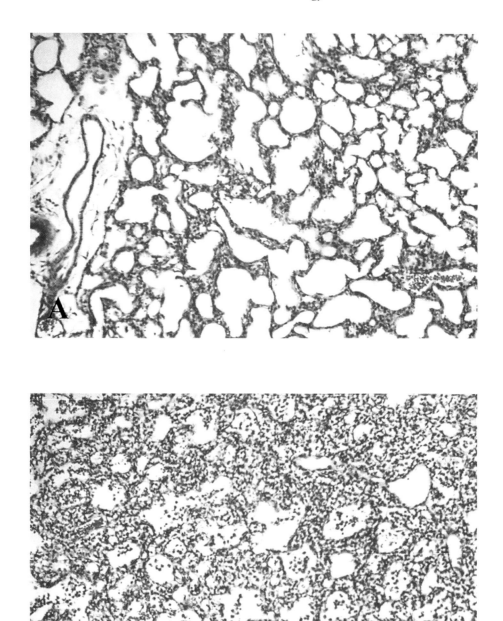

**Figure 2.** Histopathology of lungs from five days post influenza A/Bankgok/1/79 infection of a selenium-adequate (A) or selenium-deficient (B) mouse.

*Selenium as an antiviral agent* 243

Mice which were fed the selenium-deficient diet developed much more severe lung inflammation post influenza infection when compared with infected selenium-adequate mice [20]. As shown in Figure 2, mice develop interstitial pneumonitis when infected with influenza, which was exacerbated in the selenium-deficient mice. Bronchoalveolar lavage of lungs from infected selenium-deficient and selenium-adequate mice confirmed the increased pathology seen in the deficient mice. selenium-deficient mice had $8 \times 10^6$ inflammatory cells infiltrating the lung compared with $5 \times 10^6$ inflammatory cells in the lungs from selenium-adequate mice.

## Immune response of influenza-infected selenium-deficient mice

In order for the lung inflammation to occur in the infected influenza mice, a co-ordinate production of chemoattractant cytokines known as chemokines must occur. This process was altered in the selenium-deficient mice. As shown in Figure 3, mRNA levels for the chemokines RANTES, MIP-1α, MIP-β and MCP-1 were highest on days four and five post infection for the selenium-adequate mice, and then began to decrease, whereas these chemokines were highest at later time points for the selenium-deficient mice.

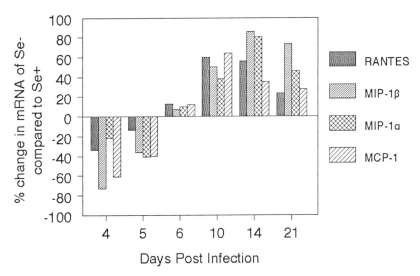

**Figure 3.** Time course of chemokine mRNA expression in influenza infected selenium-adequate and selenium-deficient mice. Data are expressed as percent change comparing selenium-deficient mice with selenium-adequate mice (selenium-deficient minus selenium-adequate divided by selenium-adequate X 100). Values above 0 indicate the percent increase of mRNA levels of selenium-deficient mice compared with selenium-adequate mice and values below 0 indicate the percent decrease of mRNA levels from selenium-deficient mice compared with selenium-adequate mice.

Clearly, selenium-deficiency leads to an increase in influenza-induced histopathology which is associated in part with altered chemokine expression. Further work is required in order to determine if the influenza virus underwent a genome change similar to what was found with the coxsackievirus infected mice.

**Selenium and other viruses**
Infection with HIV (human immunodeficiency virus) results in a loss of CD4+ helper T cells and subsequent immune dysfunction leading to increased opportunistic infections. In addition, oxidative stress increases during HIV infection. A number of studies have examined the relationship between specific nutritional factors and disease progression and survival of HIV infected individuals. Selenium status of HIV infected individuals has also been studied.

A reduction in selenium levels has been observed in symptomatic HIV patients (21-25). Low plasma and red cell levels of selenium are associated with increased disease progression in HIV-positive individuals. In a study of HIV-infected children [26], low plasma selenium levels were found to be an independent predictor of mortality and were associated with faster disease progression. Clinical studies of selenium supplementation are required in order to evaluate the effectiveness of selenium in slowing HIV disease progression.

In addition to CVB3 and HIV, selenium levels have also been associated with hepatitis B virus. Infection with hepatitis B virus is a major health problem throughout the world. In addition, chronic hepatitis B infection is thought to be a major cause of the majority of hepatocellular carcinomas, a highly malignant neoplasm with a high mortality rate. A study from Taiwan [27] demonstrated that mean selenium plasma levels were significantly lower in hepatocellular carcinoma patients, as compared with individuals testing positive for hepatitis B virus. A further study from Qidong county in China [28] demonstrated a protective effect of selenium supplementation in a population at high risk of developing primary liver cancer due to a high prevelance of hepatitis positive individuals.

**Summary**
It is clear that host selenium status can have a profound effect on a viral pathogen. A lack of selenium in the diet leads to a decrease in cellular antioxidant mechanisms, such as glutathione peroxidase. This decrease in antioxidant protection will lead to an increase in oxidative stress in the host. The oxidative stress status of the host can then lead to immune dysfunction and increased pathology post viral infection. In addition, mutations can occur in the viral genome itself as a result of the increased oxidative stress in the host, leading to a change in viral phenotype. The effects of selenium on both

the host as well as the viral pathogen are just beginning to be elucidated. The precise mechanism by which a deficiency in selenium leads to mutations in a viral genome remains to be determined.

**References**

1. BQ Gu 1983 *Chin Med J* 96:251
2. C Su, C Gong, J Li, L Chen, D Zhou, Q Jin 1979 *Chin Med J* 59:466
3. Y Li, Y Yang, H Chen 1995 *Chung Hua I Hsueh Tsa Chih* 75:344
4. J Bai, S Wu, K Ge, X Deng, C Su 1980 *Acta Acad Med Sin* 2:29
5. MA Beck, PC Kolbeck, LH Rohr, Q Shi, VC Morris, OA Levander 1994 *J Infect Dis* 170:351
6. MA Beck, PC Kolbeck, LH Rohr, Q Shi, VC Morris, OA Levander 1994 *J Med Virol* 43:166
7. JF Woodruff 1980 *Am J Pathol* 101:427
8. SA Huber, N Heintz, R Tracy 1988 *J Immunol* 141:3214
9. JF Woodruff, JJ Woodruff 1974 *J Immunol* 113:17
10. K Leslie 1989 *Clin Microbiol Rev* 2:191
11. J O'Connell, J Robinson 1985 *Postgrad Med J* 61:1127
12. A Zlotnik, O Yoshie 2000 *Immun* 12:121
13. DN Cook, MA Beck, T Coffman, SL Kirby, JF Sheridan, IB Pragnell, O Smithies 1995 *Science* 269:1583
14. MA Beck, CC Matthews 2000 *Proc Nutr Soc* 59:1
15. MA Beck, Q Shi, VC Morris, OA Levander 1995 *Nat Med* 1:433
16. MA Beck, RS Esworthy, Y-S Ho, F-F Chu 1998 *FASEB J* 12:1143
17. Centers for Disease Control and Prevention 2000 *Morb Mortal Wkly Rep* 49:13
18. BR Murphy, RG Webster 1996 Fields Virology BN Fields (Ed) Lippincott-Raven Philadelphia PA p1397
19. AC Ward 1997 *Virus Genes* 14:187
20. MA Beck HK Nelson, Q Shi, P Van Dael, EJ Schiffrin, S Blum, D Barclay. OA Levander 2000 *FASEB J* submitted
21. MK Baum, G Shor-Posner, S Lai, G Zhang, H Lai, MA Fletcher, H Sauberlich, JB Page 1997 *JAIDS* 15:370
22. A Cirelli, M Ciardi, C de Simone, F Sorice, R Giordano, L Ciaralli, S Costantini 1991 *Clin Biochem* 24:211
23. BM Dworkin, WS Rosenthal, GP Wormser, L Weiss, M Nunez, C Joline, A Herp 1988 *Biol Trace Elem Res* 15:167
24. MP Look, JK Rockstroh, GS Rao, KA Kreuzer, S Barton, H Lemoch, T Sudhop, J Hoch, K Stockinger, U Spengler, T Sauerbruch 1997 *Europ J Clin Nutr* 51:266
25. MP Look, JK Rockstroh, GS Rao, KA Kreuzer, U Spengler, T Sauerbruch 1997 *Biol Trace Elem Res* 56:31
26. A Campa, G Shor-Posner, F Indacochea, G Zhang, H Lai, D Asthana, GB Scott, MK Baum 1999 *JAIDS* 20:508
27. M-W Yu, I-S Horng, K-H Hsu, Y-C Chiang, Y-F Liaw, C-J Chen 1999 *Am J Epidemiol* 150:367
28. SY Yu, YJ Zhu, WG Li 1997 *Biol Trace Elem Res* 56:117

# Chapter 20. Role of selenium in HIV/AIDS

Marianna K. Baum, Adriana Campa, Maria José Miguez-Burbano, Ximena Burbano, Gail Shor-Posner

*Division of Metabolism and Disease Prevention, Department of Psychiatry and Behavioral Sciences, University of Miami School of Medicine, Miami, FL 33136, USA*

**Summary:** Selenium appears to have a multifactorial role in HIV-1 infection. Selenium status affects HIV disease progression and mortality [1] through various potential mechanisms. Selenium protects against oxidative stress through its function as a biological antioxidant required for the activity of gluthatione peroxidase. Adequate selenium status may also be essential in controlling viral emergence and evolution [2]. In addition, selenium may enhance resistance to infection through modulation of both cellular and humoral immunity. Plasma selenium levels affect interleukin production and subsequent changes in Th1/Th2 cytokine responses [3]. Interacting with selenium status, other nutritional factors are important in HIV-1 progression and mortality. The causative factors, including disease stage, nutritional status at the onset of the disease, types of treatment and compliance, and secondary infections may be acting independently or in combination. Treatment of malnutrition, and the accompanying micronutrient deficiencies, thus, requires a carefully individualized approach. This chapter will review the current research on the relationship of nutritional status to HIV-1 disease progression and mortality, as well as the factors that may affect this relationship, with emphasis on the role of selenium.

## Background

The total number of persons living with Human Immuno-Deficiency Virus (HIV) infection and Acquired Immuno-deficiency Syndrome (AIDS) worldwide has been estimated to be nearly 36 million [4]. Approximately 22 million adults and children have already died from AIDS. The developing world, where protein-energy malnutrition (PEM) is already an overwhelming health problem and the main cause for immune disturbances, accounts for 94% of the global HIV-1 infections. The most alarming statistic, however, is that, despite changes in incidence worldwide, 16,000 new individuals are still infected daily, a figure that lessens hopes for a rapid solution to this pandemic.

**HIV-1 infection and wasting**
Although selenium deficiency has been strongly and independently associated with mortality in HIV/AIDS [1], there are other nutritional abnormalities that may contribute to HIV progression and mortality. As the HIV disease advances to AIDS, the prevalence of selenium deficiency increases, from 2% to 4% in asymptomatic individuals, to 75% in Stage IV AIDS, a finding that suggests some degree of interaction between the characteristic AIDS wasting and selenium deficiency [1,5,6]. In the early years of the HIV-1 epidemic, the infection was known as "slim disease" [7] because marked malnutrition and wasting were the characteristic symptoms in the advanced stages of the disease. Malnutrition is the most striking symptom in individuals with AIDS. Weight loss of more than 10% has been found to be an independent risk factor for survival in HIV-1 infected individuals [8,9]. The immunological consequences and clinical manifestations of AIDS are highly reminiscent of those associated with protein-energy malnutrition (PEM) [10-12].

The immune dysregulations associated with HIV infection are extremely complex phenomena, that despite their superficial similarities involve many different aspects of host defenses and immunological functions [13]. The wasting syndrome in AIDS is accompanied by loss of a large proportion of fat free mass, a process that is more closely associated with starvation [14] than with the cachectic/catabolic pathways demonstrated in other conditions. The similarities in immunological abnormalities observed in protein-energy malnutrition and HIV-1 infection include decreased T lymphocyte and CD4 cells, reduced secretory IgA, and impaired primary and secondary delayed cutaneous hypersensitivity responses [15-17].

In HIV-1 infection, altered levels of lipid and lipoproteins have been also documented in asymptomatic and symptomatic individuals. Hypocholesterolemia and hypertriglyceridemia are the most frequently reported lipid disturbances [18-20]. These lipid levels are associated with viral activation and the accompanying immune response, especially cytokine production. The pro-inflammatory cytokines associated with HIV-1 infection include interleukin-1, tumor necrosis factor alpha, interleukin-6 and interferon [21]. Dietary supplementation with omega-3 fatty acid has been reported to depress production of tumor necrosis factor and interleukin-6 in cancer patients, a treatment that shows potential for preventing or ameliorating cancer cachexia [22]. Similar studies in AIDS patients, however, have not reached statistical significance [23]. Increased production of cytokines in AIDS interferes with metabolic processes in the liver, leading to increased gluconeogenesis, proteolysis, lipolysis, and consequently, loss of body fat stores. Lipid status in HIV-1 infection may also be influenced by opportunistic diseases and anorexia, and, in turn, may contribute to disease progression, wasting and increased mortality [24].

Nausea and anorexia are frequent in HIV-1 infected patients. A recent report suggests that imbalances between cholecystokinin and beta-endorphins, both modulators of patterns of consumption, may play a role in reduced intake and wasting in HIV-1 infected subjects [25]. Type of therapy, alteration in hormonal regulation of appetite caused by the HIV-1 virus, humoral host response, and psychological changes produced by socioeconomic factors and the disease process can significantly contribute to decrease in caloric consumption, and therefore, to the onset of HIV-1 wasting. In HIV-1 infection, though, the main cause of decreased food consumption is painful lesions in the gastrointestinal tract caused by the virus itself or secondary infections [26,27]. The contribution of reported increases in energy expenditure to wasting in HIV infection is still unclear. Measures of body composition and resting energy expenditure (REE) showed a modest but significant increase in REE in asymptomatic HIV-seropositive men, which was largely balanced by increases in caloric intake [28,29]. Other studies, however, showed that substantial hypermetabolism is a significant factor in weight loss during secondary infections [30]. Malabsorption also plays an important role in wasting and micronutrient deficiency. Diarrhea and destruction of the gastrointestinal lining by secondary infections or the virus itself are major contributors to malabsorption in HIV-1 infected patients.

**HIV-1 and nutrient deficiencies**
Although the effect of the HIV-1 virus on the immune system is the primary cause of the immunodeficiency associated with this disease, malnutrition further contributes in a significant and reversible manner to the observed immune dysregulation [31]. Moreover, independent of immune function, nutrient deficiencies may accelerate disease progression and mortality by influencing the degree of oxidative stress and viral expression, contributing to nutrition-related morbidity.

Chronic oxidative stress has been reported during the early and advanced stages of HIV-1 infection [32]. The accumulation of Reactive Oxygen Species (ROS) may contribute in several ways to disease progression. Oxidative stress has been linked to pathogenesis of HIV programmed cell death (apoptosis) of T-lymphocytes [33], to alterations in the HIV-1 promoter that may produce progression to AIDS in patients with latent HIV [34], and to the development of AIDS Kaposi sarcoma [35]. Oxidative stress has been also identified as one of the factors that may cause neural damage [36]. In addition, oxidative stress may induce alterations in the interleukin profile, contributing to the immune dysregulation and increased viral replication observed during the progression of HIV-1 infection to AIDS [37,38]. Coupled with the increased oxidative damage, and potentially aggravating it, perturbations in the antioxidant defense system, including changes in levels of ascorbic acid, tocopherols, carotenoids, selenium and glutathione, have been observed in plasma as well as in various tissues [31, 32, 34].

The interaction between nutrition and HIV-1 infection is reciprocally disturbing, creating a fatal vicious cycle. Alterations in nutritional status are widespread among various HIV-1 infected cohorts including asymptomatic and symptomatic homosexual men, drug users, and children [5,39,40]. Beyond the devastating effects of the disease on nutritional status, each population has characteristic socioeconomic, psychological and physiological factors that modulate and aggravate these effects. In turn, nutritional status has the potential for influencing the rate of disease progression, and mortality in ways that are closely related to the specific dietary practices and nutrition-related risk factors of each population. Examples of some of those population-specific nutritional risk factors are gender, type of drug of abuse, competing nutritional demands placed by growth, pregnancy, breastfeeding or co-morbidity, and life style [41-43].

Many specific nutrients have been documented to influence the immunological responses of the host. In HIV-infected individuals, who suffer from an underlying immunodeficiency disorder, and may have multiple nutritional deficiencies, cause and effect, are difficult to separate. By following relatively large numbers of HIV-1 infected subjects over time, and carefully documenting nutritional, immunological, and health status, our studies have evaluated the influence of specific nutrient deficiencies upon immune function, disease progression and mortality. Interactions between immune function and specific nutrient deficiencies in HIV-1 infection have been documented to occur with parameters of protein (albumin and prealbumin) [44] and lipid status [45] trace elements (selenium and zinc) [39,46,47], and vitamins A, E, $B_6$, and $B_{12}$ [46- 49].

In addition to the well recognized relationships between wasting and disease progression and mortality in HIV-1 infection, several studies have explored the relation of specific nutrient deficiencies to HIV-1 related morbidity and mortality [1,50]. Whereas development of biochemical deficiency of vitamins A, $B_6$ and $B_{12}$ has been associated with faster disease progression, normalization of vitamin A, vitamin $B_{12}$, and zinc levels, has been linked to slower disease progression [47]. Moreover, Tang and colleagues [48] have reported a nearly twofold increase in risk of progression to AIDS in HIV-1 infected subjects with low serum vitamin $B_{12}$

concentrations, and both vitamin A deficiency and wasting have been associated with increased mortality in HIV-1 seropositive drug users [42].

Because multiple nutrient deficiencies tend to occur simultaneously and are highly correlated, all factors that could affect immune function and survival need to be evaluated, including CD4 cell count at baseline, and over time, as well as nutrients that have been independenly shown to significantly affect HIV-1 related mortality (vitamins A, $B_{12}$, selenium and zinc) [42,51]. When these factors were considered together in a multivariate model, only a deficiency of selenium, an essential trace element that is part of the antioxidant defense system, was profoundly associated with decreased survival in HIV-1 disease [1].

A significant effect for selenium deficiency was shown, even when controlling for deterioration of overall nutritional status utilizing plasma levels of albumin and specific nutrients, and disease status with CD4 cell count. Despite the importance of other antioxidants in immune function as protection against oxidative stress, only selenium deficiency was an independent predictor of survival.

The dramatic, increased risk of mortality (10.8, p=0.002) with selenium deficiency may be related not only to its role in maintaining immunocompetency, but also in modulating viral expression and protecting against the oxidative damage caused by the HIV infection [52-56].

**Selenium and the immunity in HIV**
An important interaction between nutrition and immune function has long been recognized [57]. Selenium, among other micronutrients, has been demonstrated to affect the immune process. *In vivo* and *in vitro* studies suggest that selenium may act at different levels of immune functioning. In animal models, selenium deficiency impairs the ability of phagocytic neutrophils and macrophages to destroy antigens. In HIV infected patients, selenium deficiency has been significantly correlated with total lymphocyte counts. Plasma selenium levels have been positively correlated with CD4 cell counts and CD4/CD8 ratio, and inversely correlated with $B_2$-microglobulin and thymidine-kinase activity [58].

Although the effect of adequate selenium status on humoral immune response has not been well explored in HIV-infection, Spallholz and collaborators [59] demonstrated a close relationship between selenium status and humoral immune response. Our preliminary studies in a cohort of drug users suggest that selenium status is significantly correlated with manifestations of herpes and candida infections. This is of particular significance since both diseases have been associated with HIV-1 disease progression [60].

Immunity to infectious agents depends on the activation of T helper (Th) cells secreting the appropriate pattern of cytokines [61]. Th1 cells are especially effective when cellular response to antigens, such as viruses, is needed. Th2 cells are helpers for B cells, and appear to be adapted to support antibody response and defense against parasites. In HIV-1 infection, when a strong cellular response is essential, a change to a Th2 pattern of cytokine production may be deleterious.

In in vitro models, selenium regulates levels of interleukin-2, the cytokine responsible for the earliest and most rapid expansion of T lymphocytes. Selenium enhances interleukin-2 production in a dose-dependent manner. The mechanism of selenium action appears to occur through the increased expansion of high-affinity receptors [62].

In addition, selenium seems to affect the production of another cytokine, Tumor Necrosis Factor-$\alpha$ (TNF-$\alpha$), which is prominent in the pathogenesis of anorexia and cachexia in chronic diseases [63]. Selenium supplementation has been shown to suppress TNF-induced HIV replication. The mechanism of action of selenium on the virus-TNF interaction is not fully understood, but seems related to selenoprotein synthesis, especially in the glutathione and thioredoxin systems [64]. Look et al. [58] demonstrated in HIV-infected patients that plasma selenium levels are inversely correlated with levels of soluble TNF type II receptors. Selenium supplementation, therefore, may have the potential to reduce TNF receptors and prevent some of the adverse effects of high TNF circulating levels, such as wasting and Kaposi's sarcoma. Evidence for a selenium-cytokine mechanism of action has also been described in studies indicating the potential of selenium to decrease neuropathogenesis through suppression of interleukin-induced HIV-1 replication, neuronal apoptosis, and blood brain barrier damage [58,64-66].

Selenium levels have been inversely correlated to yet another cytokine, Interleukin-8 [58]. Two possible mechanisms have been advanced. The first proposes that the increased oxidative stress in HIV infection is caused by elevated IL-8 levels, which exhausts the available selenium to protect cells against the inflammatory response. The second, supported by in vitro studies, proposes that selenium, in the glutathione peroxidase system, can inhibit IL-8 release by endothelial cells [65].

**Selenium and chemoprevention**
The high risk of HIV-related mortality associated with selenium deficiency underscores the importance of maintaining optimal selenium status in HIV-1 infected men and women. Chemopreventive research in selenium and nutrition has resulted in striking findings [67-69]. Supplementation with selenium may help to increase the enzymatic defense systems in HIV-1

infected patients [70], and has been linked to increased blood selenium levels and an improvement in general health in HIV/AIDS patients [71-73].

Cardiac complications, similar to some of the features of Keshan Disease in China, [74] a disease associate with selenium deficiency, have been described in a child with HIV/AIDS, and, upon supplementation with selenium (4 µg/kg), the symptoms improved [75].

Rather than decreasing the importance of antioxidant chemoprevention, chronic highly active antiretroviral therapy (HAART) is creating new research challenges for the role of antioxidants in HIV-1 disease. In our preliminary research in HIV-1 infected chronic drug users receiving HAART, low selenium levels have been significantly associated with hyperglycemia, whereas thrompocytopenia has been significantly related to selenium deficiency, but not to HAART [76]. Lipodystrophy, hyperlipidemias and insulin resistance in patients receiving HIV protease inhibitors [77] may increase the long-term risk of oxidative damage associated with development of atherosclerosis and coronary hearth disease [78], turning antioxidant research in HIV/AIDS into an even more pressing matter.

In cancer research, long term selenium supplementation has produced profound affects on the incidence of, and mortality from, carcinomas of several sites [79]. In a double-blind, placebo controlled study, a nutritional dose (200 µg/daily) of selenium was used as a chemopreventive agent in healthy volunteers, regardless of their plasma selenium status. The blinded phase of the trial was stopped early, as the selenium-treatment group exhibited a 51% reduction in total cancer mortality and 41% reduction in total cancer incidence as compared to the placebo group. In other studies, nutritional supplementation of selenium has been shown to significantly reduce the incidence of primary liver cancer in China [80], and provide significantly greater resistance to Aflatoxin B1-induced carcinogenic damage in lymphocytes isolated from healthy human subjects administered daily selenium [81]. The chemoprevention trials in the United States and China have demonstrated that prolong supplementation with nutritional doses of selenium (50 to 200 µg/daily) is safe, with a low risk of toxicity.

These findings suggest that administration of selenium, as a chemopreventive agent in HIV-1 seropositive men and women, may be an effective method to slow disease progression and enhance survival. Selenium supplementation trials in HIV-1 infected men and women are now underway.

**References**

1. MK Baum, G Shor-Posner, S Lai et al 1997 *J Acquir Immune Defic Syndr Hum Retrovirol* 15:370
2. MA Beck, A Lavander 1998 *Annu Rev Nutr* 18:93
3. MK Baum, MJ Miguez-Burbano, A Campa et al 2000 *Journal of Infectious Diseases* 182(Suppl 1):S69
4. AIDS Epidemic Update: December 2000. UNAIDS/WHO 2000 www.unaids.org/
5. A Campa, G Shor-Posner, F Indacochea et al 1999 *J Acquir Immune Defic Syndr Hum Retrovirol* 20:508
6. C Avellana, B Dousset, T May et al 1995 Biol Trace Elem Res 47:1-3,133
7. D Serwadda, RD Murgewa, NK Sewankamboo, et al 1985 *Lancet* 2:849
8. WD DeWys, C Begg, PT Lavin et al 1980 *Am J Med* 68:683
9. U Suttmann, J Ockenga, O Selber et al 1995 *J Acquir Immune Defic Syndr* 8:239
10. RS Beach, PF Laura 1985 *Ann Int Med* 99:565
11. RH Gray 1983 *Am J Publ Health* 73:1332
12. VK Jain, RK Chandra 1984 *Nutr Res* 4:537
13. C Grundfeld, K Feingold 1992 *N Engl J Med* 327(5):329
14. NIJ Paton, DC Macallan, SA Jebb et al 1997 *J Acquir Immune Defic Syndr Hum Retrovirol* 14:119
15. CG Neumann, GJ Lawlor, ER Strehm et al 1975 *Am J Clin Nutr* 28:89
16. S Cunningham-Rundless 1982 *Am J Clin Nutr* 35:102
17. AS Fauci 1988 *Science* 239:717
18. C Grunfeld, D Kotler, R Hamdeh. 1989 *Am J of Med* 86:27
19. G Shor-Posner, A Basit, Y Lu 1993 *Am J of Med* 94:515
20. R Zangerle, M Sarcletti, H Gallati 1994 *Journal of Acquired Immune Deficiency Syndrome* 7:1149
21. U Keller 1993 *Supportive Care in Cancer* 1(6):290
22. S Endres, R Ghorbani, VE Kelley et al 1989 *N Engl J Med* 320:265
23. MK Hellerstein, K Wu, M McGrath et al 1996 *J Acquir Immune Defic Syndr* 11(3):258
24. J Gonzalez-Clemente, J Miro, M Navarro 1993 *AIDS* 7(7):1022
25. F Arnalich, P Martinez, A Hernanz et al 1997 *AIDS* 11:1129
26. A Schwenk, B Berger, D Wessel et al *AIDS* 7(9):1213
27. ED Schuartz, JB Greene 1992 *Seminar in Liver Disease* 12(2):142
28. RD Sharpstone, CP Murray, HM Ross et al *AIDS* 10:1377
29. MJT Hommes, JA Romijin, E Endert 1991 *Am J Clin Nutr* 54:311
30. C Grunfeld, M Pang, L Shimizu et al 1992 *Am J Clin Nutr* 55:455
31. RK Chandra (Ed) 1992 *Nutrition Immunology* ARTS Biomedical Publishers and Distributors, St. John's, Newfoundland, Canada pp 241
32. A Favier, C Sappey, P Leclerc et al 1994 *Chem Biol Interact* 91(2-3):165
33. TS Dobneyer, S Findhammer, JM Domeyer 1997 *Free Radical Biology & Medicine* 22(5):775
34. Pace GW, Leaf CD 1995 *Free Radical Biology & Medicine* 19(4):523
35. SR Mallery, RT Bailer, CM Hohl et al 1995 *Journal of Cellular Biochemistry* 59(3):317
36. S Dewhrst, HA Gelbord, SM Fine 1996 *Molecular Medicine Today* 2(1):16
37. F Muller, AM Svardal, P Aukurust 1996 *Am J of Clin Nutrit* 63(2):242
38. P Aukurist, AM Svardal, F Muller 1995 *Blood* 86(1):258
39. RS Beach, E Mantero-Atienza, G Shor-Posner et al 1992 *AIDS* 6:701
40. G Shor-Posner, MK Baum 1996 *Nutrition* 12:555
41. E Smit, NMH Graham, A Tang et al 1996 *Nutrition* 12:496

42. RD Semba, WT Caiaffa, NMH Graham et al 1995 *Journal of Infectious Diseases* 171:1196
43. MK Baum, G Shor-Posner, G Zhang et al 1997 *J Acquir Immune Defic Syndr Hum Retrovirol*, 16:272
44. CM Huang, M Ruddel, JE Ronald 1988 *Clinical Chemistry* 34(10):1957
45. Grunfeld C et al 1992 *J Clin Endocrinol and Metab* 74(5):1045
46. MK Baum 1996 *Nutrition* 12(21):124
47. MK Baum, G Shor-Posner 1998 *Nutrition Reviews* 56(S1-S2):58
48. AM Tang, NMH Graham, RD Semba et al 1997 *AIDS* 11:613
49. RD Semba, NMH Graham, T Waleska et al 1993 *Arch Intern Med* 153:2149
50. MK Baum, G Shor-Posner, Y Lu et al 1995 *AIDS* 9:1051
51. AM Tang, NMH Graham, AJ Saah 1996 *Am J Epidemiol* 143(12):1244
52. EW Taylor, CS Ramanathan, RG Nadimpalli et al 1995 *Antiviral Res* 26: A271
53. M Witted (Ed) *Computational Medicine, Public Health, and Biotechnology*, Vol. 1, Singapure: World Scientific, pp 285
54. EW Taylor, CS Ramanathan 1996 *J Orthomol Med* 10:131
55. EW Taylor, RG Nadimpalli, CS Ramanathan 1997 *Biol Trace Elem Res* 56:63
56. BM Dworkin, WS Rosenthal, GP Wormser, et al 1988 *Bio Trace Elem Res* 20:86
57. NS Scrimshaw, JP 1997 *Am J Clin Nutr* 66(2):464S
58. MP Look, JK Rocstroh, GS Rao et al 1997*Biol Trace Elem Res* 56:31
59. JE Spallholz 1981 *Advances in Experimental Medicine & Biology* 135:43
60. MJ Miguez-Burbano, A Campa, G Shor-Posner et al *STI and the Millennium, A Joint Meeting of the ASTDA and the MSSVD*, Baltimore, Maryland, May 3-6, 2000.
61. TR Mosman, S Sad 1996 *Immunol Today* 17:138
62. M Roy, L Kiremidjian-Schumacher, HI Wishe et al 1993 *Proc Soc Exp Biol Med* 202:295
63. PA Haslett 1998 *Semin Oncol* 25:53
64. K Hori, D Hatfield, F Maldarelli et al 1997 *AIDS Res Hum Retroviruses* 13:1325
65. M Moutet, P d'Alessio, P Mlette et al 1998 *Free Radic Biol Med* 25:270
66. C Sappey, S Legrand-Poels, M Best-Belpomme et al 1994 *AIDS Res Human Retrovir* 10:1451
67. LC Clark, GS Combs, BW Turnbull 1996 *FASEB* 10;550
68. JY Li, PR Taylor, B Li et al 1993 *J Natl Cancer Inst* 85:1492
69. WJ Blot, JY Li, PR Taylor et al 1993 *J Natl Cancer Inst* 85:1483
70. MC Delmas-Beauvieux, E Peuchant, A Coucouron et al 1996 *Am J Clin Nutr* 64:101
71. A Cirelli, M Ciardi, C De Simone et al 1991 *Clin Biochem* 24:211
72. L Olmsted, GN Schrauzer, M Flores-Arce et al 1989 *Biol Trace Elem Res* 20:59
73. GN Scharauzer, J Sacher 1994 *Chemico-Biological Interaction* 91:199
74. Keshan Disease Research Group *1979 Chin Med J* 92:471
75. AL Kavanaugh-McHugh, A Ruff, E Perlman et al 1991 *JPEN J Parenter Enteral Nutr* 15:347
76. MJ Miguez-Burbano, X Burbano, A Campa et al 2001 Abstract No. 8662, accepted at *FASEB 2001*, Orlando, Florida March 31-April 4.
77. A Carr, K Samaras, S Burton et al 1998 *AIDS*12:F51
78. B Halliwell 1995 *Am J Clin Nutr* 61:670S
79. LC Clark, GF Combs, BW Turnbull et al 1996 *JAMA* 276:1957
80. SY Yu, YJ Zhu, WG Li et al 1991 *Biol Trace Elem Res* 29:289
81. SY Yu YJ Zhu, WG Li 1998 *Biol Trace Elem Res* 15:231

# Chapter 21. Effects of selenium on immunity and aging

Roderick C. McKenzie, Teresa S. Rafferty

*Department of Medical and Radiological Sciences, University of Edinburgh, Edinburgh, Scotland, EH3 9YW, UK*

Geoffrey J. Beckett

*Department of Clinical Biochemistry, University of Edinburgh, Edinburgh, Scotland, EH3 9YW, UK*

John R. Arthur

*Division of Cell Integrity, Rowett Research Institute, Bucksburn, Aberdeen, Scotland AB21 9SB, UK*

**Summary**: It has been recognized from the 1970s that an adequate selenium intake is necessary for the optimum function of both cellular and humoral immune processes. This chapter reviews the effects of selenium on immunity and possible mechanisms for their occurrence. The influence of selenium on eicosanoid metabolism is discussed, followed by the modulation of adhesion molecule and cytokine expression. Turning to the cellular level, effects of selenium on humoral and cell-mediated immunity are reviewed along with the possible mechanisms. The effects of selenium on immune-mediated diseases are discussed and finally, the effects of selenium on aging and cancer are considered.

## Introduction
The immune system relies on: timely growth and death of appropriate cell populations in orderly sequence, the deployment of cells generating oxidative stress as a defense against microbial pathogens, co-ordinated regulation of adhesion molecule and the expression of soluble mediators such as eicosanoids and cytokines and their receptors. As we shall see, selenium has the potential to influence immunity at several points. The first studies with selenium were carried out in a variety of mammals and in poultry, using animals on either selenium-deficient or selenium-supplemented diets. The effects on immune processes or disease resistance were then compared.

The ability of the host immune cells to generate oxidative stress is a double-edged sword. The immune system relies upon the generation of reactive oxygen species for microbiocidal activity [1], for example, in the burst reaction of the neutrophil. In limited doses, release of reactive oxygen species generates inflammation and destroys microbial invaders, but chronic production of these species can lead to oxidative damage to the host. The host defense against this "friendly fire" is the antioxidant system. Selenium was originally shown to be immunologically important when the principal destination in dogs of incorporated radioactive selenium was found to be the leukocytes. Further studies showed that selenium was incorporated into a protein, which later proved to be the selenoenzyme glutathione peroxidase (GPx) [2,3]. Further indirect evidence for a role of selenium in immune function was the observation that this element was principally incorporated into immune-important organs such as the spleen, liver and lymph nodes [4]. Throughout the 70's and the 80's, research into the immunostimulatory effects of selenium increased dramatically after a series of papers from Berenshtein's group and Spallholz's group showing that injection of this element into rodents and rabbits enhanced the host's antibody and complement responses. This period in the selenium story has been well summarized by Spallholz et al. [5], who reviewed the studies between 1972-1989 showing immunoprotective roles for selenium. Selenium augments host antibody and complement responses to both natural and experimental immunogens such as tetanus toxoid, typhoid toxin, sheep red blood cells, and immunoglobulins [see 5 and references therein]. Deleterious effects on immunity were also demonstrated for selenium deficiency. These effects were demonstrated in selenium-deficient hosts. These effects in animals include the following: defective neutrophil function, stimulation of non-specific immunity in rabbits, increases in antibody titres to bacterial and mycotic antigens, increased $H_2O_2$ release during neutrophil phagocytosis, decreased neutrophil numbers, decreased antibody response to sheep red blood cells, decreased neutrophil fungicidal activity, inactivation of NADPH-dependent generation of superoxide by granulocytes, reduced natural killer cell activity, and increased mortality due to candidiasis.

Evidence that the immune system was being directly affected by selenium status was also suggested by studies showing that selenium injection or supplementation resulted in enhanced vaccine immunity against malaria, increased antibody-producing B-cell numbers, increased T-cell–dependent antibody production (T-cell help), increased selenium concentration in lymph nodes and immune tissue, an increase in neutrophils, increased in lymphocyte GPx activity, and increased concentration of selenium in neutrophils [see 5 and references therein]. However, the stimulatory effects of selenium on immunity were not recorded by all investigators [5] and in some cases high levels of selenium supplements actually decrease immunity.

*Effects of selenium on immunity and aging*  259

This may reflect the need for optimal forms and levels of selenium intake and we will return to this point later. Having established that selenium has immunostimulatory properties, it then remained to elucidate the mechanisms for these effects. As one would imagine, many of these effects stem from the antioxidant protection afforded by selenium compounds and by selenoenzymes such as cytosolic GPx, the phospholipid GPx (PLGPx) and thioredoxin reductase (TR) which metabolize organic peroxides. This research is still an on-going process. However, the principal mechanisms stimulated by selenium can be summarized as follows:
1) Detoxification of organic hydroperoxides and hydrogen peroxide.
2) Regulation of the balance of activity in the eicosanoid synthesis pathways, leading to preferential synthesis of leukotrienes and prostacyclins over thromboxanes and prostaglandins.
3) Down regulation of cytokine and adhesion molecule expression.
4) Upregulation of on interleukin-2 receptor expression, leading to enhanced activity of lymphocytes, natural killer and lymphokine activated killer cells.

**Seleno-enzymes as peroxynitrite reductases**
Protection against oxidative damage is the mechanism by which selenium is likely to exert its protective effects. Nitric oxide (NO) has microbiocidal effects, yet under oxidative conditions in which superoxide ($O_2^-$) is produced concomitantly (by neutrophils and mononuclear phagocytes), the highly reactive and destructive oxidant, peroxynitrite ($ONOO^-$) is produced. This free radical damages proteins (through tyrosine- nitrosylation), lipids (peroxidation) and DNA (single strand breaks). In vitro experiments, addition of selenocysteine and selenomethionine protected plasmid DNA from $ONOO^-$-mediated damage [6]. Specificity in the reaction was exhibited by later studies from the same group which showed that selenomethionine, selenocysteine and the GPx mimic ebselen were more effective protectants than selenite [7]. However, selenoenzymes also protect from $ONOO^-$-mediated damage. Again, in vitro studies have shown that GPx protects against nitrotyrosine formation in fibroblast lysates [8]. TR may also protect against $ONOO^-$ [9]. The role of the GPx family and TR as a peroxynitrite reductase needs to be further elucidated by studies in intact cells.

**Selenium and eicosanoid metabolism**
The eicosanoids are the 20-carbon- containing family of metabolites of arachidonic acid. This includes the leukotrienes, the thromboxanes, the prostaglandins and the lipoxins. The thromboxanes have, predominantly, coagulatory effects causing platelet degranulation, aggregation and thrombosis [see 10 for review]. The leukotrienes have pro-inflammatory effects and the prostacyclins prevent clotting. Selenium (mediated through

the PLGPx and the GPx) probably has anti-inflammatory effects, preventing the release of inflammatory mediators and thus inflammation, and also by preventing blood clotting. Selenium deficiency has profound effects on this potent family of mediators. Several steps require reduction of organoperoxides by a peroxidase in order to allow production of the active leukotrienes. This step is inhibited by selenium deficiency.

The cyclo-oxygenase enzyme catalyses the conversion of arachidonic acid to prostaglandin $GG_2$. This enzyme requires a minimal level of peroxide to function, however, if peroxide levels in the cell are high, cyclo-oxygenase activity is inhibited [11]. Thus, selenium deficiency, resulting in a decrease in GPx activity will inhibit further conversion of arachidonic acid, particularly to prostacyclins [12]. However, despite the fact that reduction of the hydroperoxyeicosatetraenoic acids to hydroxyeicosatetraenoic acids requires the reductive power of peroxidases, the resultant products are generally pro-inflammatory. Anti-inflammatory activity of selenium may be explained by the ability of selenoenzymes to inhibit the 5- and 15- lipoxygenase enzymes which convert arachidonic acid to the 5-hydroperoxyeicosatetraenoic acid precursor of the leukotrienes [13,14]. The conversion of selenite to selenide (which inhibits lipoxygenase) appears to be catalysed by a reaction of NADPH with TR [15]. Selenium deficiency also leads to decreased leukotriene B4 synthesis since the synthase is selenium-dependent [16], and lack of leukotriene B4 impairs the functions and mobility of phagocytes.

Another important effect of selenium deficiency is the disturbance in the balance of production of the pro-coagulant thromboxanes and the anti-clotting prostacyclin family of metabolites. There is preferential production of thromboxanes over prostacyclins in selenium-deficient endothelial cells [17] or selenium-deficient rats [18]. This could explain the prevalence of atherosclerosis in populations which have low dietary selenium intake [see 19 for review]. Apart from the promotion of blood clotting in selenium deficiency, the favoring of inflammatory events will occur more readily in a low selenium environment. The disagreement in the findings of selenium protection studies in atherosclerosis has been explained by Huttenen`s "threshold hypothesis" which states that protective effects of selenium will only be seen in populations whose selenium intake is below the recommended daily allowance [20]. Platelet GPx is particularly high in humans and is thus extremely sensitive to the effects of selenium deficiency. The platelets of selenium-low subjects show increased aggregation, thromboxane B2 production and the synthesis of lipoxygenase-derived products. In such people, selenium supplementation increases platelet GPx activity and concomitantly decreases hyperaggregation. This suggests that selenium supplements may help prevent thrombosis and coronary heart disease [see 11 for review].

**Effect of selenium on adhesion molecules and cytokines**
Up regulation of adhesion molecules is a means by which endothelial cells can regulate the influx of inflammatory cells into damaged tissue. The effects of selenium on cytokine production will be considered first, since pro-inflammatory cytokines, such as tumor necrosis factor-α and interleukin-1, induce many of the adhesion molecules which are upregulated in inflammation. A direct inhibitory effect of selenium on cytokine production, of course does not preclude a direct effect of selenium on adhesion molecules expression *per se*. However, existing evidence is more consistent with a selenium effect on adhesion molecule expression through regulation of cytokine release.

The endothelium or endothelial cells in culture have been the systems most widely used to study these effects. In general, selenium-deprived cells or endothelium from selenium-low individuals have higher constitutive expression of adhesion molecules, and selenium supplementation decreases expression of adhesion molecules. Thus, the selenium-induced down regulation of adhesion molecules expression acts as a mechanism to inhibit inflammation. For example, Horvatha et al [21] found that in endothelial cells obtained from asthmatics, that the constitutive expression of P-selectin, vascular adhesion molecule-1, E-selectin and intercellular adhesion molecule-1 was significantly higher than on cells from normal subjects. However, after 3 months of selenium supplements, there was a significant decrease in vascular adhesion molecule-1 and E-selectin expression. This was also confirmed by treatment of an in vitro culture of endothelial cells with 6 nM to 48 nM selenium. Similarly, bovine endothelial cells in culture, grown under selenium-deficient conditions and stimulated with tumor necrosis factor-α, had higher levels of E-selectin, P-selectin expression and intercellular adhesion molecule-1, which was manifested by greater adherence of neutrophils [22]. More direct evidence that the down regulation of adhesion molecules on endothelial cells by selenium supplementation is due to effects on selenoproteins was provided by the observation by D`Alessio et al. [23] that a GPx mimic inhibited the expression of intercellular adhesion molecule-1 and vascular adhesion molecule-1. Following up on these observations, the same group demonstrated that GPx analogs prevented tumor necrosis factor-α-stimulated expression of P-selectin and E-selectin, as well as tumor necrosis factor-α and interleukin-1-stimulated interleukin-8 release in human endothelial cells [24].

Turning to the effects of selenium on cytokine expression, oxidative stress induces several pro-inflammatory cytokines. These include: interleukin-1, interleukin-6, interleukin-8 and tumor necrosis factor-α and may be through seleniums effects on the activation of the transcription factors AP-1 and NF-κB [25], binding sites for which are present in several cytokine gene promoters. Pre-incubation of keratinocytes with selenium abrogated upregulation of the mRNAs for interleukin-6 and interleukin-8 (see Figure 1) in response to ultraviolet radiation B, which is a potent environmental oxidative stress. Interleukin-6 induces the acute phase response and interleukin-8 is a potent chemoattractant for neutrophils and T-cells. Thus, these cytokines mediate inflammation.

However, the source of selenium seems important. For example, selenomethionine seems a better protectant than selenite against ultraviolet radiation B-induction of cytokines. Since GPx depletes reduced glutathione, it was found that cytokine release into the culture fluid of endothelial cells increased in endothelial cells treated with selenite [26]. Also, selenite (but not selenomethionine) supplementation of BALB/c mice exposed to 9 mg/kg of selenite increased the release of interleukin-1 and tumor necrosis factor-$\alpha$ from phytohaemagluttinin-P stimulated splenic macrophages [27]. This could result from the pro-oxidant effects of high doses of selenite. Moreover, Stewart et al. showed that selenite and selenocystamine, but not selenomethionine, increased oxidative stress, oxidative DNA damage and apoptosis in keratinocytes [28]. The nutritional significance of selenomethionine and the ability of selenium compounds to cause pro-oxidant toxicity have been recently reviewed [29; see 30 for further discussion].

There are few reported studies on the effects of cytokines on selenium metabolism. Mostert et al. [31] demonstrated that treatment of liver cell line Hep G2 with interleukin-1$\beta$, tumor necrosis factor-$\alpha$, or interferon-gamma had no effect on selenoprotein P expression. However, incubation with transforming growth factor-$\beta$ (100 pM for 48 h) led to a 21% decrease in expression of selenoprotein P mRNA, which was corroborated by a decrease in expression of a luciferase gene under the control of a selenoprotein P promoter in a transfected cell line incubated with transforming growth factor-$\beta$. Incubation with transforming growth factor-$\beta$ also down regulated mRNA for GPx and catalase, as well as enzyme activities for both antioxidants [31]. Taken together, this may indicate a role for transforming growth factor-$\beta$ in oxidative damage to cells.

**Effects of selenium on cell-mediated and humoral immunity (Post 1990)**
Spallholtz has written an excellent summary of the early studies on the effects of selenium on cell–mediated immune cells and on antibody production [see 5 for review of studies carried out prior to 1990]. In this section, significant advances which have been reported since 1990 are considered.

Many of these studies were performed by veterinary researchers. This recognizes the fact that if animal husbandry is to be effective and profitable, that care must be taken - especially in "factory farming" - that animals have optimal nutrient intake to protect them from diseases and to maximize growth. Many of the animal studies have used simultaneous supplements of selenium and vitamin E, since the nutrients can act synergistically, and to a limited degree, can substitute for each other. This is a factor that needs to be borne in mind when interpreting these studies.

In rats, selenium deficiency decreased IgG production slightly, but had no effects on IgA production. However, IgM production was greatly decreased and compounded by vitamin E deficiency. Selenium supplementation was partially able to compensate for the vitamin E deficiency-induced decreases in IgA and IgG [32]. The effects of selenium deficiency and supplementation on the production of colostral immunoglobulins was examined in cattle by Swecker et al. [see 33 and refs therein]. IgG levels were higher in cows given 120 µg/kg selenium and calves from these cows had higher post suckle serum IgG levels. Thus, maintaining optimal selenium intake may promote health of offspring as well as of the mothers. The effect of vitamin E and selenium supplements on the immune responses of domestic animals was reviewed by Finch and Turner in 1996 [33]. Since then it has been demonstrated that selenium-enriched diets given to poultry improve their antibody responses to salmonella and aflatoxin vaccination [34]. A combination of vitamin E and selenium supplements in diet increased antibody titres to Newcastle disease virus and give maximum gain in body weight [35]. In sheep vaccinated against *Chlamydia psittaci*, which causes abortion, injection of selenium (0.1mg/kg) alone increased the *Chlamydia* antibody response. But this was decreased, oddly enough, if co-administered with vitamin E [36].

Selenium supplementation improves responses in most studies of cell-mediated immunity. Selenium deficiency in rats resulted in impaired superoxide production and GPx activity in neutrophils. This was associated with decreased candidiacidal activity, which was partially reversed after selenium supplementation for 30 days [37]. Polymorphonuclear cells also provide defense against mastitis in cattle, supplementation in vitro with selenium and vitamin E increased superoxide production and migration of polymorphonuclear cells after stimulation with phorbol esters in vitro [38].

Selenite enhanced chemotaxis of macrophages. Murine infection with the parasite *Trypanosoma cruzi* was treated with selenium at 0 ppm, 2 ppm, 4 ppm, 8 ppm, or 16 ppm as sodium selenate in drinking water [39]. Sixty four days after infection, the mice without selenium supplements had all died. But 60 % of the animals in groups supplemented with 4 and 8 ppm selenium survived. Interestingly, survival was much less in the group fed 16 ppm, again suggesting a bell-shaped response to selenium protection, possibly due to toxicity at higher doses and subsequent immune suppression.

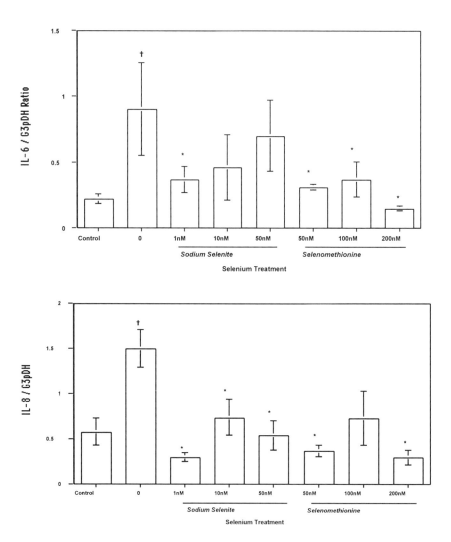

**Figure 1.** Selenium suppresses ultraviolet radiation-induced expression of interleukins-6 and -8. Human primary keratinocytes were supplemented with either sodium selenite or selenomethionine for 24 hours prior to cells being exposed to ultraviolet radiation (200 $J/m^2$). Fresh media was returned to the cells and they were incubated for 6 hours before being harvested for interleukin-6 (and for 24 h for interleukin-8) mRNA assay by RT-PCR. The RNA was extracted and RT-PCR analysis was carried out on the samples using $^{32}$P-labeled primers. The PCR products were resolved on gels and quantitated using a phosphoimager. The cytokine levels were normalized using the housekeeping gene G3PDH. Control cells had no selenium added and were mock irradiated. Results are expressed as the mean ratios ± S.E.M, n=3. Significant difference from the control cells, † = $P<0.05$, significant difference from the irradiated cells with no selenium added * = $P<0.05$.

**Effects of selenium on interleukin-2 receptor and lymphocytes**
Supplementation with selenium augments the performance of both T- and B-lymphocytes and perhaps the common effect is through the upregulation of the interleukin-2 receptor α and β subunits which results in a greater number of high affinity interleukin-2 receptors in mice [40] and humans [41] and is accompanied by enhanced proliferation and differentiation into cytotoxic effector cells [42]. Selenium supplementation in humans (200 μg/day for eight wks) also upregulates the activity of cytotoxic T-cells (118%), natural killer cells (82%) [42] and downregulates the activity of suppressor T-cells. The lytic capabilities of natural- and lymphokine activated- killer cells in humans, was increased, purportedly by upregulation of interleukin-2 receptors [43]. However, the stimulatory effects of selenium on cell-mediated and humoral immunity seem to be dose-dependent increasing, then decreasing with respect to increasing selenium intake. For example, Koller et al. examined the effects of selenium supplementation in a selenium–replete population of rats. The animals were given doses of selenite of 0.5 ppm, 2.0 ppm and 5.0 ppm in their water supply. The response of natural killer cells was boosted in the group receiving 0.5 and 2.0 ppm selenite, but the natural killer cell activity in animals receiving 0.5 ppm was similar to that seen in unsupplemented animals. Antibody synthesis was not significantly increased, but fell in the group given 5.0 ppm selenium. Production of prostaglandin E$_2$ was decreased at all selenium doses [44]. Similarly, an inhibitory effect of selenite on natural killer cell activity and lymphokine activated killer cell activity was seen in human lymphocytes supplemented in culture with 0.8 μg/ml selenite. However, this is a very large dose and may be toxic. In the same study, lymphocyte proliferation to T-cell mitogens was suppressed by selenium in the range 0.5-1.0 μg/ml [45]. This underlines the importance of considering the dose of selenium, the chemical form and possibly the species under investigation when drawing conclusions on the effects of selenium.

The requirement for selenium in the immune response is corroborated by studies showing that activated T-cells have increased activity of the enzyme selenophosphate synthetase, which is essential for the synthesis of selenocysteine, an obligatory step in selenoprotein synthesis [46]. This supports the earlier findings that selenium is concentrated in immune-active tissues such as the spleen, lymph nodes and the liver [4]. The importance of selenium in maintaining cellular immunity was exemplified by a study on patients from Italy and Germany in which uremic patients were studied [47]. These subjects were shown to have lower selenium plasma levels than controls. Supplementation with selenium at 500 μg/day thrice weekly for three months was followed by 200 μg/day for the next three months. Another group was given placebos. Although no change in lymphocyte numbers or subpopulations was observed, delayed–type hypersensitivity responses (to phytohaemoagluttinin) were significantly higher in the selenium-supplemented group after 6 months, compared with their own pre-experimental levels and compared to the placebo group. The augmented responses dropped to pre-supplementation values 3 months after ceasing selenium supplementation. The overall conclusion was that selenium supplementation could be beneficial in uremic patients. On the other hand, Girodon [48] found no improvement with selenium and zinc supplements on the delayed–type hypersensitivity responses in elderly patients, despite an improved humoral immune response to influenza vaccination. A recent scan of the literature suggests that the effects of selenium on mast cells or basophils have not been explored.

**Selenium and immune-mediated disease in humans**
From the previous discussion, one would intuitively expect that selenium would have therapeutic effects on inflammatory, and in particular, on chronic inflammatory conditions such as rheumatoid arthritis. However, there is little evidence of this in the literature. Epidemiological studies have been performed showing that serum selenium levels are correlated with disease state, but little in the way of intervention studies exist. Reports do not agree on the activity levels of GPx in rheumatoid arthritis patients, but this could be because sub-forms of the disease were not categorized in earlier studies. Moreover, in some forms of the disease there appears to be an inability of the neutrophil to increase GPx activity despite dietary selenium supplementation. This may explain the reported lack of effect of selenium to alleviate arthritic symptoms in some studies. The results of several studies have been summarized [see 19,49 for review]. Nevertheless, some studies have shown that low selenium status may be a risk factor for rheumatoid factor-negative (but not rheumatoid factor-positive) arthritis [50]. Because selenoproteins can be acute phase reactants, a decrease in plasma selenium is not necessarily associated with loss of antioxidant function. More work and trials with better controls are needed in this area before one can draw clear conclusions.

Reimund et al. [51] examined the relationship between selenium and immune parameters in Crohn's disease patients - a disease in which immune activation is mediated by reactive oxygen species. There was a negative correlation between plasma selenium and soluble interleukin-2 receptor and erythrocyte sedimentation rate. The soluble interleukin-2 receptor concentration is positively correlated with the degree of immune activation.

Oxidative stress and micronutrient deficiencies have been identified along with selenium deficiency and decreased red cell GPx activity as risk factors for the development of asthma [see 52 for review]. In a limited study on juvenile asthma patients with intrinsic asthma, it was found that giving 100 µg selenite/day improved their clinical symptoms [53]. Protection against asthmatic wheeze has been found in English adults [19] asthma patients and in asthmatic children in New Zealand [54]. The reasons for this are not clear,

*Effects of selenium on immunity and aging*                                   267

but atopic asthmatics have been shown to have low platelet and red blood cell GPx activities. However, a study by Misso et al. [55], showed that GPx activity was higher in eosinophils in both normal and asthmatic subjects (but was not different between these groups) than in neutrophils from both groups. It was suggested that the higher GPx activity of eosinophils prolonged their survival at inflammatory sites, thus driving the inflammatory process. An increase in the incidence of asthma has recently been noted in the Western World and more studies into the therapeutic utility of selenium supplements in asthmatics are required.

Clinical trials on sepsis and systemic inflammatory response syndrome patients suggest that these patients have low plasma selenium and GPx activity. Two independent prospective studies showed a beneficial therepeutic response in patients given selenium supplements [56].

**Selenium and the skin**
The skin is the body`s largest organ which of course interfaces with the environment. Skin is continually exposed to oxidative stress due to the products of commensal organisms on the surface as well as the oxidative stress and cell damage caused by exposure to ultraviolet radiation. Several groups have shown that both selenomethionine and selenite can protect keratinocytes, melanocytes and fibroblasts from UV-induced cell death and apoptosis at nanomolar concentrations. The processes behind these effects have been partially elucidated and these include inhibition of oxidative DNA damage, lipid peroxidation, apoptosis, suppression of inflammatory (see Figure 1) and immune suppressive cytokine release, and modulation of p53 activity [see 57 for review]. In 1985, Overvad et al. demonstrated that selenium could prevent UVB-induced skin tumors in hairless mice, which was confirmed in studies by Burke and Pence [see 57 for discussion]. It remains to be seen if the protective effect against skin cancer in mice also operates in humans.

**Selenium and aging**
Several studies have demonstrated that aging cells accumulate oxidative damage (the oxidative theory of aging - [see 58 for review] and that the aging lymphocyte population fails to expand as effectively on antigenic challenge, and damage to both mitochondrial and nuclear DNA occurs. Apart from lipid peroxidation, there is an accumulation of carbonyl moieties on protein, both types of lesions are produced by oxidative stress. Because of the presence of oxidative metabolism, mitochondria accumulate age-related damage, which release more reactive oxygen species, exacerbating the process [see 58 and refs therein]. The treatment of fibroblasts with non-lethal doses of hydrogen peroxide seems to activate a senescence program, which

leads to growth cessation. Thus, a role for GPx and other selenoproteins in slowing cellular damage and the aging process are credible.

The efficiency of the immune system declines with age and the elderly are more prone to infections than young or middle aged adults. The thiobarbituric reactive substances content of plasma - which is a measure of degree of lipid peroxidation - from elderly subjects was inversely related to plasma selenium. These higher levels of oxidative stress in the elderly were most prominent in those at risk for cognitive decline [59]. Generally, the response to antigen challenge and the ratio of effector to naïve T-cells decreases, along with a decrease in the ratio of CD4 to CD8 T-cells and decreased ratio of $CD5^-$ to $CD5^+$ B-cells [60]. There is also decreased ability of macrophages and monocytes to destroy microbes. For example, aged mice produced weak interferon-$\gamma$ and interleukin-2 responses to the parasite *Trypanosoma musculi* [61]. Roy et al. showed that the decreased proliferation of spleen lymphocytes to allogeneic or mitogen stimulation from aged mice could be restored by dietary supplementation with selenium [62]. The mechanism appeared to be via upregulation of the interleukin receptor. The in vitro effects of selenium and vitamin E on polymorphonuclear cell chemotaxis and monocyte chemo attractant protein release were studied in well-nourished elderly donors. Selenium supplementation was able to enhance the previously depressed chemotactic and cytokine release capabilities of their polymorphonuclear cells [63].

In humans, low selenium status (low selenium and erythrocyte GPx activity) in the elderly [64] was correlated with lower triiodothyrinine to thyroxine ratios, due to raised thyroxine concentrations, and was seen with advancing age [65]. Selenium supplementation decreased the serum thyroxine concentration. The age-related decline in triiodothyrinine content was ascribed to the requirement for iodothyronine-5`-deiodinase, which is a selenoprotein, to catalyze the conversion of thyroxine to triiodothyrinine. A deficiency in thyroxine to triiodothyrinine conversion will affect general metabolism, including immunity. Longevity in areas of the world which had selenium-rich soils was noted by Foster and Zhang in 1995 [66]. It was found that less people over 80 years of age were found in areas where the selenium-deficiency diseases Kashin-Beck and Keshan disease were endemic [66]. More recently, an interesting hypothesis has been put forward by Foster which proposes that the areas of the world which have higher lifespan than (national) average are areas where soil selenium is high, but mercury content, which sequesters selenium, is lowest [67].

However, some investigators have found increases in antioxidant enzyme production with aging which may be induced by stress itself [see 58 for discussion]. In humans, cancer is a disease associated principally with old age, pointing to an age-dependent decrease in the efficiency of the immune system to detect and destroy tumors. This may be due to a decrease in the effectiveness of natural killer cells and nutritional deficiencies. In a study of free-living elderly Italians (ages 90-106 years of age), Ravaglia et al. showed that in the women in this group that the percentage of natural killer cells in the circulation was related to selenium serum content and that low selenium status was noted in 50 % of the total cohort [68]. An adequate selenium intake may also be important in maintaining GPx activity in aging cells. Increasing the intracellular hydrogen peroxide content by blocking GPx activity with buthionine sulfoximine and inhibiting catalase activity by aminotriazol treatment raised the levels of collagenase mRNA [69]. Increased collagenase-1 levels contribute to connective tissue damage, which is a feature of tumor expansion, inflammatory disease and photo-aging. Evidence that GPx activity is lower in neutrophils from human volunteers over 65 years of age has been produced [70]. A decreased V max for the enzyme was seen in the elderly group compared with neutrophils from younger donors (21-34 years of age) after in vitro stimulation with formyl peptide. Furthermore, in the neutrophils of the young group, the affinity (Km) of the enzyme for its substrate increased on neutrophil activation. This did not occur in the elderly group.

Finally, telomere length decreases with age in peripheral leukocytes and is accelerated by oxidative stress in fibroblasts [see 71 and refs therein]. A role for GPx in the maintenance of telomere length has been proposed. The rate of telomere shortening and carbonyl group accumulation was inversely correlated with GPx activity in fibroblasts [71]. Furthermore, experiments with human breast cells which were transfected with DNA constructs to produce cell lines that were GPx-replete or GPx-deficient revealed an important role for the enzyme in the protection against oxidative–induced mitochondrial DNA damage. The GPx-replete lines contained 100 times greater GPx activity than the GPx-depleted lines. Exposure to 25 μM menadione for 1 hour caused approximately three-fold more single-strand breaks and 8-oxo-deoxyguanosine residues in the GPx-depleted cell lines [72]. As well as GPx, the mitochondrial TR [73] may play an important role in limiting oxidative damage in immune cells.

**Selenium and cancer**
Associated with aging is a decline in the effectiveness of the immune system, which increases the chances of neoplasia. Chronic exposure to oxidative stress also predisposes to cancer. We will now briefly discuss the role of selenium in cancer. It is important to note that some of the putative anticancer effects of selenium may be unrelated to its role in antioxidant enzymes.

A large number of animal studies show an association between selenium and the processes which lead to, or may prevent cancers [see 74-76 for discussion]. Dietary selenium deficiency in animals leads to a greater susceptibility to chemical carcinogens which can affect a range of tissues. This susceptibility can be prevented by supplementation of deficient animals or sufficient animals with selenium in various inorganic and organic forms [77-81]. There are numerous studies which show that selenium intakes above those that are considered the normal nutritional range have chemoprotective effects. The effectiveness of these higher selenium intakes has been attributed to methylated selenium compounds which can kill rapidly growing cancerous cells [see 74,82,78 and Chapter 23]. Many different selenium compounds which have similar effects have been synthesized and tested and it is relevant that many of these compounds do not provide selenium that is readily available for selenoprotein synthesis. Thus, both selenium intake and selenium status have the potential to influence cancer initiation and proliferation at many different levels. At lower selenium intakes these effects are likely to be associated with impairment of antioxidant functions or redox regulatory effects of the GPxs and TRs [75, 83]

In humans there are many epidemiological studies which associate higher cancer incidence with low dietary selenium intake [see 19 and Chapter 17]. In addition, in case-control studies, low selenium status, as determined by plasma selenium or GPx has been associated with increased cancer incidence and mortality. However, in many of these studies changes in plasma selenium were small and may be attributed to an effect, rather than the cause, of the disease. Recently, however, work incidence and mortality were compared with placebo treatment in a double blind trial [see 83,84 and Chapter 17]. Although the trial was set up to examine the effect of selenium on skin cancer (but produced negative results in this aspect of the study), the effect on cancer mortality from other causes was a 50% protection. The authors of the study also emphasized that the greater protective effects were observed in subjects who started with the lowest blood selenium levels. These lowest blood selenium values found in the USA are still much higher than those of most European populations. Since this cancer prevention trial was carried out in a population with a dietary selenium intake that would saturate selenoenzyme activities, it is unlikely that selenoproteins are associated with this protective effect of selenium. The chemopreventive mechanism(s) may therefore be the production of methyl selenol-like compounds that are thought to be effective in the animal trials with high selenium intakes. While similar mechanisms could contribute to potential chemopreventive effects of selenium in populations with an initially low selenium status, beneficial effects of optimizing selenoprotein activity could also be involved.

In conclusion, selenium deficiency is likely to impair immunity, promote the aging process and some types of cancers. Therefore, an optimum selenium intake may improve immunity and protect against a variety of diseases.

## References

1. JC Fantone JC, PA Ward 1982 *Am J Pathol* 107:397
2. JT Rotruck, AL Pope, HE Ganther, AB Swanson, DG Hafeman, Hoekstra 1973 *Science* 179:588
3. L Flohe, WA Gunzler, HH Schock 1973 *FEBS Lett* 32:132
4. RC Dickson, RH Tomlinson 1967 *Clin Chim Acta* 16:311
5. JE Spallholz, LM Boylan, HS Larsen 1990 *Ann NY Acad Sci* 587:123
6. I Roussyn, K Briviba, H Masumoto, H Sies 1996 *Arch Biochem. Biophys* 330:216
7. K Briviba, I Roussyn I, VS Sharov, H Sies 1996 *Biochem J* 319:1315
8. H Sies, VS Sharov, LO Klotz, K Briviba 1997 *J Biol Chem* 272:27812.
9. GE Arteel, K Briviba, H Sies 1999 *Chem Res Toxicol* 12:264
10. GR Davies, DS Rampton 1997 *Euro J Gastroenterol Hepatol* 9: 1033
11. D Vitoux, P Chappuis, J Arnaud, M Bost, M Accominotti, AM Roussel 1996 *Annales De Biologie Clinique* 54:181
12. MJ Parnham, E Graf 1987 *Biochem Pharmacol* 36:3095
13. Werz, D Steinhilber 1996 *Eur J Biochem* 242:90
14. C Schewe, T Schewe, A Wendel 1994 *Biochem Pharmacol* 48:65
15. M Bjornstedt, B Odlander, S Kuprin, HE Claesson, A Holmgren 1996 *Biochem* 35:8511
16. C Gairola, HH Tai 1985 *Biochem Biophys Res Commun* 132:397
17. YZ Cao, CC Reddy, LM Sordillo 2000 *Free Radical Biol Med* 28:381
18. M Meydani 1992 *Biol Trace Element Res* 33:79
19. MP Rayman 2000 *Lancet* 356:233
20. JK Huttenen 1997 *Biomed Environment Sci* 10:220
21. M Horvathova, E Jahnova, F Gazdik 1999 *Biol Trace Element Res* 69:15
22. JF Maddox, KM Aherne, CC Reddy, LM Sordillo 1999 *J Leuco Biol* 65:658
23. P D'Alessio, M Moutet, E Coudrier, S Darquenne, J Chaudiere 1998 *Free Radical Biol Med* 24:979
24. M Moutet, P D'Alessio, P Malette, V Devaux, J Chaudiere 1998 *Free Radical Biol Med* 25: 270
25. G Powis, JR Gasdaska, A Baker 1997 *Adv Pharmacol* 38:329
26. R Tolando, A Jovanovic, R Brigelius-Flohe, F Ursini, M Maiorino 2000 *Free Radical Biol Med* 28:979
27. VJ Johnson, M Tsunoda, RP Sharma 2000 *Arch Environ Contamination Toxicol* 39:243
28. MS Stewart, JE Spallholz, KH Neldner, BC Pence 1999 *Free Radical Biol Med* 26:42
29. GN Schrauzer 2000 *J Nutr* 130:1653
30. JE Spallholz 1994 *Free Radical Biol Med* 17:45
31. V Mostert, I Dreher, J Kohrle, J Abel 1999 *FEBS Letts* 460:23
32. S Bauersachs, M Kirchgessner, BR Paulicks 1993 *J Trace Elements Electrolytes Health & Dis* 7:147.
33. JM Finch, RJ Turner 1996 *Res Vet Sci* 60:97
34. SM Hegazy, Y Adachi 2000 *Poultry Sci* 79:331
35. BK Swain, TS Johri, S Majumdar 2000 *Brit Poultry Sci* 41:287
36. N Giadinis, G Koptopoulos, N Roubies, V Siarkou, A Papasteriades 2000 *Comp Immunol Microbiol Infec Dis* 23:129
37. R Kukreja, A Khan 1994 *Indian J Biochem Biophys* 31:427
38. N Ndiweni, Finch JM 1996 *Vet Immunol Immunopathol* 51:67
39. CD Davis, L Brooks, C Calisi, BJ Bennett, DM McElroy 1998 *J Parasitol* 84:1274
40. M Roy, L Kiremidjianschumacher, Hi Wishe, MW Cohen, G Stotzky 1992 *Proc Soc Exp. Biol Med* 200:36
41. M Roy, L Kiremidjianschumacher, Hi Wishe, MW Cohen, G Stotzky 1994 *Biol Trace Element Res* 41:103

42. L KiremidjianSchumacher, M Roy, HI Wishe MW Cohen, G Stotzky 1996 *Biol Trace Element Res* 41:115
43. L KiremidjianSchumacher, M Roy, HI Wishe MW Cohen, G Stotzky 1996 *Biol Trace Element Res* 52:227
44. LD Koller, JH Exon, PA Talcott, CA Osbourne GM Henningsen 1986 *Clin Exp Immunol* 63:570
45. MP Nair, SA Schwartz 1990 *Immunopharmacol* 19:177
46. MJ Guimaraes, D Peterson, A Vicari et al. 1996 *Proc Natl Acad Sci USA* 93:15986
47. M Bonomini, S Forster, F Derisio et al. 1995 *Nephrol Dialysis Transpl* 10:1654
48. F Girodon, P Galan, AL Monget et al 1999 *Arch Internal Med* 159:784
49. U Tarp 1995 *Analyst* 120:877
50. P Knekt, M Heliovaara, K Aho, G Alfthan, J Marniemi, A Aromaa 2000 *Epidemiology* 11:402
51. JM Reimund, C Hirth, C Koehl, R Baumann, B Duclos 2000 *Clinical Nutr* 19:43
52. LS Greene LS 1995 *J Am Coll Nutr* 14:317
53. L Hasslemark, R Malgren, U Zetterstorm 1993 *Allergy* 48:30
54. R Shaw, K Woodman, J Crane et al. 1994 *N Zealand J Med* 107:387
55. NLA Misso, DJ Peroni, DN Watkins, GA Stewart, PJ Thompson 1998 *J Leuk Biol* 63:124
56. R Gartner, M Angstwurm 1999 *Medizinische Klinik* 94:54
57. RC McKenzie 2000 *Clin Exp Dermatol* 25:1
58. T Finkel NJ Holbrook 2000 *Nature* 408:239
59. C Berr, B Balansard, J Arnoud et al. 2000 *J Am Geriatric Soc* 48:1285
60. BM Lesourd 1997 *Medizinische Klinik* 66:S478
61. JW Albright, JF Albright 1998 *Exp Gerontol* 33:13
62. M Roy, L Kiremidjianschumacher, Hi Wishe, MW Cohen, G Stotzky 1995 *Proc Soc Exp Biol Med* 209:369
63. MT Ventura, E Serlenga, C Tortorella, S Antonaci 1994 *Cytobios.* 77:225
64. Olivieri, D Girelli, Azzini M et al 1995 *Clinical Sci* 89:637
65. Olivieri, D Girelli, AM Stanzial, L Rossi, A Bassi, R Corrocher 1996 *Biol Trace Element Res* 51:31
66. HD Foster, LP Zhang 1995 *Sci of Total Environ* 170:133
67. HD Foster 1997 *Medical Hypoth* 48:355
68. G Ravaglia, P Forti, F Maioli et al. 2000 *Am J Clin Nutr* 71:590
69. P Brenneisen, KBriviba, M Wlashek, J Wenk, K ScharfetterKochanek 1997 *Free Radical Biol Med* 22:515
70. Y Ito, O Kajkenova, RJ Feuers et al 1998 *J Gerontol Series A- Biol Sci Med Sci* 53:M169
71. V Serra T Grune, N Sitte, G Saretski T Von Zglinicki 2000 *Ann NY Acad Sci* 908:327
72. J Legault, C Carrier, P Petrov et al 2000 *Biochem Biophys Res Com* 272: 416
73. S Watabe. Y Makino K Ogawa K et al 1999 *Eur J Biochem* 264:74
74. C Ip 1998 *J Nutr* 128:1845
75. HE Ganther 1999 *Carcinogenesis* 20:1657
76. C Ip, DK Sinha 1981 *Cancer Res* 41:31
77. C Redman, JA Scott, AT Baines et al 1998 *Cancer Lett* 125:103
78. C Ip, DJ Lisk, HE Ganther 1998 *Anticancer Res* 18:4019
79. C Ip, HE Ganther1992 *Carcinogenesis* 13:1167
80. RL Nelson, H Abercarian, TM Nelson et al 1996 *Am J Surg* 172:85
81. C Ip, K El-Bayoumy, P Upadhyaya et al 1994 *Carcinogenesis* 15:187
82. C Ip, HJ Thompson, ZJ Zhu, HE Ganther 2000 *Cancer Res* 60:2882
83. G Combs 2000 *Cancer Res* in press
84. LC Clark, GF Combs, BW Turnbull et al 1996 *JAMA* 276:1957

# Chapter 22. Selenium and male reproduction

Leopold Flohé

*Department of Biochemistry, Technical University of Braunschweig, Mascheroder Weg 1, D-38124 Braunschweig, Germany*

Regina Brigelius-Flohé

*Department of Vitamins and Atherosclerosis, German Institute for Research of Nutrition (DIfE), Arthur-Scheunert-Allee 114-116, D-14558 Potsdam, Germany*

Matilde Maiorino

*Dipartimento di Chimica Biologica, Università di Padova, Viale G. Colombo 3, I-35121 Padova, Italy*

Antonella Roveri

*Dipartimento di Chimica Biologica, Università di Padova, Viale G. Colombo 3, I-35121 Padova, Italy*

Josef Wissing

*Department of Biochemistry, Technical University of Braunschweig, Mascheroder Weg 1, D-38124 Braunschweig, Germany*

Fulvio Ursini

*Dipartimento di Chimica Biologica, Università di Padova, Viale G. Colombo 3, I-35121 Padova, Italy*

**Summary:** Selenium deficiency has long been documented to result in impaired male fertility of rats, mice and boars. The prominent feature of selenium-deficient spermatozoa is a distorted architecture of the midpiece, where normally the mitochondria are embedded into a keratinous matrix called the mitochondrial capsule. This material, which contains most of the selenium of sperm, is composed of oxidatively cross-linked proteins, the major component being the selenoprotein phospholipid hydroperoxide glutathione peroxidase (PHGPx). PHGPx is abundantly synthesized in round spermatids under indirect control of testosterone. In late phase of spermatogenesis, the active soluble peroxidase is transformed into an enzymatically inactive structural protein by an oxidative process that is not yet understood in detail. The dual role of PHGPx is considered to be pivotal for spermatogenesis in mammals and to explain selenium dependency. The relevance of other selenoproteins in the male reproductive system remains to be established. Preliminary data indicate that PHGPx deficiency could also be a cause of human fertility problems.

## Introduction

The potential relevance of selenium to the reproductive system in livestock, laboratory animals and humans has been considered for at least five decades [1]. Impaired reproductive abilities due to selenium deficiency were reported for both sexes. In cows, cystic ovarian disease [2] and retained placenta [3-5] appear to respond to selenium supplementation; infertility of ewes may be associated with selenium deficiency [6,7]; and a selenium-deficient diet resulted in reduced egg production and embryonic survival in hens that could be normalized by selenium supplementation [8]. The biochemical basis of these disturbances in female reproductive ability remains elusive. In contrast, a molecular basis for the impaired spermatogenesis, as was first reported for rats [9-12], mice [13,14], and boars [15], is emerging.

## Function and morphology of sperm in selenium deficiency

Data on the influence of selenium on human reproductive performance are scarce and contradictory [compiled in 1,16] and do not reveal the essential role of selenium in spermatogenesis or sperm function. Thus, the role of selenium in human reproduction must be inferred from studies in laboratory animals.

Interestingly, testes have the ability to accumulate selenium and to retain this trace element even during substantial selenium deficiency [17,18]. Specific alterations of sperm were first seen in rats depleted of selenium for two generations [9,10]. In mice, these alterations increased through successive generations of selenium deprivation [13,14]. Therefore, impairment of male fertility cannot reasonably be expected to result from transient variations of selenium supply.

In severe and prolonged selenium deficiency, male rats and mice become sterile as spermatogenesis is arrested. The seminiferous epithelium is degenerated and the lumen of the testicular tubules has the appearance of being more or less devoid of sperm [13,19]. Clearly, this kind of azoospermia or aspermia mimic a block in cell division.

The functional and morphological alterations of spermatozoa, observed in less severe selenium deprivation, are more discrete. In rats, the prominent feature is reduced sperm motility leading to an impaired fertilization capacity [11]. Sperm motility is less affected by selenium deprivation in mice [13]. In both species, abnormal sperm morphology is observed [11,12,14]. Characteristically, the midpiece of the spermatozoon, that harbors the helix of mitochondria embedded in a keratin-like matrix, appears structurally disturbed and fuzzy or broken. The sperm tails, consequently, appear distorted, and isolated sperm heads and tails are often seen. Interestingly, the midpiece is precisely the part of the spermatozoon where Brown and Burk [17] found most of the selenium accumulated when $^{75}$Se was injected into rats. Within the midpiece, selenium proved to be primarily associated with the keratin-like material surrounding the mitochondrial helix. In this material, selenium was reported to be associated with a cysteine- and proline-rich protein in rodents [20,21] and bulls [22]. Cloning of this "mitochondrial capsule selenoprotein (MCS)", however, revealed that, in rats and mice at least, it was not a selenoprotein [23,24]. It therefore was renamed "sperm mitochondria-associated cysteine-rich protein (SMCP)" [23]. After this investigation, the search for the real selenoprotein(s) in sperm became revitalized, since selenium, irrespective of its chemical identity, evidently contributes to the structural integrity of the sperm midpiece.

## Selenoproteins in the male genital system

Pulse-labeling experiments with $^{75}$Se in selenium-deprived rats show specific selenium incorporation into a variety of testicular and epididymal proteins. They cover a wide range of apparent molecular masses when electrophoresed on SDS-polyacrylamide gels [25,26]. Some of bands on gels could be tentatively identified as their molecular weights correspond to those of cytosolic glutathione peroxidase (cGPx), phospholipid hydroperoxide GPx (PHGPx), thioredoxin reductases and selenoprotein P. A band with an apparent MW of 34kDa, which is only seen in testis [26] and reportedly contained in sperm nuclei [27], does not correspond in molecular mass to any known selenoprotein sequence. A band corresponding to a 15kDa selenoprotein is detected in prostate epithelium [27]. The selenoproteins clearly identified in the genital system by measuring activities, by isolation, by immunochemistry or in situ hybridization, are cGPx, PHGPx and selenoprotein P.

Selenoprotein P is a secreted extracellular protein with multiple selenocysteine residues. It was surprisingly found to be expressed in Leydig cells of mice by means of in situ hybridization [28]. The promoter region of the SelP gene contains putative SRY sites that are presumed to bind the sex-determining region gene product of the Y chromosome [28]. The specific role of SelP in testis remains elusive.

Glutathione peroxidase activities have been repeatedly measured in testis and epididymis. Yet most of the early investigations did not differentiate between the different types of GPx [reviewed in 29] and thus have been practically useless in elucidating a role of selenium in fertility. The data may be compromised by summing up GPx activities of the selenoperoxidases, of the androgen-responsive cysteine homologs of extracelluar GPx [30] and even of GSH-S-transferases [31].

Cytosolic GPx is present in testis as shown by Maiorino et al. [32]. These investigators measured cGPx activity after its separation from PHGPx and by in situ hybridization [32,33]. cGPx has been implicated in antioxidant defense in Leydig cells that are presumed to produce $H_2O_2$ during steroid hormone synthesis [34]. There appears to be a general feeling that the seminiferous epithelium and mature sperm also require a particularly efficient protection against oxidative stress [33,35,36]. cGPx, as the selenoperoxidase most efficient in $H_2O_2$ reduction, would indeed be the enzyme of choice to meet this demand [37]. There is, however, no indication of any predominance of cGPx in particular sites of the genital system. In situ hybridization studies display an uncharacteristic low level of cGPx mRNA in rat testis [32] and cGPx activities are accordingly low [29,32]. Finally, a specific role of cGPx in male reproduction, can be ruled out as cGPx knock-out mice develop and reproduce normally [38].

In contrast, PHGPx is abundantly present in rat testis [36,39,40,41], but only after puberty [39,40]. The peripuberal increase in testicular PHGPx can be prevented by hypophysectomy and restored by application of chorion gonatotropin [39] indicating a hormonal control of PHGPx gene expression. Attempts to verify the hormonal control of PHGPx gene transcription by means of reporter gene constructs, however, have been unsuccessful [32]. In addition, testosterone or forskolin did not activate transcription directly and inhibition by 17-β-estradiol could not be detected in hormone-responsive T47D and MCF7 cells [32]. Testosterone, human chorion gonadotropin and forskolin did not enhance PHGPx activity either when added to decapsulated testes [32]. An explanation for these seemingly contradictory results was provided by in situ hybridization: PHGPx mRNA was seen to be predominantly expressed in a cell layer of the seminiferous epithelium representing round spermatids [32]. Later studies with isolated spermatogenic cells confirmed the preferential expression in round spermatids [42]. The thickness of the spermatid layer reflecting proliferation of the germ epithelium is controlled by testosterone that is provided by gonadotropin-stimulated Leydig cells. When the Leydig cells are selectively destroyed, for example, by ethane dimethane sulfonate, the spermatid layer shrinks with some delay and the PHGPx content of whole testis decreases in parallel to the disappearance of spermatids [32]. The hormonal control of

PHGPx in testis, thus, is an indirect one. The PHGPx gene is not regulated by hormone action, but the cell type abundantly transcribing the gene is dependent on testosterone.

The burst of PHGPx gene transcription in spermatids is reflected by a high content of PHGPx protein detected by immune histochemistry [39] and high PHGPx activity measured with the specific substrate phosphatidyl choline hydroperoxide [32,39,43,44]. PHGPx mRNA declines with elongation of spermatids and is no longer detectable in spermatozoa. Immunostained PHGPx protein declines similarly, but remains faintly visible in spermatozoa. In contrast, PHGPx activity becomes almost undetectable in mature epididymal spermatozoa [43]. The PHGPx protein, however, is still present in spermatozoa, but as an enzymatically inactive, densely packed material constituting a considerable portion of the mitochondrial capsule in the midpiece of spermatozoa [43,44]. The PHGPx protein can be solubilized out of this keratinous material by strong reduction and chaotropic agents and detected by MALDI-TOF mass spectrometry or western blotting [43]. Prolonged preincubation with 0.1 M DTT or mercaptoethanol even leads to a partial recovery of enzymatic activity [43].

## Impact of PHGPx moonlighting on sperm maturation

The puzzling switch of PHGPx from an active peroxidase in spermatogenic cells to an enzymatically inactive protein in spermatozoa raises the question of what this new example of "moonlighting" [45,46] might mean in the context of male fertility.

One of the first attempts to quantify the relative PHGPx protein content in sperm was carried out by scanning Coomassie-Blue stained, 2D electrophoresis gels. About 50% of the protein was found in the mitochondrial capsule [43]. The remaining 50% was contributed by minor amounts derived from different cellular compartments. The capsule material had been prepared by trypsination of spermatozoa followed by density gradient centrifugation, as described by Calvin et al. [21]. Trypsination could have removed part of the capsule proteins, which would lead to an overestimation of the PHGPx content. On the other hand, the capsule material, as prepared, was not clean, which became evident from the appearance of mitochondrial ghosts after solubilization and the detection of flagellar, mitochondrial and cytoplasmatic proteins [43]. In short, therefore, there is no question that PHGPx represents the major constituent of the sperm mitochondrial capsule in rats and that it is the real mitochondrial capsule selenoprotein. The presumed role of MCS or SMCP [20] may now be attributed to PHGPx and PHGPx may be considered the raw building material of the midpiece structure of spermatozoa that guarantees appropriate function.

The chemical process leading to the transformation of the soluble active peroxidase to a structural protein, although unknown in detail, is an oxidative one. The keratinous capsule material resists conventional solubilizers like guanidine or sodium dodecyl sulfate, unless thiols or dithiols are added. Only upon such reductive treatment, monomeric PHGPx can be recovered. Inversely, when total sperm proteins are reductively solubilized and exposed to $H_2O_2$ in the absence of low molecular weight thiols, they readily form high molecular weight aggregates containing PHGPx, which can again be reductively depolymerized. Interestingly, highly purified PHGPx is not equally polymerized by $H_2O_2$ [43]. Taken together, these observations indicate that PHGPx, when oxidized by hydroperoxides in the absence of GSH, reacts with exposed thiols of other proteins and thereby becomes crosslinked probably via Se-S bonds. In enzymological terms, the proposed reaction simply means that the selenolate function of the ground state PHGPx is oxidized by ROOH to a first intermediate, which is a selenenic acid derivative (E-SeOH) [see Chapter 14]. The first intermediate then reacts with protein-SH instead of GSH to form the second intermediate (E-Se-S-Prot). Regeneration of the ground state enzyme (E-Se$^\ominus$), that is commonly carried out by a second GSH molecule, proceeds only slowly with this intermediate which could not occur in the absence of suitable thiols. The inactive, insoluble oxidation products of PHGPx might thus be considered as alternate substrate dead-end intermediates. Such reactions likely occur in late spermatogenesis. The transition from round to elongated spermatids is paralleled by a decrease of GSH and protein thiols [35,47-49]. Evidently, the loss of GSH in this phase of spermatogenesis forces PHGPx into the alternate substrate pathway.

Beyond these basics, little else can be said at present about the transformation process. The mechanism leading to the pivotal disappearance of GSH in late spermatogenesis is obscure. Interestingly, not only GSH, but also GSSG and mixed disulfides derived from GSH become undetectable. This could result from increased GSH metabolism or, more likely, from GSH oxidation followed by release of GSSG [50,51]. Again, the source of the required oxidation equivalents that becomes up-regulated at the specific point of sperm differentiation remains elusive. Also the proteinaceous reaction partners of PHGPx that are evidently required to build-up the mitochondrial capsule remain to be identified. The abundance of PHGPx in the capsule and the failure to identify any other major capsule protein in this material [43] suggest a co-reacting protein that can accommodate many attached PHGPx molecules. SMCP, containing about 30% cysteine residues, could be envisioned as an ideal candidate to meet this requirement. Gene disruption in mice, however, indicates that SMCP is not absolutely required for male fertility as it depends on the genetic background of the mice and whether the fertility of the corresponding SMCP gene knockout mice was affected [IM Adham, personal communication]. Attempts to generate a knockout of the PHGPx gene have so far failed, but recently reported preliminary results [52] appear nevertheless to be revealing. Nine male chimeric mice with genotype +/- were raised having more than 50% PHGPx. They proved to be fertile, but among 190 offspring not a single homozygously or hemizygously mouse that was deficient in PHGPx could be identified. This observation suggests that not even hemizygous cells could contribute to the male germ line. The chimeric mice did not display any obvious altered phenotype apart from a mosaic-like disturbance of the testes. While parts of the testicular tissue looked perfectly normal, others, obviously derived from hemizygous cells, were markedly altered. In the tubules of affected areas, only a few morphologically disturbed spermatozoa were detectable. Some of the alterations, for example, distorted tails, fuzzy and broken midpieces, and isolated heads strongly mimic pathologies seen in severe selenium deficiency. Thus far, the results comply with our presumption that PHGPx is indispensible for the integrity of the midpiece architecture in spermatozoa. Another observation, however, suggests that PHGPx must indeed have a dual function in spermatogenesis. In the affected testicular tissue of chimeric mice, the seminiferous epithelium looked degenerated and severely disorganized. Also, in the early phases of spermatogenesis, PHGPx, which serves as an active peroxidase, may therefore contribute to differentiation processes in a still unknown manner.

## Conclusion and outlook

Of the known selenoproteins, cGPx appears irrelevant to fertility. One or the other thioredoxin reductases, because of the link of thioredoxin to nucleic acid metabolism, may be indispensible to sustain proliferation of the seminiferous epithelium. Thioredoxin reductase deficiency may therefore account for the complete arrest of spermatogenesis observed in rodents deprived of selenium for several generations. PHGPx proved to be pivotal for rodent spermatogenesis. In early spermatogenic cells, it is present as an active peroxidase and may regulate proliferation and/or differentiation. In mature spermatozoa, PHGPx represents a structural component of the mitochondrial matrix. The relevance of other selenoproteins to male fertility remains to be established.

The dual role of PHGPx during sperm maturation, as determined with rodents, appears to apply to other mammals including man, but not to non-mammalian vertebrates or other metazoa [53]. The clinical relevance of sperm PHGPx content to fertility is supported by preliminary studies on a limited number of subjects with fertility problems. The PHGPx activity of the sperm samples, as measured after reductive reactivation of the enzyme, correlated positively with functional parameters indicating the fertilization potential [53]. Large scale and well designed studies will be required to determine how frequently PHGPx deficiency accounts for impaired fertility and whether such deficiencies are due to insufficient selenium supply, defects in the PHGPx gene, or an altered regulation of PHGPx biosynthesis.

**Acknowledgements:** The preparation of this article was supported by the Deutsche Forschungsgemeinschaft (grants Fl61/12-1 and Br778/5-1).

**References**

1.  Subcommittee on Selenium, Committee on Animal Nutrition, Board on Agriculture, National Research Council 1983 *Selenium in Nutrition*, National Academy Press, Washington, DC
2.  JA Harrison, DD Hancock, HR Conrad 1984 *J Dairy Sci* 67:132
3.  N Trinder, CD Woodhouse, CP Renton 1969 *Vet Rec* 85:550
4.  WE Julien, HR Conrad, JE Jones, AL Moxon 1976 *J Dairy Sci* 59:1954
5.  EC Segerson, GJ Riviere, HL Dalton, MD Whitacre 1981 *J Dairy Sci* 64:1833
6.  WJ Hartley, AB Grant 1961 *Federation Proc* 20:679
7.  ED Andrews, WJ Hartley, AB Grant 1968 *N Z Vet J* 16:3
8.  AH Cantor, ML Scott 1974 *Poultry Sci* 53:1870
9.  KEM Mc Coy, PH Weswig 1969 *J Nutr* 98:383
10. ASH Wu, JE Oldfield, OH Muth, PD Whanger, PH Weswig 1969 *Proc West Soc, Am Soc An Sci* 20:85
11. ASH Wu, JE Oldfield, PD Whanger, PH Weswig 1973 *Biol Reprod* 8:625
12. ASH Wu, JE Oldfield, LR Shull, PR Cheeke 1979 *Biol Reprod* 20:793
13. E Wallace, HI Calvin, GW Cooper 1983 *Gamete Res* 4:377
14. E Wallace, GW Cooper, HI Calvin 1983 *Gamete Res* 4:389
15. CH Liu, YM Chen, JZ Zhang, MY Huang, Q Su, ZH Lu, RX Yin, GZ Shao, D Feng, PL Zheng 1982 *Acta Vet Zootech Sinica* 13:73
16. GN Schrauzer 1998 Selen *Neue Entwicklungen aus Biologie, Biochemie und Medizin* Johann Ambrosius Barth Verlag, Hüthig GmbH, Heidelberg, Leipzig p 84
17. DG Brown, RF Burk 1973 *J Nutr* 103:102
18. D Behne, T Hofer-Bosse 1984 *J Nutr* 114:1289
19. D Behne, H Weiler, A Kyriakopoulos 1996 *J Reprod Fert* 106:291
20. HI Calvin, GW Cooper 1979 *The Spermatozoon*, 135
21. HI Calvin, GW Cooper, E Wallace 1981 *Gam Res* 4:139
22. V Pallini, E Bacci 1979 *J Submicr Cytol* 11:165
23. L Cataldo, K Baig, R Oko, MA Mastrangelo, KC Kleene 1996 *Mol Reprod Dev* 45:320
24. IM Adham, D Tessmann, KA Soliman, D Murphy, H Kremling, C Szpirer, W Engel 1996 *DNA Cell Biol* 15:159
25. HI Calvin, K Grosshans, SR Musicant-Shikora, SI Turner 1987 *J Reprod Fert* 81:1
26. D Behne, H Hilmert, S Scheid, H Gessner, W Elger 1988 *Biochim Biophys Acta* 966:12
27. D Behne, A Kyriakopoulos, M Kalcklösch, C Weiss-Nowak, H Pfeifer, H Gessler, C Hammel 1997 *Biomed Environm Sci* 10:340
28. P Steinert, D Bächner, L Flohé 1998 *Biol Chem* 379:683

29. L. Flohé 1989 *In: (Dophin, D., Poulson, R., and Avramocic, O. (Eds): Glutathione: Chemical, biochemical, and medical aspects - Part A*. John Wiley & Sons, Inc., New York p 643
30. C Jinemez, NB Ghyselinck, A Depeiges, JP Dufaure 1990 *Biol Cell* 68:171
31. RF Burk, RA Lawrence 1978 *Functions of glutathione in liver and kidney* H Sies and A Wendel (Eds) Springer Verlag, Berlin p 114
32. M Maiorino, JB Wissing, R Brigelius-Flohé, F Calabrese, A Roveri, P Steinert, F Ursini, L Flohé 1998 *FASEB J* 12:1359
33. A Zini, PN Schlegel 1997 *Int J Androl* 20:86
34. V Peltola, I Huhtaniemi, T Metsa-Ketela, M Ahotupa 1996 *Endocrinology* 137:105
35. F Bauché, MH Fouchard, B Jégou 1994 *FEBS Letters* 349:392
36. F Tramer, F Rocco, F Micali, G Sandri, E Panfili 1998 *Biol Reprod* 59:753
37. F Ursini, M Maiorino, R Brigelius-Flohé, K-D Aumann, A Roveri, D Schomburg, L Flohé 1995 *Meth Enzymol* 252:38
38. Y-S Ho, JL Magnenat, RT Bronson, J Cao, M Gargano, M Sugawara, CD Funk 1997 *J Biol Chem* 272:16644
39. A Roveri, A Casasco, M Maiorino, P Dalan, A Calligaro, F Ursini 1992 *J Biol Chem* 267:6142
40. A Giannattasio, M Girotti, K Williams, L Hall, AJ Bellastella 1997 *Endocrinol Invest* 20:439
41. F Weitzel, F Ursini, A Wendel 1990 *Biochim Biophys Acta* 1036:88
42. K Mizuno, S Hirata, K Hoshi, A Shinohara, M Shiba 2000 *Biol Trace Elem Res* 74:1112
43. F Ursini, S Heim, M Kiess, M Maiorino, A Roveri, JB Wissing, L Flohé 1999 *Science* 285:1393
44. M Maiorino, L Flohé, A Roveri, P Steinert, JB Wissing, F Ursini 1999 *BioFactors* 10:251
45. J Piatigorski 1998 *Prog Ret Eye Res* 17:145
46. CJ Jefferey 1999 *Trends Biol Sci* 24:8
47. R Shalgi, J Seligman, NS Kosower 1989 *Biol Reprod* 40:1037
48. J Seligman, NS Kosower, R Shalgi 1992 *Biol Reprod* 46:31
49. HM Fisher, RJ Aitken 1997 *J Exp Zool* 277:390
50. H Sies, PM Akerboom 1984 *Meth Enzymol* 105:445
51. SC Lu, WM Sun, J Yi, M Ookhtens, G Sze, N Kaplowitz 1996 *J Clin Invest* 97:1488
52. M Conrad, U Heinzelmann, W Wurst, GW Bornkamm, M Brielmeier 2000 Abstract submitted to the $7^{th}$ *International Symposium on Selenium in Biology and Medicine*, October 1-5, 2000, Venice, Italy
53. M Maiorino, A Roveri, L Flohé, F Ursini 2000 Abstract submitted to the $7^{th}$ *International Symposium on Selenium in Biology and Medicine*, October 1-5, 2000, Venice, Italy

# Chapter 23. Role of low molecular weight, selenium-containing compounds in human health

Henry J. Thompson

*Center for Nutrition in the Prevention of Disease, AMC Cancer Research Center, Denver, CO 80214, USA*

**Summary:** Most ingested forms of selenium ultimately are metabolized to low molecular weight inorganic and organic compounds that play a central role in human health either via incorporation into selenoproteins or binding to selenium binding proteins. Less attention has been paid to other effects of low molecular weight selenium compounds on human health. This chapter explores the role of intermediates derived from the intracellular biotransformation of selenium on cellular and molecular mechanisms that could potentially affect the development of cancer. A hypothesis is presented that may account, in part, for the cancer chemopreventive activity of selenium, and criteria that candidate mechanisms should fulfill are delineated. It is anticipated that the cellular and molecular effects of low molecular weight selenium compounds discussed in relationship to cancer will also be applicable to other disease processes.

## Introduction

Much of our current understanding about biological activities of selenium has come from studies in a diverse array of species in both the plant and animal kingdom. However, knowledge of the role of low molecular weight selenium containing compounds in human health is limited. Consequently, the approach used in writing will be translational, many times drawing on current understanding of selenium's effects on health in various mammalian species and then making inferences relative to possible effects of selenium in human health. In order to proceed logically with an exploration of the role of low molecular weight, selenium containing compounds in human health, operational definitions of human health and low molecular weight selenium compounds will be given as a framework for the chapter. A goal in writing this chapter is to underscore the role in human health of the products of the intracellular biotransformation of selenium, and to emphasize the importance of an understanding of selenium metabolism in efforts to elucidate the mechanisms by which selenium exerts these effects.

## Human health

For the purposes of this chapter, the term human health is considered to have two components, health promotion and disease prevention. Discussion of the role of selenium in health promotion will be limited to a consideration of the provision of selenium in amounts and chemical forms sufficient to meet cellular needs for synthesis of the twelve mammalian selenoproteins that have been identified [1]. While provision of selenium in sufficient amounts for selenoprotein synthesis is tantamount to disease prevention, the definition of disease prevention adopted for the purposes of this chapter will be the use of low molecular weight selenium compounds to influence a specific disease process directly, and apart from the potential of the administered selenium to provide hydrogen selenide for selenoprotein synthesis. The consideration of disease prevention will also include issues related to the occurrence of selenium toxicity.

## Low molecular weight selenium-containing compounds

It is well known that the chemical form and dose of selenium to which an organism is exposed are key determinants of its biological activities [2]. It is also recognized that the metabolism of selenium produces a wide array of low molecular weight products, many of which have distinct biological activities [3]. Given that the majority of selenium containing compounds, both inorganic and organic, to which a mammalian organism is exposed are likely to undergo biotransformations that ultimately result in the selenium from the compound entering a common pathway of metabolism [2,3], the chemical forms of selenium produced during its intermediary metabolism will provide the focus for the chapter.

## Selenium metabolism

Within a cell, selenium undergoes biotransformation and experiences one of two metabolic fates [2]. Selenium compounds either enter a common pathway, albeit at different points of entry, of reductive metabolism with the selenium ultimately entering the hydrogen selenide pool, or the selenium enters the methylation pathway and is eliminated from the body primarily as trimethylselenonium in the urine and/or is exhaled as dimethyselenide [3]. The metabolic fate of ingested selenium is shown in Figure 1. It is important to note that forms of selenium that enter the hydrogen selenide pool are either used to form selenophosphate which is a precursor for selenocysteine synthesis and subsequent incorporation into selenoproteins [4], or the hydrogen selenide not used for this purpose enters the methylation pathway and is eventually eliminated from the body [5].

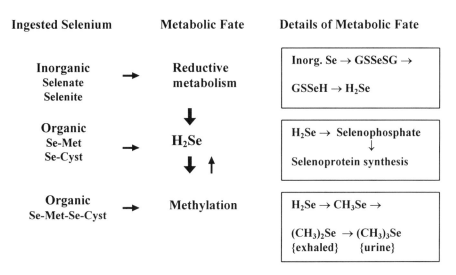

**Figure 1.** Metabolic fate of ingested selenium. Most organic and inorganic forms of ingested selenium undergo biotransformation and enter the hydrogen selenide pool directly or following reductive metabolism that involves the formation of intermediates with glutathione. Hydrogen selenide that is formed either undergoes activation to selenophosphate with subsequent incorporation into selenoproteins, or it is further metabolized via methylated intermediates. Selenium metabolized via methylation is eliminated from the body in urine or is exhaled.

## Selenium Metabolism and Health

### Health promotion

Mammalian species studied to date synthesize selenoproteins from hydrogen selenide, a key intermediate in selenium metabolism [5]. Hydrogen selenide is formed by the reduction of inorganic selenium salts or via the release of selenium from organoselenium compounds by the scission of carbon-selenium bonds. For example, the metabolism of selenomethionine involves conversion to selenocysteine via the trans-sulfuration pathway, followed by selenocysteine metabolism via its lyase to release inorganic selenium into the hydrogen selenide pool. Hydrogen selenide provides selenium for selenoprotein synthesis after activation to selenophosphate [6]. The process by which hydrogen selenide is made available for selenoprotein synthesis is clearly the best documented and most studied example of the central role that low molecular weight selenium compounds play in human health. This topic has been the subject of a recent review [7]. As discussed in the following sections, metabolic intermediates leading up to the formation of hydrogen selenide as well as the selenium intermediates formed during the further metabolism of hydrogen selenide via methylation are likely to be involved in disease prevention as well as selenium toxicity.

**Disease prevention**
While it is becoming clear that the approximately 20 identified mammalian selenoproteins [see Chapter 9] play critical roles in many vital cellular functions and are therefore essential for disease prevention [7], the only disease for which there has been an extensive amount of research in the use of low molecular weight selenium compounds for its prevention is cancer [5,8]. Therefore, cancer prevention by selenium will be the focus of the subsequent discussion with the expectation that low molecular weight selenium compounds will likely be found to impact other disease processes. Since an understanding of the cellular and molecular effects of selenium will provide a basis for translating what is determined to be underlying the cancer inhibitory activity of selenium to its effects on other disease states, mechanistic considerations will dominate the discussion of this topic.

**Cancer chemoprevention**
A combination of epidemiological, ecological, case-control and most recently prospective clinical trials support the hypothesis that an inverse relationship exists between selenium exposure and the incidence of and/or mortality from cancer in a number organ sites. This topic has been extensively reviewed [9-12]. While most data from studies with animals provides little evidence of an effect of chemopreventive levels of selenium on selenoprotein activity, it has been argued that in humans, long term exposure to supplemental levels of selenium does elevate selenoprotein activities and prevent cancer. This topic has been discussed in a number of reviews [5,7,8,13-15]. Clearly, a clarification of the role of selenoproteins is an important topic for investigation in prospective clinical trials of chemopreventive levels of selenium.

Nonetheless, this situation prompts the question, what else could account for the cancer inhibitory effects of selenium if selenoproteins are not involved? While sometimes overlooked, a significant body of literature indicates that inhibitory activity is directly attributable to specific low molecular weight selenium compounds as reviewed in [8]. Research in this area can be divided into two categories: 1) identification of the selenium species accounting for cancer inhibition, and 2) elucidation of the affected cellular and molecular pathways.

**Identification of the selenium species accounting for cancer inhibition**
It has been recognized for sometime that whenever cancer chemoprevention has been observed in a particular tumor model system, that the levels of selenium required to achieve the effect are considerably above those levels of

selenium at which maximal selenoprotein activity is observed [5,13]. Since most early studies were conducted using readily available inorganic forms of selenium, and given the extensive literature indicating that inorganic forms of selenium require biotransformation for their biological activity, efforts were directed at determining whether the cancer inhibitory species was a metabolite formed during reductive metabolism of selenium, or whether it was a product of the methylation pathway. In an elegant series of experiments from the laboratories of Ip and of Ganther, it was shown that cancer inhibitory activity is derived from a product of the methylation pathway. The reader is referred to two reviews for a detailed discussion of this topic [5,8]. The current hypothesis is that a monomethylated selenium species, Me-Se-X, formed as a metabolic product of the intracellular metabolism of hydrogen selenide, a toxic compound if intracellular levels accumulate, is responsible for selenium's cancer inhibitory activity. This concept is supported by studies that show that naturally occurring or synthetically produced selenium compounds that enter the monomethyl selenium pool directly also inhibit the development of cancer. The reactions mechanisms by which Me-Se-X, the presumed active species, is likely to work have been summarized [5]. It is noteworthy that di- and tri-methylated forms of selenium have limited to no cancer inhibitory activity.

One other candidate molecule formed during the reductive metabolism of selenium merits comment. Selenodiglutathione is an intermediate that has been shown to influence cell growth in vitro and the development of transplantable tumors in vivo [16]. A number of factors argue that it is unlikely to play a primary role in selenium cancer chemoprevention. They include: rapid metabolism to other intermediates, likelihood of a mode of action via a reactive oxygen species in which cellular damage is induced, and selenium chemoprevention can be achieved by selenium compounds that do not form selenodiglutathione as an intermediate during metabolism. As will be discussed in subsequent sections, the potential toxicity of intermediates formed during the reductive metabolism of selenium further detracts from their consideration as candidate molecules for cancer chemoprevention. Rather, the argument can be made that an appropriate strategy for health promotion and disease prevention is to minimize the formation of reactive intermediates of selenium during its metabolism while insuring an adequate supply of the element for selenoprotein synthesis and the generation of monomethylated selenium species.

**Other low molecular weight cancer chemopreventive compounds**
El-Bayoumy and coworkers have developed a series of selenium compounds, p-methoxybenezeneselenol, benzylselenocyanate, and 1,4-phenylene-bis (methylene) selenocyanate (p-XSC). The reader is referred to a recent review for a detailed discussion of these agents [17]. Of the compounds

evaluated, p-XSC has the highest chemopreventive index, i.e. it has the greatest cancer inhibitory activity in relationship to the amount of the compound that suppresses growth rate [8]. While it is clear that the selenium from the compounds in this series can be released into the hydrogen selenide pool during their metabolism and elimination from the body, it has been argued that p-XSC itself has chemopreventive activity apart from the activity that may result when selenium is released [18]. p-XSC has been shown to inhibit both the initiation and the promotion/progression stages of the carcinogenic process. While effects on initiation appear to be mediated by alterations in carcinogen metabolism, the mechanism(s) that account(s) for its effect on promotion/progression has yet to be identified. Recently, this team of investigators has developed glutathione conjugates of the parent compounds mentioned above in an effort to improve chemopreventive efficacy, and the conjugates appear to have significant cancer inhibitory activity [19]. This class of selenium compounds holds significant potential for evaluation in clinical trials.

Another type of low molecular weight selenium compound that has been studied for anticancer activity has the selenium bonded directly to a benzene ring. The selenium bonding to the aromatic ring is very stable and less of the element is released from such compounds during metabolism. Of the three compounds tested, methylphenylselenide, diphenylselenide, and triphenylselenonium (TPSe), TPSe had the greatest cancer chemopreventive activity [20,21]. This compound is not capable of regenerating glutathione peroxidase activity in selenium deficient animals. It appears to inhibit cell proliferation both in vitro and in vivo via a cytostatic mechanism [22]. Available evidence indicates that the primary effect of TPSe is on the stage of carcinogenic promotion and progression. The chemopreventive index of this compound is high, meaning that the range between the amount that is cancer inhibitory and that which is poorly tolerated is very large. Thus, this compound has considerable promise for clinical development.

**Elucidation of the cellular and molecular pathways affected by low molecular weight cancer chemopreventive selenium compounds**
When the body of literature reporting the chemopreventive activity of selenium is considered in aggregate, the majority of evidence indicates that the predominant effect of compounds that release their selenium into the common pathway of metabolism described above, is on the process of carcinogenic promotion/progression, although effects on the process of initiation also have been reported. There are a number of conditions that should be considered in the search for the mechanism(s) by which selenium inhibits the promotion/progression stage of the carcinogenic process. In general, these criteria have not been considered in discussions of selenium's cancer inhibitory activity, yet available animal data indicate that these

conditions should be satisfied by any candidate mechanism. These conditions include 1) the mechanism does not involve the induction of cytotoxicity, since selenium cancer chemoprevention has been reported both in animal models and in humans in the absence of toxic side effects; 2) the mechanism explains the specific effect of selenium against transformed cells since normal cells and tissues in general appear to be unaffected by a chemopreventive regime of selenium; and 3) the mechanism accounts for the selectivity of selenium against pre-invasive stages of the disease process since in vivo, selenium has little effect against established cancers in the absence of cytotoxicity. If a candidate mechanism meets these conditions, it merits serious consideration.

## Candidate cellular mechanisms

In a practical sense, cancer is the result of a failure in tissue size homeostasis as discussed in [23]. Cancer can occur in a tissue only if the rate of cell proliferation exceeds the rate of cell death that is characteristic of a tissue when it is in size equilibrium. Thus, selenium could inhibit carcinogenesis either by 1) decreasing the rate of cell proliferation, 2) increasing the rate of cell death by apoptosis, and/or 3) by concomitant effects on both processes. The results of in vitro experiments document that many forms of selenium can inhibit cell proliferation as well as induce apoptosis [see 8 for review]. Of particular interest is that monomethylated selenium compounds have been shown to inhibit cell proliferation and induce apoptosis in the absence of cytotoxicity, while forms of selenium which enter the reductive pathway of selenium metabolism also have comparable effects, but they are accompanied by a loss of cell viability and induction of DNA damage [24]. This is an important distinction since DNA damage, irrespective of the agent that induces it, can result in cell cycle arrest and apoptosis. Thus, it appears that forms of selenium that undergo reductive metabolism have the potential to affect cell proliferation and cell death secondary to cytotoxicity; whereas, forms of selenium that do not undergo reductive metabolism and enter the methylation pathway exert similar effects, but in the absence of DNA damage.

Few studies of the effect of selenium on tissue size homeostasis have been conducted in vivo. One reason for this has been the lack of animal models in which to study the effects of selenium on the occurrence of premalignant lesions, which as noted above and initially reported in [25], appears to be the stage of the carcinogenic process against which selenium exerts its dominant effect(s). Using one of the animal models for breast cancer in which an extensive amount of selenium chemoprevention work has been conducted, it was recently shown that a monomethyl precursor form of selenium does not alter rates of cell proliferation in either histologically normal appearing mammary epithelium or in premalignant mammary gland lesions, the development of which selenium appears to inhibit [26]. If these data are considered in conjunction with the in vitro data discussed above, they point to the need to investigate in vivo, the potential role of apoptosis in accounting for the chemopreventive effects of selenium. Nonetheless, it must be recognized that this will not be a trivial task. Apoptosis is a very rapid and effective mechanism of tissue size regulation. Even small changes, e.g. 1%, in rates of apoptosis could have dramatic effects on the process of clonal expansion of neoplastic foci of cells. Currently, there are many issues that need to be resolved including 1) the timeframe over which to study induction of apoptosis since the entire process lasts approximately three hours, and 2) whether current assays are sensitive enough to detect the small changes in rates of apoptosis that are biologically plausible given the power of this process to rapidly reduce tissue size.

**Candidate molecular mechanisms**
Molecular aspects of the potential regulation of the clonal expansion of neoplastic foci by selenium have been studied in vitro. In general, molecular studies have focused on one of two areas: the regulation of cell cycle progression or the induction of apoptosis.

**Regulation of cell cycle progression**
Much of the evidence regarding cell cycle regulation by low molecular weight selenium containing compounds has been obtained under experimental conditions (form and dose of selenium) that have also been shown to induce alterations in DNA integrity. This is an important point since studies of Medina and coworkers indicate that mechanisms of cell cycle arrest by genotoxic, inorganic forms of selenium are likely to be different than those exerted by monomethylated forms of selenium [27]. Based on those observations, the work that is likely to be most relevant to chemoprevention suggests that monomethylated selenium compounds retard the progression of cells through the G1 phase of the cell cycle [28]. Specifically, published data show that selenium reduces the kinase activity of either the cyclin D-cdk4 complex or the cyclin E-cdk2 complex. The kinase activity of both of these complexes is necessary for the initiation of and for sustaining the phosphorylation of the retinoblastoma protein, an event that is essential for cells to pass the $G_1$-S transition [29]. Data from the laboratory of Thompson and coworkers indicates that selenium compounds reduce levels of phosphorylated retinoblastoma protein [35,36]. These observations provide a focus from which to design additional experiments to identify the upstream effector events that selenium regulates and that are likely to account for the decreased phosphorylation of retinoblastoma. One candidate target of selenium's effect are protein kinases such as protein kinase C. Based on hypothetical mechanisms by which monomethyl species of

selenium may act [5] and reports that selenium compounds inhibit the activity of protein kinase C [28,30,31], further investigations of this protein and the signal transduction pathway of which it is a component appear warranted.

### Regulation of apoptosis
Given both direct in vitro evidence [32-34] and the implications of the in vivo experiments cited above [22], a potentially rich and virtually unexplored area of mechanistic inquiry is the induction of apoptosis by selenium. Recent evidence shows that a nongenotoxic form of selenium can induce p53 expression in cells that undergo both cell cycle arrest and the induction of apoptosis, raising the possibility that the events are interrelated [35,36]. What is clearly needed at this time is a careful exploration of the mechanisms by which non-genotoxic forms of selenium induce apoptosis. Due to an explosion of work, the major pathways accounting for the induction of apoptotic cell death have been defined and the assays and reagents are currently available for a detailed investigation of apoptosis induction by selenium [37]. Currently, only one study of the effects of selenium on caspase induction has been reported and this was done using a genotoxic form of selenium and focused on an effector caspase, caspase 3, rather than on delineating which initiator caspase pathway is involved [38]. Significant insights about the signal transduction pathways induced by selenium are likely to result from efforts to identify whether initial effects of selenium are mediated via caspase 2, 8, or 9. Such studies will indicate whether the effects of selenium are likely to be mediated via cell survival mechanisms, predicted if procaspase 9 is initially activated, or via death receptor mediated pathways, predicted if procaspases 2 or 8 are initially activated. This effort would also permit assessment of the merit of studies of selenium's effects on mitochondrial function and activities of the Bcl-2 family of proteins.

## Mechanisms: Other considerations

### Inhibition of neoangiogensis by selenium
New blood vessel formation is essential to the process of the clonal expansion of transformed populations of cells [39]. This fact led Lu and coworkers to determine if the specificity of selenium in inhibiting the development of early premalignant lesions might be do to the inhibition of this process. As reported in [40], treatment with a monomethylated form of selenium inhibited the process of neoangiogenesis. Preliminary evidence indicates that the inhibition of neoangiogenesis by selenium might be do to an effect of selenium on endothelial cells that are involved in new blood vessel formation. This area of inquiry is virtually unexplored and merits consideration.

**Effects of selenium on insulin like growth factor metabolism**
One outcome common to many chemopreventive agents, including low molecular weight selenium compounds, is that at cancer inhibitory doses, these compounds slow animal growth rate. In general, rates of growth are suppressed by 5 to 10%, and the reduction in growth rate is not considered to contribute significantly to cancer inhibitory activity. Rather than disregarding the effects of selenium on weight gain, the question can be asked, does this effect give insights about selenium's mechanism of action? A number of reports indicate that low molecular weight selenium compounds can affect both circulating levels of growth hormone and insulin-like growth factor-1 (IGF-1) [41-43]. Interestingly, some evidence indicates that selenium may have an effect on IGF-1 metabolism that is distinct from its effect on growth hormone [42]. This observation is of interest since there is a rapidly accumulating literature indicating that factors that regulate IGF-1 metabolism are associated with a reduction in risk for cancer in multiple organ sites [44-47]. Further studies of how selenium may regulate the IGF pathway appear warranted, especially when it is recognized that changes in this pathway are involved in the regulation of cell cycle progression as well as the induction of apoptosis.

**Selectivity and specificity: a new hypothesis**
How can nongenotoxic levels of selenium selectively and specifically inhibit the development of premalignant clones of cells without affecting nontransformed populations of cells? A hypothesis about selenium chemoprevention that could lead to specific tests for a novel mechanism or mechanisms explaining such effects is presented.

In general, neoplastic transformation is the result of mutation [48]. These mutations can cause subtle changes in the structure and function of the proteins they encode. When such initiating events occur in a protooncogene(s) of a cell, cell transformation, promotion, and progression can occur if sufficient levels of the abnormal protein(s) accumulate(s) in the cell. Such events do not occur concurrently in non-mutated cells in the same tissue. The hypothesis being advanced is that chemopreventive levels of selenium induce higher cellular levels of a monomethylated selenium species that interacts with and counteracts the effects of oncoproteins that drive early stages of malignant transformation while having no effect on the oncoprotein's wild-type counterpart. Thus, the hypothesis is that selenium can alter the phenotype of a transformed cell without affecting its genotype. There are several conceivable ways this effect could be achieved. One possibility is that a point mutation results in substitution of amino acid residue "Y" by cysteine in the oncoprotein, causing some chemical functionality to be gained or lost. The binding of selenium to the cysteine via covalent adduct formation, i.e., conversion of the cysteine (Cys-SH) to Cys-

S-Se-CH$_3$ by mechanisms outlined in reference [5] would abolish this effect by disrupting the activity of the oncogenic protein. This concept is illustrated in Figure 2. Consequently, only mutations involving specific amino acid substitutions would be affected by selenium. That such a mechanism is plausible is supported by reports that mutations in known oncogenes do result in the substitution of glycine or arginine by cysteine [49,50]. Another way that selenium could alter the activity of an oncoprotein is that introduction of a new cysteine could allow a chemical functionality to occur that was precluded in the native protein, such as formation of a disulfide bond between two cysteines, either stabilizing an active conformation or, alternatively, destabilizing the native conformation of the protein. As postulated above, adduct formation by a monomethylated form of selenium with cysteine could abolish the change in chemical functionality of either the oncoprotein or the tumor suppressor protein.

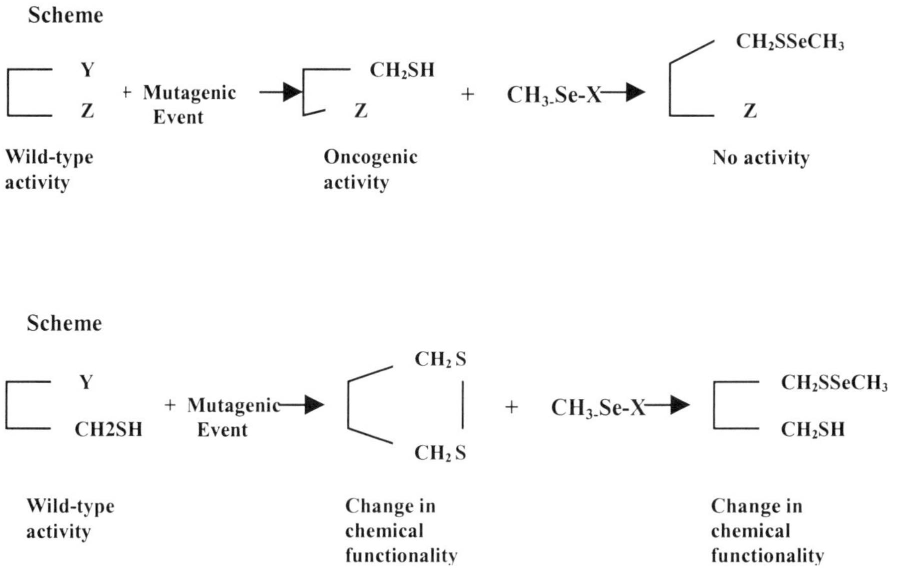

**Figure 2.** Hypothetical scheme of the mechanism of cancer chemoprevention by selenium. In scheme A, a mutagenic event results in the substitution of a cysteine for residue Y in a wild-type proto-oncogene resulting in an activated oncoprotein. CH$_3$-Se-X forms an adduct with the cysteine of the oncoprotein causing it to lose its oncogenic activity. In scheme B, a mutagenic events causes a substitution of a cysteine for residue Y in a wild-type proto-oncogene and formation of a disulfide bond with another cysteine in the molecule. This changes the chemical functionality of the protein giving it oncogenic activity, or alternatively, resulting in loss of function. CH3-Se-X forms an adduct with one of the cysteines which inhibits disulfide bridge formation and the change in chemical functionality caused by the mutagenic event.

This postulated activity of selenium in selectively inhibiting the activity of specific oncoproteins would suppress the development of initiated clones of cells while having no effect on non-transformed cells. Such interactions would occur stoichiometrically. If levels of active oncoprotein exceeded a threshold, in excess of the selenium available for adduct formation, transformation would progress and selenium would no longer exert a detectable influence on the neoplastic process. While only a hypothesis, this idea does fit several observations from animal studies, namely, 1) chemopreventive forms and levels of selenium do not appear to have an overall effect on an animal, 2) selenium appears to preferentially affect early stages in the process of malignant transformation, and 3) selenium does not appear to have an effect on established tumors apart from acute effects caused by nonspecific cytotoxicity.

**Toxicity**
Given the history of selenium as a naturally occurring toxicant, it is important to consider the potential of low molecular weight selenium compounds to induce toxicity [51]. Most of this discussion requires that elements of this chapter that focused on health promotion and disease prevention be revisited, but with the intent of identifying the origins of toxicity. While the literature on selenium toxicity is extensive, an attempt is made in the following paragraphs to simplify the story and to view toxicity from the same metabolic perspective from which the mechanisms of disease prevention were considered.

Ingested selenium can result in toxic side effects when excessive amounts are consumed from either inorganic or organic sources [52-54]. Estimates of safe versus potentially detrimental intakes of selenium have recently been published [55]. While many factors can influence the onset of toxicity, most evidence indicates that at least two metabolic fates of selenium can be linked to the potential for toxicity. Intermediates in the reductive metabolism of selenium including selenodiglutathione and hydrogen selenide can, if they occur at elevated levels within the cell, result in the formation of reactive molecular species that can damage cellular macromolecules including DNA [56]. Consistent with this are numerous reports that excessive exposure of cells to inorganic forms of selenium induces DNA strand breaks as well as an increase in the rate of other clastogenic events [57]. Low molecular weight organic forms of selenium such as selenocysteine and selenomethionine are also likely to induce a similar spectrum of toxic effects since they undergo degradation and ultimately lyase-mediated scission of carbon-selenium bonds to release hydrogen selenide [52]. However, the propensity of selenoamino acids, in particular selenomethionine, to induce toxicity is reduced by their nonspecific incorporation into cellular protein and slower release into the inorganic pool as a result of protein turnover.

Another potential source of cellular stress and toxicity is excess levels of monomethylated selenium compounds that also are hypothesized to account for the cancer chemopreventive activity of selenium. As outlined in [5], these compounds have the potential to react with protein sulfhydryls and alter protein activity, an effect that could adversely influence cell function. That monomethylated species are relatively well tolerated is due to their facile methylation to dimethylselenide and trimethylselenonium which have decreased biological activity and are rapidly eliminated from the body [3]. Hence, the adage that dose makes the poison clearly applies to selenium, but also requires that consideration be given to the form of selenium ingested. The form of selenium ingested, that affects both the point of entry of selenium into the pathway of metabolism and the rate at which a particular ingested dose of selenium will be released into the cascade of events, eventually leads to the excretion of the element.

**Translational considerations**
In 1996, Clark et al. reported a significant reduction in the incidence of mortality due to cancer of the lung, prostate, and colon that was achieved by ingestion of a supplement containing 200 µg selenium (as high selenium brewer's yeast) in addition to the amount of selenium consumed via the diet and drinking water [11]. While this observation has prompted a great deal of activity and discussion in the research community, little effort has been directed at a comparative analysis of these clinical findings with what is known from animal studies.

Adequate selenium nutrition for the synthesis of selenoproteins can clearly be achieved via the biotransformation of any number of low molecular weight selenium compounds ingested at nutritionally recommended levels. The fact that Clark and coworkers found no change in plasma glutathione peroxidase activity despite an increase in serum selenium levels in excess of 50%, raises the question of whether an effect of the high selenium yeast supplement on selenoprotein activity is likely to explain the cancer chemopreventive activity observed. This finding is consistent with the results of animal experiments discussed above. A second consideration relates to the effects of the high selenium yeast supplement on selenium metabolism. Available data indicate that humans metabolize selenium in a manner analogous to other mammalian species. Thus, while it has not been documented, it can be predicted that supplementation would ultimately increase the flow of selenium into the methylation pathway, generating the putative monomethylated active form of selenium. This would be consistent with the selenium metabolic profile observed in animals protected against cancer by selenium [58]. Third, it is consistently observed in human and animal work that chronic ingestion of chemopreventive levels of selenium has no detectable adverse side effects. However, there is one observation from the clinical trial that does not appear consistent with expectations from animal studies. The effects of selenium on incidence rates were detected at an early point following the initiation of supplementation. This implies that the effect of selenium was not limited to early stages of the carcinogenic process in lung, prostate, and colon as would be expected from work done in animal models of cancer in other organ sites. Inadequate information exists to speculate on possible reasons for this difference. However, it is hoped that recently initiated clinical trials that are being conducted to confirm the inhibitory effect of selenium supplementation against cancer will shed further light on the cellular and molecular mechanisms involved.

## Concluding comments

An attempt has been made to use the intermediary metabolism of selenium as a basis for understanding the potential effects of low molecular weight selenium containing compounds on human health. It is hoped that this presentation will promote further discussion of candidate mechanisms that account for many of the biological effects of selenium apart from its role in selenoproteins and other proteins that bind the element. This manuscript's focus was largely limited to cancer chemoprevention. It is hoped that investigators working with models of other human diseases will consider the implications of the information presented to those cellular processes, and that new lines of inquiry will emerge.

**Acknowledgements:** I would like to thank Howard E. Ganther for reviewing this manuscript and for his helpful comments and suggestions. This work was supported by PHS grant CA 84059 from the National Cancer Institute.

**References**

1. CB Allan, GM Lacourciere, TC Stadtman 1999 *Annu Rev Nutr* 19:1
2. HE Ganther 1986 *J Am Coll Toxicol* 5:1
3. HE Ganther JR Lawrence 1997 *Tetrahedron* 53:12299
4. TC Stadtman 1991 *J Biol Chem* 266:16257
5. HE Ganther 1999 *Carcinogenesis* 20:165
6. Z Veres, IY Kim, TD Scholz, TC Stadtman 1994 *J Biol Chem* 69:10597
7. MP Rayman 2000 *Lancet* 356:233
8. C Ip 1998 *J Nutr* 128:1845
9. M Vinceti, S Rovesti, M Bergomi, G Vivoli 2000 *Tumori* 86:105
10. GF Combs, Jr. ,WP Gray 1998 *Pharmacol Ther* 79:179
11. LC Clark, GF Combs, Jr., BW Turnbull, EH Slate, DK Chalker, J Chow, LS Davis, RA Glover, GF Graham, EG Gross, A Krongrad, JL Lesher, Jr., HK Park, BB Sanders, Jr., CL Smith , JR Taylor 1996 *JAMA* 276:1957
12. GF Combs, Jr., LC Clark 1999 *Selenium and cancer. Nutritional Oncology* D. Heber (Ed) Academic Press New York pp 215
13. D Medina 1986 *Adv Exp Med Biol* 206:465
14. D Medina, H Lane, F Shepherd 1981 *Anticancer Res* 1:377
15. C Ip, HE Ganther 1992 *Cancer Chemoprevention* L Wattenberg, M Lipkin, CW Boone, GJ Kelloff, (Eds) CRC Press Boca Raton p 479

16. J Lanfear, J Fleming, L Wu, G Webster, PR Harrison 1994 *Carcinogenesis* 15:1387
17. BS Reddy, A Rivenson, K El Bayoumy, P Upadhyaya, B Pittman, CV Rao 1997 *J Natl Cancer Inst* 89:506
18. C Ip, K El Bayoumy, P Upadhyaya, H Ganther, S Vadhanavikit, H Thompson 1994 *Carcinogenesis* 15:187
19. T Kawamori, K El Bayoumy, BY Ji, JG Rodriguez, CV Rao, BS Reddy 1998 *Int J Oncol* 13:29
20. C Ip, H Thompson, H Ganther 1994 *Carcinogenesis* 15:2879
21. C Ip, HJ Thompson, HE Ganther 1998 *Anticancer Res* 18:9
22. C Ip, HJ Thompson, HE Ganther 2000 *Cancer Epidemiol Biomarkers Prev* 9:49
23. HJ Thompson, R Strange, PJ Schedin 1992 *Cancer Epidemiol Biomarkers Prev* 1:597
24. D Medina, HJ Thompson, HE Ganther, C Ip 2000 *Cancer Lett* in press
25. D Medina, CJ Oborn 1981 *Cancer Lett* 13:333
26. C Ip, HJ Thompson, HE Ganther 2000 *Cancer Epidemiol Biomarkers Prev* 9:49
27. R Sinha, TK Said, D Medina 1996 *Cancer Lett* 107:277
28. R Sinha, SC Kiley, JX Lu, HJ Thompson, R Moraes, S Jaken, D Medina 1999 *Cancer Lett* 146:135
29. CJ Sherr 2000 *Cancer Res* 60:3689
30. R Gopalakrishna, ZH Chen, U Gundimeda 1997 *Arch Biochem Biophys* 348:37
31. R Gopalakrishna, U Gundimeda, ZH Chen 1997 *Arch Biochem Biophys* 348:25
32. J Lu, M Kaeck, C Jiang, AC Wilson, HJ Thompson 1994 *Biochem Pharmacol* 47:1531
33. HJ Thompson, A Wilson, J Lu, M Singh, C Jiang, P Upadhyaya, K el Bayoumy, C Ip 1994 *Carcinogenesis* 15:183
34. MR Kaeck, S Briggs, HJ Thompson 1993 *Anal Biochem* 208:393
35. W Jiang, Z Zhu, HE Ganther, C Ip, HJ Thompson 2000 *Cancer Lett* in press
36. Z Zhu, W Jiang, HE Ganther, C Ip, HJ Thompson 2001 *Molecular Carcinogenesis* in press
37. SW Lowe AW Lin 2000 *Carcinogenesis* 21:485
38. HS Park, SH Huh, Y Kim, J Shim, SH Lee, IS Park, YK Jung, IY Kim, and EJ Choi 2000 *J Biol Chem* 275:8487
39. J Folkman, K Watson, D Ingber, D Hanahan 1989 *Nature* 339:58
40. C Jiang, W Jiang, C Ip, H Ganther, J Lu 1999 *Mol Carcinog* 26:213
41. O Thorlacius-Ussing, A Flyvbjerg, H Orskov 1988 *Toxicology* 48:167
42. O Thorlacius-Ussing, A Flyvbjerg, KD Jorgensen, H Orskov 1988 *Acta Endocrinol* 117:65
43. O Thorlacius-Ussing, A Flyvbjerg, J Esmann 1987 *Endocrinology 120*:659
44. M Pollak 2000 *Eur J Cancer* 36:1224
45. A Grimberg 2000 *Genet Metab* 70:85
46. R O'Connor, C Fennelly, D Krause 2000 *Biochem Soc Trans* 28:47
47. E Giovannucci 1999 *Horm Res 51 Suppl* 3:34
48. D Hanahan, RA Weinberg 2000 *Cell* 100:57
49. F LePage, A Margot, AP Grollman, A Sarasin, A Gentil 1995 *Carcinogenesis* 16:2779
50. A M Pace, YH Wong, HR Bourne 1991 *Proc Natl Acad Sci USA* 88:7031
51. JE Oldfield 1987 *J Nutr* 117:2002
52. JE Spallholz 1994 *Free Radic Biol Med* 17:45
53. W Mertz 1995 *Nutr Rev* 53:179
54. TC Stadtman 1994 *Adv Inorg Biochem* 10:157
55. OA Levander 1987 *Annu Rev Nutr* 7:227
56. JE Spallholz 1997 *Biomed Environ Sci* 10:260
57. RJ Shamberger 1985 *Mutat Res* 154:29
58. S Vadhanavikit, C Ip, HE Ganther 1993 *Xenobiotica* 23:731

# Chapter 24. Evolution of human dietary standards for selenium

Orville A. Levander

*Beltsville Human Nutrition Research Center, U.S. Department of Agriculture, Agricultural Research Service, Beltsville, MD 20705, USA*

**Summary:** An overview of the evolution of human dietary standards for selenium in the United States is presented beginning with the Estimated Safe and Adequate Daily Dietary Intakes (ESADDIs) in 1980, followed by the Recommended Dietary Allowances (RDAs) in 1989, and concluding with the Dietary Reference Intakes (DRIs) in 2000. Also included is a discussion of the 1996 standards of the World Health Organization (WHO). The overall trend in the selenium recommendations for adults in the U.S. during the past 20 years has been downward and the WHO standards are even lower. This is not surprising since nutritionists initially tend to propose somewhat elevated standards that gradually decline as the knowledge base for a given nutrient increases. Also, WHO recommendations have historically been lower than their U.S. counterparts for a variety of nutrients. Whether it will be desirable to halt or reverse the trend toward lower dietary selenium standards in the future depends upon the ability to demonstrate significant cancer chemopreventive or other benefits of an increased selenium intake.

## Introduction

Initially, the role of selenium in biology was considered only in terms of its toxic properties and the focus on selenium in human and animal disease was entirely on selenosis, or selenium poisoning [1]. Despite the observation of Pinsent in 1954 that selenium had beneficial properties in bacteria [2], few in the nutrition community were prepared for the announcement by Schwarz and Foltz in 1957 [3] that traces of dietary selenium could protect vitamin E-deficient rats from liver necrosis. The fundamental significance of this seminal discovery in trace element nutrition was grasped quickly, and this increased knowledge was rapidly and widely applied throughout agriculture around the globe. Many animal diseases of formerly uncertain etiology were found to respond to selenium supplementation, including white muscle disease in sheep, horses and cattle, exudative diathesis in poultry, and hepatosis dietetica and cardiomyopathy (mulberry heart disease) in swine [4]. Never before had a single breakthrough in basic trace element nutrition research provided such an immediate and substantial economic benefit to farmers and ranchers worldwide.

Meanwhile, progress in clarifying the role of selenium in human nutrition was advancing more slowly. In fact, 22 years elapsed between the original report of Schwarz and the publication by the Beijing group in 1979 linking selenium deficiency with Keshan disease, a cardiomyopathy that affects primarily infants, children, and women of child-bearing age [5]. The Chinese scientists found that Keshan disease occurred only in those areas of China that had selenium-poor soils. Low blood and hair selenium levels correlated with the increased incidence of the disease and, most importantly, it was possible to protect against the disease by giving selenium supplements to the general population [6].

Less than fifty years have passed since selenium was discovered to be an essential element. Most nutrition scientists would agree that the advances in our understanding of the role of selenium in human nutrition during that time have been remarkable. Prior to the Ninth Edition of the Recommended Dietary Allowances (RDAs), trace elements in general were accorded scant space in the discussion of dietary standards. For example, in the Seventh Edition (1968), selenium and molybdenum were lumped together in a single paragraph of six lines, given three literature citations each, and put under the heading "Chromium, Cobalt, Manganese, Molybdenum, Selenium, and Zinc" [7]. By the time of the Eighth Edition (1974), selenium warranted its own nine-line paragraph but still got only three literature citations and was placed under the heading "Other Trace Elements" along with seven other minerals [8]. No attempt was made to estimate quantitative dietary requirements for selenium in either of these editions.

The latest three editions of the RDAs, published approximately every ten years during the period 1980 to 2000, have seen selenium gain more and more attention as a factor of importance in human nutrition. This review will examine the state of our knowledge about selenium during each of those three editorial cycles as a way of demonstrating the growth in our understanding of this trace mineral needed for health and well-being.

**RDAs—Ninth Edition (1980)**
In the 1970's, research in trace element nutrition was accelerating. Selenium was shown to be a constituent of the active site of glutathione peroxidase in 1973 and Keshan disease was shown to respond to selenium supplementation in 1979. The latter finding, however, was too late for inclusion in the Ninth Edition of the RDA in 1980 [9] because of the long lead time involved in reviewing and approving the document. The discussion about selenium now occupied almost two full pages and contained much more definitive statements concerning its role in nutrition such as "the essential function of selenium has been proven conclusively in many animal species" and "in

view of the well defined selenium requirement of many animal species and the role of selenium in an important enzyme system, it can be stated that selenium must be essential for man as well".

The Ninth Edition also proposed an innovative dietary standard by introducing the so-called "Estimated Safe and Adequate Daily Dietary Intakes" (ESADDI). In the past, RDAs were established only when there were sufficient data to justify a separate quantitative value that could be listed in the "RDA Table" for all the various individual age, weight, and gender groupings. The ESADDI, on the other hand, were presented as a range of values that were to be regarded as "more tentative and evolutionary than the RDA". The ESADDI were presented for only seven age categories, whereas RDAs were developed for 17 age/weight/gender designations. The 1980 ESADDI for selenium are listed below:

**Table 1.** The 1980 Estimated Safe and Adequate Daily Dietary Intakes for Selenium

| Life Stage (µg/day) | Age (years) | Selenium Intake |
|---|---|---|
| Infants | 0-0.5 | 10-40 |
|  | 0.5-1 | 20-60 |
| Children | 1-3 | 20-80 |
|  | 4-6 | 30-120 |
|  | 7-10 | 50-200 |
|  | 11+ | 50-200 |
| Adults |  | 50-200 |

Adapted from [9].

It should be emphasized that in 1980 there were no quantitative human data available from which to calculate a dietary standard for people. Rather, the ESADDI were proposed on the basis of extrapolation from animal experiments. First of all, it was pointed out that "in all mammalian species examined, a selenium concentration of 0.1 µg/g of diet is adequate for optimal performance of growth and reproduction." Then it was stated that "by extrapolation from the selenium requirement of mammalian animal species, it can be assumed that this dietary concentration will also meet the human requirement." The lower limit of the selenium ESADDI for adults was calculated by assuming a consumption of 500 g of a mixed diet (dry matter basis) containing 0.1 µg of selenium per gram of diet to yield 50 µg of selenium daily. The upper limit of the selenium ESADDI for adults was based on that maximal intake "...which should not be exceeded habitually if the risk of long-term chronic overexposure is to be avoided." Recommendations for other age groups were extrapolated from the adult range on the basis of expected food consumption.

A major concern at the time of the Ninth Edition was to avoid the overconsumption by the general population of any of the six minerals accorded ESADDI status, including selenium. Although the upper limit of the ESADDI was never meant to be a toxicological standard, it quickly became a *de facto* ceiling which helped control abuse of mineral supplements.

**RDAs---Tenth Edition (1989)**
Interest in the possible benefits of selenium in human health continued to grow well into the 1980's and this was reflected in the doubling of the text pages devoted to selenium in the Tenth Edition [10]. Literature citations grew also, increasing six-fold. This was due not only to the increasing number of papers dealing with selenium but also to the expressed desire of the RDA Committee to make the RDA book more "scientific" and one in which every step of the derivation of the RDA was "transparent" so that the logic and reasoning behind the derivation of the dietary standard was clear and open for everybody to see.

Fortunately for selenium researchers, several excellent studies from China allowed the RDA Committee to pinpoint human selenium requirements with increased precision such that it was possible to advance selenium from ESSADI to full RDA status. One group of studies examined the dietary selenium intake needed to prevent Keshan disease in certain endemic regions of China [11]. The disease did not occur in those areas where the selenium intake by adults was 17 µg/day or more. Thus, 17 µg/day was suggested as a kind of minimum daily requirement.

In another study, a "physiological" selenium requirement was determined by following increases in GPX activity in the plasma of men living in a Keshan disease area who were given graded doses of selenomethionine as a supplement over a period of several months [12]. At total intakes of 40 µg/day or more (10 µg from diet plus supplement), the plasma GPX activities all maximized at about the same level. Therefore, it was concluded that the Chinese men had a physiological requirement of 40 µg/day.

In order to convert this figure to an RDA for North American males, it was necessary to apply a correction factor for differences in body weight (79/60) and to apply a safety factor to allow for individual differences in requirement (1.3). Thus, the calculation for adult males became:

$$40 \times 79/60 \times 1.3 = 70 \text{ μg/day (rounded)}.$$

For adult North American females, the calculation was:

$$40 \times 63/60 \times 1.3 = 55 \text{ μg/day (rounded)}.$$

**Table 2.** The 1989 Recommended Dietary Allowances for Selenium (μg/day)

| Life Stage (μg/day) | Age (years) | RDA for Selenium |
|---|---|---|
| Infants | 0-0.5 | 10 |
| | 0.5-1.0 | 15 |
| Children | 1-3 | 20 |
| | 4-6 | 20 |
| | 7-10 | 30 |
| Males | 11-14 | 40 |
| | 15-18 | 50 |
| | 19-24 | 70 |
| | 25-50 | 70 |
| | 51+ | 70 |
| Females | 11-14 | 45 |
| | 15-18 | 50 |
| | 19-24 | 55 |
| | 25-50 | 55 |
| | 51+ | 55 |
| Pregnant | | 65 |
| Lactating | | 75 |

Adapted from [10].

A more detailed explanation of the RDA calculations for adults was presented elsewhere [13]. Because of the lack of data, the RDAs for young adults also served as the basis of RDAs for the elderly. Likewise, because of the lack of data, RDAs for infants and children were based on adult values with extrapolations downward on the basis of body weight plus a factor arbitrarily allowed for growth. The RDA during pregnancy was calculated using a factorial technique based on the fetal accretion of selenium. The RDA during lactation provided sufficient selenium to avoid depletion of the mother and permit a satisfactory selenium content in breast milk. A critique of the assumptions and calculations used to derive the 1989 RDA for selenium has been offered [14].

The Tenth Edition discussed selenium toxicity only in general terms. An episode of human selenosis in China was described in which hair loss and fingernail changes were observed on intakes approximating 5 mg/day (i.e., 5000 µg/day). It was pointed out that sensitive and specific biochemical indices of selenium overexposure were not available and no attempt was made to establish an upper limit of selenium intake.

**World Health Organization (1996)**
In 1996, the World Health Organization (WHO) published its dietary standards for several trace elements, including selenium [15]. WHO has the responsibility for setting recommendations that must apply to many different countries around the globe (United Nations members) that have highly varied national diets. For that reason, the Organization tends to suggest nutrient intakes that are often somewhat lower than those set in the U.S.A This also turned out to be true for selenium since large parts of the U. S. Great Plains, a major wheat production area, have soils that are rich in selenium and relatively generous amounts of the trace element are incorporated into the food chain. Intakes of selenium exceeding 100 µg/day are not uncommon in the U.S. and so meeting an RDA of 55 to 70 µg/day is not difficult.

On the other hand, meeting the 1989 RDA for selenium could be quite a challenge for some other countries. Many parts of China, for example, routinely consume much lower amounts of selenium in their diet [11] and New Zealanders rarely ingest such RDA levels [16]. Likewise, Finland had a low-selenium food supply before deciding in 1985 to add selenium to its fertilizers [17]. In fact, dietary surveys indicate that several European countries would have problems achieving intakes as high as the 1989 RDA, including Belgium, Denmark, France, Germany, United Kingdom, Slovakia, and Sweden [reviewed in 18]. The selenium intake in Switzerland was somewhat higher because of the common use of North American wheat rich in selenium. So it is not surprising that WHO was reluctant to set a dietary standard that so many of its member states could not attain, especially in the absence of any evidence of overt signs of human selenium deficiency outside of China.

The reader will recall that the rationale used by the 1989 RDA Committee for its selenium recommendation was full expression of GPX activity. The WHO Committee decided that such full activity was probably not necessary for human health and that only two-thirds full activity of GPX still afforded sufficient protection against oxidative stress. This conclusion was based on clinical observations that blood cells metabolized hydrogen peroxide satisfactorily until their GPX activity fell to one-quarter or less than normal. Of course, if one selects a lower target GPX activity for the biochemical criterion of adequate nutriture, this allows a lower dietary standard to be proposed also. In this case, the WHO Committee (formal designation: Joint FAO/IAEA/WHO Expert Consultation on Trace Elements in Human Nutrition) came up with 40 and 30 µg/day for the lower limit of the safe range of population mean dietary selenium intake that would meet the normative requirement of most adult males and females, respectively.

As defined by WHO, the *normative requirement* referred to the "level of intake that serves to maintain a level of tissue storage or other reserve that is judged by the Expert Consultation to be desirable" [15]. WHO also defined a *basal requirement* that referred to the "intake needed to prevent pathologically relevant and clinically detectable signs of impaired function attributable to inadequacy of the nutrient". For selenium, the basal requirement was taken from the quantity needed to protect against Keshan disease. The lower limit of the safe range of population mean dietary selenium intake that would meet the basal requirement of most adult males and females was calculated to be 21 and 16 µg/day, respectively, after adjusting for body weight.

The WHO Committee also attempted to deal with the question of tolerances of high dietary selenium intakes. On the basis of considerable fieldwork with human selenosis in China, Yang and associates proposed 750-850 µg as a marginal level of daily safe dietary selenium intake [19], defined as "the level of selenium intake at which few individuals have functional signs of excessive intake and above which the tendency to exhibit functional signs is apparent and symptoms may first appear among ... susceptible individuals [whose] selenium intake [is] further increased". The Committee took one-half of the average of this range because of the uncertainty surrounding the harmful dose of selenium for people to suggest a maximal daily safe dietary selenium intake of 400 µg.

**Dietary Reference Intakes (2000)**
The new millennium saw a host of changes in the way that dietary standards for selenium (and many other nutrients) were handled in the U.S.A. [20]. First of all, selenium now was grouped with a variety of so-called "dietary antioxidants" (vitamins C and E and the carotenoids) instead of with the trace elements where it had traditionally been put. This change made sense because selenium, due to its multitude of roles protecting against oxidative stress, really had more in common with the nutritional antioxidants than it did with a collection of various microminerals.

Another substantial change was in the dietary standards themselves [20]. The general term "Dietary Reference Intakes" was used to describe not only the RDA, but also Adequate Intake (AI), Tolerable Upper Intake Level (UL), and Estimated Average Requirement (EAR).

Table 3. The 1996 WHO Lower Limits of the Safe Ranges of Population Mean Intakes of Dietary Selenium

| Life Stage | Age (years) | Se Basal Intakes (µg/day) | Se Normative Intakes (µg/day) |
|---|---|---|---|
| Infants | 0-0.25 | 3 | 6 |
| | 0.25-0.5 | 5 | 9 |
| | 0.5-1.0 | 6 | 12 |
| Children | 1-3 | 10 | 20 |
| | 3-6 | 12 | 24 |
| | 6-10 | 14 | 25 |
| Males | 10-12 | 16 | 30 |
| | 12-15 | 19 | 36 |
| | 15-18 | 21 | 40 |
| | 18+ | 21 | 40 |
| Females | 10-12 | 16 | 30 |
| | 12-15 | 16 | 30 |
| | 15-18 | 16 | 30 |
| | 18+ | 16 | 30 |
| Pregnancy | | 18 | 39 |
| Lactation | | | |
| 0-3 months | | 21 | 42 |
| 3-6 months | | 25 | 46 |
| 6-12 months | | 26 | 52 |

Adapted from [15].

Each of these terms has a particular role in describing the dietary standards of a nutrient and since they are relatively new terms (even the RDA was newly defined), it might be worthwhile to repeat here their meanings as presented by the Panel on Dietary Antioxidants and Related Compounds:

*"Recommended Dietary Allowance (RDA)*: the average daily dietary intake level that is sufficient to meet the nutrient requirements of nearly all (97 to 98 percent) healthy individuals in a particular life stage (which considers age, and when applicable, pregnancy and lactation) and gender group.

*Adequate Intake (AI)*: a value based on observed or experimentally determined approximations or estimates of nutrient intake by a group (or groups) of healthy people that are assumed to be adequate—used when an RDA cannot be determined.

*Tolerable Upper Intake Level (UL)*: the highest level of daily nutrient intake that is likely to pose no risk of adverse health effects to almost all individuals in the general population. As intake increases above the UL, the risk of adverse effects increases.

*Estimated Average Requirement (EAR)*: a daily average nutrient intake value that is estimated to meet the requirement of half the healthy individuals in a life stage and gender group."

Thus, the recent definition of the RDA echoes that of the 1989 version [10], which states that they are "... the levels of intake of essential nutrients that ... are judged ... to be adequate to meet the known nutrient needs of practically all healthy persons." The AIs are reminiscent of the ESADDIs that were dietary standards to be used when insufficient data were available to posit an RDA. The UL represents the first formal attempt by an "RDA Committee" to establish a ceiling of intake for all the nutrients being considered by the group. In the "Dietary Reference Intakes" (what the handbook on dietary standards in North America is now called—the title is no longer "Recommended Dietary Allowances") an entire chapter is devoted to describing a model for the development of ULs for nutrients.

The EAR occupies a critical place in the new dietary standards for without it, there can be no RDA. These two entities are related by the equation:

$$RDA = EAR + 2SD$$

where SD is the standard deviation of the EAR.

If the SD is unknown, the 2000 Committee generally assumed a coefficient of variation of 10% for the EAR so that

$$RDA = 1.2 \times EAR.$$

The 2000 Committee based its EAR on 2 intervention trials designed to estimate selenium requirements by determining the intake needed to maximize plasma GPX activity. The first trial was carried out in China [12] and in fact was the same study that served as the basis for the 1989 RDA [13]. The selenium intake needed to maximize GPX in that work was 41 µg/day that came to 52 µg/day after adjustment for Western body weight. The second intervention trial was from New Zealand [16] and the 2000 Committee interpreted that research as suggesting an EAR of 38 µg/day. Although other interpretations of the New Zealand trial may be possible [18], the average of both the New Zealand and Chinese trials, 45 µg/day, was selected as the EAR. The RDA for adult males then was calculated as 45 × 1.2 to yield 55 µg/day.

Thus, by using a lower base requirement figure than the 1989 Committee (45 vs. 52 µg/day after adjustment for body weight) and a smaller correction factor for individual variation (1.2 vs. 1.3), the 2000 Committee arrived at a lower RDA figure for adult males than the 1989 Committee (55 vs. 70 µg/day). Given the reported greater susceptibility of women to develop Keshan disease, their RDA was also set at 55 µg/day despite their smaller body weight.

The 2000 Committee could find no data available to calculate an EAR for children or adolescents, so the RDAs for these age groups were extrapolated from young adult values. Similarly, there were no data that specifically addressed the selenium requirement for elderlies and the 2000 Committee found no information that suggested that the aging process impaired selenium absorption or utilization, so their RDA was the same as young adults.

A major philosophical shift occurred in the way that requirements were presented for infants up to one year of age. Since "no functional criteria of selenium status have been demonstrated that reflect response to dietary intake in infants", the 2000 Committee rescinded the 1989 RDA, so to speak, and replaced it with an AI based on the "mean selenium intake of infants fed principally with human milk." This fundamental change in viewing infant requirements was not limited to selenium. In fact, all nutrients to date have been accorded only AI status for infants, including calcium, magnesium, vitamin D, thiamin, riboflavin, etc. The 2000 Committee selenium AI for infants for the first and second six months of life are 15 and 20 µg/day, respectively, up 50 and 33% from their 1989 counterparts, respectively.

Using somewhat different assumptions, the 2000 Committee came up with RDAs for pregnancy and lactation that were slightly less than those set by the 1989 Committee (60 vs. 65 µg/day and 70 vs. 75 µg/day, respectively).

Another innovation in the 2000 DRI was the establishment of an upper limit of intake. The UL for selenium was based on the criteria of hair and nail brittleness and loss due to dietary overexposure in a high selenium region in China. Intakes of selenium from food sources were inferred from blood levels. A No-Observed-Adverse-Effect-Level (NOAEL) was calculated to be 800 μg/day. An uncertainty factor of 2 was chosen to protect sensitive individuals thereby leading to a UL of 400 μg/day for adults 19 years and older, a figure in agreement with upper limits set by others [15]. Finding no evidence of teratogenicity or selenosis in infants of mothers consuming high but not toxic amounts of selenium, the 2000 Committee kept to 400 μg/day UL for pregnant and lactating women. The UL for infants 0 to 6 months old consuming human breast milk exclusively was set at 45 μg/day based on the lack of any adverse effects (NOAEL) reported in such infants consuming breast milk containing 60μg selenium/L. The ULs for older infants, children, and adolescents were extrapolated on the basis of body weights.

Table 4. The 2000 Dietary Reference Intakes (DRI) for Selenium

| Life Stage | Age | DRI for Selenium (μg/day) |
|---|---|---|
| Infants | 0-6 mo | 15* |
| | 7-12 mo | 20* |
| Children | 1-3 y | 20 |
| | 4-8 y | 30 |
| Males | 9-13 y | 40 |
| | 14-70 y | 55 |
| | >70 y | 55 |
| Females | 9-13 y | 40 |
| | 14-70 y | 55 |
| | >70 y | 55 |
| Pregnancy | | 60 |
| Lactation | | 70 |

Adapted from [20]; values with asterisk are AI, others are RDA.

## Future human dietary selenium standards

Comparison of the human dietary standard for selenium over the past 20 years reveals that there has been a definite downward trend in the recommendations. The mean value for the adult ESADDI in 1980 was 125 µg/day, followed by the RDA of 70 µg/day for adult males in 1989 and then the 2000 RDA of 55 µg/day for adults of either sex. Such a decline is not unusual for a nutrient that is newly introduced into the RDA process. Faced with incomplete knowledge, nutrition scientists are more prone to err on the side of safety to prevent deficiency and adopt standards that tend to be somewhat inflated. However, as knowledge of requirements becomes more complete, nutritionists are more comfortable with accepting lower standards since the likelihood of doing harm through inadequate intake becomes correspondingly less.

What does the future hold for selenium? If past is prologue, the next set of U.S. dietary standards should be out in about 2010. Will the decline in the selenium standard continue as dieticians home in more exactly on the amount needed to prevent deficiency? Or will there be a reversal of this trend in anticipation of the possible beneficial effects of higher intakes of selenium against various viral infections and chronic human diseases such as cancer? Only future research can answer these questions and the next decade of selenium research promises to be as exciting as the last one.

### References

1. AL Moxon, MA Rhian 1943 *Physiol Rev* 23:305
2. J Pinsent 1954 *Biochem J* 57:10
3. K Schwarz, CM Foltz 1957 *J Am Chem Soc* 79:3292
4. Subcommittee on Selenium, Committee on Animal Nutrition, Board on Agriculture, National Research Council 1983 *Selenium in Nutrition,* Revised Edition National Academy Press Washington pp 174
5. Keshan Disease Research Group 1979 *Chin Med J* 92:477
6. Keshan Disease Research Group 1979 *Chin Med J* 92:471
7. Food and Nutrition Board, National Research Council 1968 *Recommended Dietary Allowances,* 7th Revised Edition National Academy Press Washington pp 101
8. Committee on Dietary Allowances, Food and Nutrition Board, National Research Council 1974 *Recommended Dietary Allowances,* 8th Revised Edition National Academy Press Washington pp 128
9. Committee on Dietary Allowances, Food and Nutrition Board, National Research Council 1980 *Recommended Dietary Allowances,* 9th Revised Edition National Academy Press Washington pp 185
10. Subcommittee on the Tenth Edition of the RDAs, Food and Nutrition Board, National Research Council 1989 *Recommended Dietary Allowances,* 10th Edition National Academy Press Washington pp 284
11. GQ Yang, KY Ge, J Chen, X Chen 1988 *Wld Rev Nutr Diet* 55:98
12. GQ Yang, LZ Zhu, SJ Liu, LZ Gu, PC Qian, JH Huang, MD Lu 1987 *Selenium in Biology and Medicine* Part B GF Combs Jr, JE Spallholz, OA Levander, JE Oldfield (Eds) Van Nostrand Reinhold, New York p589
13. OA Levander 1991 *J Am Diet Assoc* 91:1572

14. GF Combs Jr 1994 *Risk Assessment of Essential Elements* W Mertz, CO Abernathy, SS Olin (Eds) ILSI Press, Washington p167
15. *Trace Elements in Human Nutrition and Health. Report of a Joint FAO/IAEA/WHO Expert Consultation* 1996 World Health Organization, Geneva pp 343
16. AJ Duffield, CD Thomson, KE Hill, S Williams 1999 *Am J Clin Nutr* 70:896
17. A Aro, G Alfthan, P Varo 1995 *Analyst* 120:841
18. MP Rayman 2000 *Lancet* 356:233
19. GQ Yang, S Yin, RH Zhou, L Gu, B Yan, Y Liu, Y Liu 1989 *J Trace Elem Electrolytes Hlth Dis* 3:123
20. Panel on Dietary Antioxidants and Related Compounds, Food and Nutrition Board, Institute of Medicine 2000 *Dietary Reference Intakes for Vitamin C, Vitamin E, Selenium, and Carotenoids* National Academy Press Washington pp 506

# Chapter 25. Selenium in biology and human health: controversies and perspectives

Vadim N. Gladyshev

*Department of Biochemistry, University of Nebraska, Lincoln, NE 68588, USA*

**Summary:** Important unresolved questions raised by the contributors of this book and addressing roles of selenium in biology and human health are discussed. Resolving major scientific controversies in the field should further highlight a bright future for selenium in fundamental science, biotechnology and medicine.

## Introduction
The chapters of this book make a compelling case for the remarkable progress in selenium research on the role of this element in biology and medicine in recent years. Still, many questions remain unanswered and contradictions unresolved. Research in these areas should provide investigators with many exciting years in the field.

In this concluding chapter, some of the areas, which were selected by the contributors of this book to be important unresolved questions, are highlighted. Answers to these questions should lead to major advances in the field. These unresolved issues are supplemented with topics, which may constitute important challenges for future research or enhance applications of selenium biology in medicine and biotechnology.

## Mechanism of selenocysteine incorporation
Recent findings yielded long-sought components of the eukaryotic Sec insertion machinery, the SECIS-binding protein [Chapter 6] and the Sec-specific elongation factor [Chapter 7]. Further characterization of these and other components of the eukaryotic Sec insertion system [Chapters 2-5 and 8] should lead to advances in our understanding of the mechanism for Sec biosynthesis and insertion into polypeptide chains.

Future experiments should also reveal whether all components responsible for Sec insertion have now been identified. One question to be resolved is the role of the kinase in phosphorylating serine that is attached to selenocysteine tRNA [Chapter 3]. The presence of phosphoserine has long been recognized, but its function remains unknown. A second uncharacterized protein that may be specific for the eukaryotic Sec insertion system is the methylase that is involved in the formation of mcmUm at the anticodon position in Sec tRNA. Eukaryotes have two Sec tRNA isoacceptors [Chapter 3] and two selenophosphate synthetases [Chapter 4] and future studies should reveal specific roles for each of these factors.

In the last decade, many studies in the eukaryotic system were built on classic experiments by Böck and his colleagues that identified and characterized the principal components and the mechanism of Sec insertion into protein in bacteria [Chapter 2]. However, recent studies revealed unexpected complexities in composition and regulation of the eukaryotic Sec insertion system, where RNA-based mechanisms, such as nonsense mediated decay [Chapter 8], the role of the SECIS element [Chapter 5], hierarchy in selenoprotein expression [Chapters 8 and 14] and alternative splicing in selenoprotein genes, as well as multifunctional protein machinery [Chapters 6 and 7] appear to play dominant roles.

Finally, it is not clear if the mechanism for Sec insertion is common to all systems. Bacterial Sec insertion has only been characterized in *E. coli* [Chapter 2]. One limitation of this system is that *E. coli* has only three selenoproteins and these exhibit extensive sequence homology. Characterization of the Sec insertion system in other bacteria as well as in archaea should provide an answer as to whether the Sec insertion mechanism is general to all organisms [Chapter 2].

**Selenoproteins and regulation of cellular processes**
Only a few physiological processes in humans are known that are mediated by selenoproteins. These include activation and inactivation of thyroid hormones [Chapter 16], antioxidant defense [Chapters 14 and 15], sperm maturation [Chapter 22] and control of cellular redox processes [Chapters 14 and 15]. Prokaryotic processes involving several selenoproteins are better understood, but major unresolved questions also remain with respect to selenoprotein function and the specific role of selenium in protein [Chapter 10]. Less than a half of known eukaryotic selenoproteins have been characterized with respect to function, although ongoing studies may result in identification of functions for several additional selenoproteins [Chapters 9 and 11-14]. Identification and functional characterization of new selenoproteins will also reveal other processes in which selenoproteins are involved. Future research may extensively use functional genomics approaches, such as cDNA microarrays, to reveal a global picture of selenium regulation.

**Human selenoproteome**
Currently, the number of known vertebrate selenoproteins is 22 [Chapter 9]. With the completion of the sequencing of the human genome, it should be possible to identify, in the near future, the majority, if not all, human

*Controversies and perspectives* 315

selenoproteins through a combination of bioinformatics and functional genomics approaches. This goal is complicated by the fact that no reliable tools exist that identify selenoprotein genes because currently available programs recognize Sec-encoding TGA codons as stop signals. However, recently developed bioinformatics tools that analyze SECIS elements and homologies between selenoproteins and their homologs should be very useful in identifying selenoprotein genes [Chapter 9]. In addition to the human genome, all or the majority of selenoprotein genes may soon be identified in several other eukaryotic genomes, such as rats, mice and zebrafish. These approaches will also be useful to determine how widespread is the use of selenocysteine in organisms on earth. Although selenoproteins were found in the three major domains of life (e.g., bacteria, archaea and eukaryotes), certain representatives of these organisms lack selenoproteins.

**Mechanism of cancer prevention by selenium**
The landmark study by Clark et al. provided strong support for the cancer chemoprevention effect of dietary selenium and is consistent with the majority of epidemiological and animal studies [Chapter 17]. Nevertheless, available data do not justify the use selenium as a cancer preventive supplement in the human population. Clearly, additional studies are necessary to support the conclusion that supplementation of the diet with selenium decreases the incidence of human cancers, and to determine which cancers are prevented by this trace element and to what extent.

Future clinical trials will be assisted tremendously if the mechanism for cancer prevention by selenium is known. The lack of information on the mechanism appears to be the major drawback for further advances in this area. A current view on how selenium prevents cancer implicates low molecular weight selenium compounds in the chemoprevention effect of selenium [Chapters 17 and 23]. The argument in favor of this hypothesis is that cancers are best prevented by selenium concentrations greatly exceeding those needed for maximal expression of glutathione peroxidases 1 and 3. These enzymes are chosen as markers because they located at the bottom of the selenoprotein hierarchy and their expression is thought to be consistent with the selenium status of an organism [Chapter 24].

However, evidence also emerges for the role of selenoproteins in cancer prevention [Chapters 13 and 17]. For example, the recently identified 15 kDa selenoprotein may be one of the proteins involved in this process. Its expression is changed in cancers relative to normal tissues, its gene is located in a chromosomal region that encodes a possible tumor suppressor gene, and the selenoprotein has polymorphisms that affect expression of the protein in an allele- and selenium-dependent manner [Chapter 13].

Questions also remain as to whether glutathione peroxidases 1 and 3 best illustrate selenium requirements under environmental and genetic stresses. It is possible that in some individuals, who genetically predisposed to cancer, requirements for selenium differ from those in the normal human population.

**Selenium and human diseases other than cancer**
The role of selenium in etiology of several diseases has been established and the number of disorders, in which selenium is implicated, has steadily expanded over the past decade. As reviewed in this book, selenium is involved or implicated in Keshan and Kashin-Beck diseases [Chapter 18], viral suppression [Chapter 19], HIV infection [Chapter 20], aging, immune function [Chapter 21], male reproduction [Chapter 22] and other disorders and pathophysiological conditions [Chapter 18]. However, molecular mechanisms for these effects remain largely unknown. Further research should define the mechanisms and identify other biomedical areas in which dietary selenium is involved.

**Selenium nutritional levels in a general population versus individual-specific requirements**
The chemopreventive mechanistic considerations appear to be linked to a more general question of whether selenium requirements in the general human population can be adapted to satisfy specific ethnic, age and genetic needs of groups of individuals or even needs of a single individual. With the ongoing advances in cDNA microarrays, microchips and parallel genomic sequencing, information on the genetic make-up of an individual is not out of reach in the next decade or so. It is possible that the use of these new technologies will lead to determination of individual requirements for dietary selenium.

**Applications of selenium in biotechnology**
Recent discoveries of new selenoproteins [Chapter 9], characterization of reaction mechanisms and properties of selenoproteins [Chapter 10] and advances in the mechanism of selenocysteine incorporation [Chapters 5-7] may lead to various biotechnological applications. Selenium in the form of selenomethionine is already extensively used in x-ray crystallography in solving the phasing problem. The targeted incorporation of selenium into proteins has also been used occasionally for protein characterization, but technically was difficult to achieve. The finding that efficiency of selenocysteine incorporation is higher than previously thought and that it may be regulated by adjusting levels of components of the selenocysteine insertion machinery [Chapters 6-8] may allow expression of high levels of proteins containing selenocysteine at a specific place in the sequence. These proteins may then be characterized in vivo in cell culture or animal models or in vitro mechanistically and spectroscopically. In particular, this method should be useful to evaluate functions of specific cysteine residues. Cysteine

and selenocysteine differ by a single chalcogen atom and exhibit similar chemical properties. Nevertheless, nucleophilicity and redox properties of selenocysteine differentiate this residue from cysteine and thus provide unique opportunities for functional characterization.

# Index

## A

Activator protein-1 (AP1), 126, 210, 212, 261
Adequate Intake (AI), 306, 307, 308
Adhesion molecules, 257, 259, 261
Aging, 1, 197, 267–269, 316
$A^{1125}/G^{1125}$ polymorphism, 152
AIDS, *see* Human immunodeficiency virus
Alkylselenocyanates, 211
Alkylseleno-cysteine, 211
Allyl-selenocysteine, 211
Allylselenol, 211
Alzheimer's disease, 173
Apoptosis, 171, 210, 211, 230, 262, 289–290, 291
*Aquifex aeolicus*, 37, 38f
Archaea
    selenocysteine incorporation into, 19–20
    selenoproteins of, 101, 102t
Aryl selenocyanates, 211
Arylselenol, 211
Asthma, 173, 266–267
*Azotobacter vinelandii*, 39, 40

## B

Bacteria
    selenocysteine incorporation into, 7–20
    selenoenzymes of, 115–121
    selenoproteins of, 100–101, 102t
Basal cell carcinoma, 206
Benzylselenocyanate, 211, 287
Biotechnology, 316–317
Bladder cancer, 206
Brain
    selenoprotein P uptake, 132
    selenoprotein W uptake, 142–143
    thyroid hormone and, 196, 197
Brain cancer, 206
Breast cancer, 208, *see also* Mammary cancer
BthD, 103, 107, 112

## C

*Caenorhabditis elegans*, 37, 38f, 72, 112, 186
Calcium, 137, 140
Cancer prevention, 1, 2, 106, 205–214, 253, 268, 269–270
    emergence of selenium link, 205–207
    glutathione peroxidase and, 154, 174, 205, 207, 208, 209, 210, 212, 270, 315–316
    low molecular weight selenium compounds and, 285–296
    mechanisms of, 207–208, 315–316
    metabolic bases for, 208–213
    selenium species accounting for, 286–287
    Sep15 and, 147–148, 153–154
Carbon monoxide dehydrogenase, 101, 116t, 117–118
Cardiomyopathy, 198, 300, *see also* Keshan disease
$\beta$-Carotene, 207
Cell-mediated immunity, 257, 262–263
Cellular glutathione peroxidase (GPx), *see* Glutathione peroxidase-1
Children, 306t, 308, 309
*Chlamydia psittaci*, 263
*Clostridium*, *see* Glycine reductase; Proline reductase; Purine hydroxylase; Xanthine dehydrogenase
*Clostridium barkeri*, 117
*Clostridium purinolyticum*, 118–119, 120
*Clostridium sticklandii*, 13, 39, 120
Colon cancer, 206, 295
Coxsackievirus, 224, 235, 236–238, 239, 240, 241t, 244
Cretinism, 192, 197–198, 219, 220–224
Crohn's disease, 229, 266
Cysteine, 8–9, 41, 109, 110, 116, 126, 149, 161, 292–293, 316–317
Cysteyl-tRNA synthetase, 8, 12
Cystic fibrosis, 197
Cytokines, 257, 259, 261–262, *see also* specific types
    HIV and, 248–249, 252
    Keshan disease and, 238–239
Cytosolic glutathione peroxidase (GPx), *see* Glutathione peroxidase-1

## D

Deiodinase-1 (DI-1), 46, 47–48, 49, 76, 103t, 105, 191, 196, 197, 198, 221
    regulation of expression, 86, 87, 88, 89, 93, 94
    selenium deficiency and, 230
    structure and functions, 192–193, 194–195
Deiodinase-2 (DI-2), 103t, 105, 193–195, 196, 197, 221
Deiodinase-3 (DI-3), 103t, 105, 193–195, 196, 221

Deiodinases (DIs), 102, 105, 109, 189, 196–197, 223
  aging and, 268
  cancer prevention and, 205, 208, 209
  functions of specific types, 192–196
  Kashin-Beck disease and, 227
Diabetes, 173, 189, 198
*Dictyostelium discoideum,* 37, 38f, 112
Dietary Reference Intake (DRI), 300, 305–309
Dietary selenium, 2, *see also* Selenium deficiency
  cancer and, 153–154
  evolution of standards for, 300–310
  general population vs. individual needs, 316
  selenoprotein W and, 142–143
Dimethyldiselenide, 213
Dimethylselenide, 210, 295
Dimethyl selenoxide, 210
*Dio1* gene, 195
*Dio2* gene, 193, 195
*Dio3* gene, 195
Diphenylselenide, 288
Diselenide bonds, 212
Disulfide bonds, 212
*Drosophila,* 105, *see also* BthD; G-rich
*Drosophila melanogaster,* 37, 38, 72, 112

### E
eEF1, 70
eEF2, 70
eEF1A, 72, 76
eEFsec, 56, 58, 63, 65, 69, 72, 76, 77, 78, 79
  interactions between SBP2, SECIS element, and, 73–75
EF, 70–71, 72–73
EF-Tu, 7, 12–13, 18, 70
Ehrlich ascites tumors, 210
Eicosanoids, 257, 259–260
Elongation factor, *see* EF
Endocrine function, 189–199
eRF1, 58, 64, 76, 78
eRF3, 64, 76, 78
*Escherichia coli,* 1, 7–8, 9, 10, 11f, 14–15, 16–17, 18, 33, 34, 55, 69, 75
  SECIS elements of, 45, 46, 47, 51
  selenoenzymes of, 115, 116, 119–120
  selenophosphate synthetase of, 35, 37–39, 40, 41
  selenoprotein W of, 142
  thioredoxin reductase of, 180–181, 184
E-selectin, 261
Esophageal cancer, 206, 207
Estimated Average Requirement (EAR), 306, 307–308
Estimated Safe and Adequate Daily Dietary Intake (ESADDI), 300, 301–302, 307, 310
*Eubacterium acidaminophilum,* 13, 100
*Eubacterium barkeri,* 119
Eukaryotes
  essential trans-acting factors in, 70–73
  SECIS-binding protein-2 of, 59
  selenocysteine incorporation into, 55, 69–79, 88, 314–315
  selenoproteins of, 102–109
  selenoprotein W of, 139
  selenoproteomes of, 108–109
Eukaryotic release factors, *see* eRF
Eukaryotic selenocysteyl-tRNA-specific elongation factor, *see* eEFsec
Extracellular glutathione peroxidase (GPx), *see* Glutathione peroxidase-3
Exudative diathesis, 300

### F
*fdhA* gene, 9, 12
*fdhB* gene, 9
*fdhC* gene, 9, 10
*fdhF* gene, 14, 16, 18, 33
*fdhF-lacZ* fusion reporter gene, 9, 15
*fdnG* gene, 14
Female reproduction, 189, 198–199
Formate dehydrogenase H, 9, 10, 33, 119–120
Formate dehydrogenase N, 9, 33
Formate dehydrogenases, 1, 100, 101, 102t, 115, 116, 119–120
Formylmethanofuran dehydrogenase, 102t, 116t
Fulvic acid, 226, 227–228

### G
*gadd* gene, 212
Gastrointestinal glutathione peroxidase (GPx), *see* Glutathione peroxidase-2 Genes
  glutathione peroxidase, 167
  selenocysteine tRNA, 25–27
  selenoprotein, 2, 107–108, 109, 110–113
  selenoprotein W, 140–141
  Sep15, 148–149
  thioredoxin reductase, 16–17
Glucose-6-phosphate dehydrogenase, 158, 169
χ-Glutamylcysteine synthetase, 158
Glutathione (GSH), 138, 143, 145, 157, 158–159, 162, 165
  clinical relevance, 173
  deiodinases and, 195
  male reproduction and, 278
  response to selenium, 169, 170

Glutathione peroxidase (GPx), 1, 10, 76, 99, 102–103, 116, 138
  aging and, 268, 269
  basic characteristics of family, 159–165
  cancer prevention and, 154, 174, 205, 207, 208, 209, 210, 212, 270, 315–316
  clinical relevance, 172–174
  dietary selenium and, 302, 304–305, 308
  immune function and, 261, 262, 263, 266, 267
  Kashin-Beck disease and, 227
  Keshan disease and, 302
  low molecular weight selenium compounds and, 288
  male reproduction and, 275–277
  overview of specific types, 104
  response to selenium, 167–172
  SECIS binding proteins and, 56, 65–66
  selenium deficiency and, 81, 83, 84, 127, 143, 167–170, 171–172, 173, 230
  selenium supplmentation and, 295
  selenoprotein P compared with, 123, 132–133
  selenoproteins of system, 157–174
  thyroid function and, 189, 197
  tissue distribution and subcellular location, 166–167
Glutathione peroxidase-1 (GPx1), 103t, 108, 157
  basic characteristics, 104, 159, 161, 162, 165
  as biological selenium buffer, 94–95
  cancer prevention and, 154, 315–316
  clinical relevance, 172–173
  immune function and, 259, 260
  Keshan disease and, 239–240
  male reproduction and, 275, 276, 279
  mRNA stability, 90–92, 167–168
  regulation of expression, 81–85, 86–87, 88, 89, 93
  response to selenium, 167–170, 171, 172
  SECIS elements and, 46, 49, 51
  selenium deficiency and, 81, 83, 84, 127, 167–170, 172
  in thyroid gland, 191–192
  tissue distribution and subcellular location, 166
Glutathione peroxidase-2 (GPx2), 103t, 104, 157
  basic characteristics, 159–161, 165
  response to selenium, 167, 170–171
  tissue distribution and subcellular location, 166
Glutathione peroxidase-3 (GPx3), 103t, 157
  basic characteristics, 104, 159, 161, 165
  cancer prevention and, 154, 315–316
  clinical relevance, 172–173
  male reproduction and, 276
  regulation of expression, 85, 86
  response to selenium, 167, 170
  selenoprotein P compared with, 124, 132
  in thyroid gland, 192
  tissue distribution and subcellular location, 166
Glutathione peroxidase-4 (GPx4), 103t, 108, 157
  basic characteristics, 104, 159, 161, 165
  immune function and, 259, 260
  male reproduction and, 172, 273–274, 275, 276–280
  regulation of expression, 81, 85–86, 87, 88, 89, 91–92, 93, 94, 95
  response to selenium, 167, 171–172
  SECIS-binding protein-2 and, 57, 60–61, 65
  in thyroid gland, 191–192
  tissue distribution and subcellular location, 166–167
Glutathione peroxidase-5 (GPx5), 104
Glutathione reductase, 158, 179, 181–182, 183, 185, 186
Glutathione-S-transferase (GSH-S-transferase), 276
Glutathione synthetase, 158
Glycine reductase, 99, 100, 115, 116, 120–121
Goiter, 220, 222
Grain contamination, 226, 228
G-rich, 103t, 104, 107, 112

# H

*Haemophilus influenzae,* 37, 38, 116t
*Halocynthia roretzi,* 112
Head and neck cancer, 206
Heart disease, 173, 260
Heparin binding, 127–128, 129–130
Hepatitis B, 206, 244
Hepatitis C, 235
Hepatosis dietetica, 300
Heterodisulfide reductase, 102t
Human immunodeficiency virus (HIV), 1, 2, 229, 235, 244, 247–253, 316
  chemoprevention, 252–253
  glutathione peroxidase and, 169, 230
  selenium and immunity, 251–252
  thioredoxin reductase and, 186
  wasting and, 248–251
Humans
  immune-mediated disease in, 266–267
  selenium deficiency and disease in, 219–231
  selenoprotein P in, 123–124, 125f, 126, 133
  selenoprotein W in, 139, 142, 143

Sep15 in, 148–149, 150f, 153
  thioredoxin reductase in, 183–184, 186
  thyroid hormone economy in, 197–198
Humoral immunity, 257, 262–263
Hydrogenase, 100, 101, 102t, 116t
*Hydrogenophaga pseudoflava,* 118
Hydrogen selenide, 205, 209–210, 211, 213, 287, 294
Hydroxylases, 117
Hypothyroidism, 194, 196, 220, 227

**I**

Immune function, 1, 257–270, 316
  cell-mediated, 257, 262–263
  HIV and, 251–252
  human disease and, 266–267
  humoral, 257, 262–263
  influenza virus and, 243–244
  Keshan disease and, 236–239
Infants, 306t, 308, 309
Influenza virus, 240–244
Insulin-like growth factor-1 (IGF-1), 292
Intercellular adhesion molecule-1, 261
Interferon (IFN), 248
Interferon-χ (IFN-χ), 239, 262, 268
Interleukin-1 (IL-1), 172, 239, 248, 261, 262
Interleukin-2 (IL-2), 239, 252, 268
Interleukin-2 (IL-2) receptors, 259, 265–266
Interleukin-4 (IL-4), 239
Interleukin-5 (IL-5), 239
Interleukin-6 (IL-6), 239, 248, 261, 262, 264f
Interleukin-8 (IL-8), 252, 261, 262, 264f
Interleukin-15 (IL-15), 239
Iodine deficiency, 192, 193, 196, 197–198, 208, 220, 221, 222, 225, 227
Iodine supplementation, 223–224
Iodothyronine deiodinases (DIs), *see* Deiodinases
Ischemia/reperfusion damage, 229

**K**

Kaposi's sarcoma, 252
Kashin-Beck disease, 173, 219, 225–228, 268, 316
Keshan disease, 137, 157, 169, 173, 219, 224–225, 235–240, 253, 268, 302, 305, 308, 316
  animal model for, 236
  viral mutations in, 239–240

**L**

Lactation, 304, 306t, 308, 309
*lacZ* gene, 9, 10

*lacZ*-luciferase fusion, 89
*Leishmania major,* 112
Leukemia, 172, 210
Leukotrienes, 259, 260
Liver cancer, 206, 207, 235, 244, 253
Low molecular weight selenium compounds, 285–296, 315
  neoangiogenesis inhibition by, 291
  selectivity and specificity hypothesis, 292–294
  toxicity, 294–295
Lung cancer, 206, 295
Lymphocytes, 265–266

**M**

Macrophage inflammatory protein 1α (MIP-1α), 237, 238, 243
Macrophage inflammatory protein 1β (MIP-1β), 238, 243
Malaria, 187, 258
Male reproduction, 1, 2, 172, 198, 273–280, 316, *see also* Sperm
Mammalian selenocysteine tRNA, 23–31
Mammalian Selenoprotein Gene Signature (MSGS), 99, 110–112
Mammalian thioredoxin reductase (TR), 180–181, 183–186
Mammary cancer, 210, 211, 212, 289, *see also* Breast cancer
MCP-1, 238, 243
Melanoma, 206
*Methanococcus jannaschii,* 11f, 19, 20, 37, 38f, 70, 116t
*Methanococcus maripaludis,* 20
*Methanococcus vannielii,* 19
*Methanococcus voltae,* 19
*Methanopyrus kandleri,* 19
Methionine, 9, 116
*p*-Methoxybenzeneselenol, 287
*p*-Methoxybenzyl-selenocyanate, 211
Methylated selenide metabolites, 209
Methylphenylselenide, 288
Methyl-selenocyanate, 210
Methylselenocysteine, 210, 211
Methylselenol, 205, 210–211, 213
Miscarriage, 198–199
Mobile Clinic Health Examination cohort study, 173
*Molluscum contagiosum,* 230
Monoselenophosphate, 7, 12, 27, 33, 34, 35
mRNA
  accumulation of selenoprotein W, 143–144
  stability of glutathione peroxidase, 90–92, 167–168

*Index* 323

stability of selenoprotein W, 144–145
Mulberry heart disease, 300
Mycotoxins, 228
Myxedematous cretinism, 192, 219, 220–224

**N**
NADPH, 170, 179, 180, 181, 182, 183, 184, 258, 260
Newcastle disease virus, 263
NFκB, 169, 171, 172, 230, 261
Nicotinic acid hydroxylase, 101, 115, 116t, 117
No-Observed-Adverse-Effect-Level (NOAEL), 309
Nutritional Prevention of Cancer (NPC) Trial, 206, 207–208

**O**
*Oligotropha carboxydovorans,* 117, 118
Ovarian cancer, 206

**P**
Pancreatic cancer, 206
Pancreatitis, 229
Panel on Dietary Antioxidants and Related Compounds, 306
Parkinson's disease, 173
Peroxiredoxin, 100, 101, 102t
Peroxynitrite reductases, 132–133, 259
1,4-Phenylene-bis (methylene) selenocyanate (p-XSC), 287–288
Phenylketonuria, 197, 229
*p*-Phenylselenocyanate, 211
6-Phosphogluconate dehydrogenase, 158
Phospholipid hydroperoxide glutathione peroxidase (PHGPx), *see* Glutathione peroxidase-4
Phosphoseryl-tRNA, 27–28
Plasma glutathione peroxidase (GPx), *see* Glutathione peroxidase-3
*Plasmodium falciparum,* 161, 165
Plateau break point, 83–84, 85–86
Pregnancy, 306t, 308, 309
Prokaryotes
 SECIS-binding protein-2 and, 59, 64
 selenocysteine incorporation into, 55
 selenoproteins of, 100–102
 thioredoxin system of, 180–181
Proline reductase, 100, 102t, 116t
Prostacyclins, 259, 260
Prostaglandins, 259, 260, 265
Prostate cancer, 153, 154, 206, 295
Protein kinase C (PKC), 210, 212, 290–291
P-selectin, 261

Purine hydroxylase, 115, 116t, 118–119

**R**
RANTES, 238, 243
Recommended Dietary Allowance (RDA), 300, 301, 302–304, 306, 307, 308, 310
Retinoblastoma, 290
Rheumatoid arthritis, 186, 187, 266

**S**
*Saccharomyces cerevisiae,* 116
*Salmonella,* 34
*Schistosoma mansoni,* 112, 139
SECIS-binding protein-1 (SBP1), 52, 57
SECIS-binding protein-2 (SBP2), 20, 51, 52, 55, 56
 eukaryote incorporation of selenocysteine, 69, 71–72, 73–75, 77–78, 79
 function, 59–61
 interactions between eEFsec, SECIS element, and, 73–75
 known properties, 57–59
 as master regulator, 65–66
 model for function, 63–65
 the ribosome and, 61–63
 selenoprotein expression and, 88
SECIS binding proteins (SBPs), 55–66, 314
SECISearch, 112–113
SECIS elements, 7, 17, 18, 19–20, 33–34, 45–52, 69, 99, 106, 107, 108, 315
 applications, 51–52
 archaeal selenoproteins and, 101
 in deiodinases, 195
 functional characterization, 46–47
 interactions between eEFsec, SBP2, and, 73–75
 Mammalian Selenoprotein Gene Signature and, 110
 SECISearch and, 112–113
 selenoprotein B interaction with, 13–16
 selenoprotein expression and, 81, 88, 89, 91, 92
 in selenoprotein W, 139
 in Sep15, 147, 149, 152
 structure and conserved features, 47–51
 thioredoxin reductase and, 183
*selA* gene, 12, 33, 69
*selB* gene, 12, 15, 33
*selC* gene, 10, 33, 69, 70
*selD* gene, 33, 34, 37–39, 69, 70
Selenide, 12, 33, 35, 38, 181, 182, 211, 260
Selenite, 28, 39
 cancer prevention and, 207, 209, 210, 211

immune function and, 262, 265
thioredoxin reductase and, 181, 182, 186
Selenium-carboxymethylselenocysteine, 115
Selenium deficiency, 81, 83, 84, 86, 91–92, 93, 94, 137
  cancer and, 269–270
  as a consequence of disease, 228–229
  endocrine function and, 189, 192, 196–197
  geography of endemic, 220
  glutathione peroxidase and, 81, 83, 84, 127, 143, 167-170, 171-172, 173, 230
  human disease and, 219-231
  immune function and, 236–239, 258, 260
  influenza virus and, 240–244
  selenoprotein biosynthesis and, 229–231
  selenoprotein P and, 123, 127, 130
  selenoprotein W and, 142–143
  sperm function and morphology in, 274–275
Selenium dioxide, 213
Selenium-metabolites, 205, 209–213
Selenium supplementation, 93, 197, 295–296
  cancer prevention and, 206–208
  HIV and, 252–253
  immune function and, 260, 265
  Keshan disease and, 235
  myxedematous cretinism and, 223–224
Selenium toxicity, 189, 219, 294–295
Selenobetaine, 210, 211
Selenocystamine, 262
Selenocystathione, 8
Selenocysteine, 1–2, 33, 34, 41, 99, 115–116, 316–317
  biosynthesis of, 11–12, 27–28
  cancer prevention and, 211
  deiodinases and, 194–195
  evolution of insertion in selenoprotein genes, 107–108
  glutathione peroxidase and, 159, 161, 165
  incorporation into archaeal proteins, 19–20
  incorporation into bacteria, 7–20
  incorporation into eukaryotes, 55, 69–79, 88, 314–315
  incorporation into prokaryotes, 55
  Mammalian Selenoprotein Gene Signature and, 110
  mechanism of incorporation, 314–315
  selenoprotein P and, 126, 127, 129
  selenoprotein W and, 138, 139, 141
  Sep15 and, 148, 149
  thioredoxin system and, 181, 182, 183, 184–185
Selenocysteine Insertion Complex (SIC), 55–56
Selenocysteine insertion sequence elements,
  see SECIS elements
Selenocysteine lyase, 39–40

Selenocysteine-specific elongation factor, see eEFsec
Selenocysteine synthase, 12, 69
Selenocysteine synthetase, 33, 34
Selenocysteine tRNA, 108, 314
  gene copy number, transcription and maturation, 25–27
  mammalian, 23–31
  mutant, 24–25, 26–27, 30–31
  over- and under-expression, 29–30
  population in cells and tissues, 28–29
  primary and secondary structures, 23–25
  SECIS-binding protein-2 and, 58, 63, 64, 65, 66
  selenoprotein expression and, 81, 88, 92, 93
Selenocysteyl-tRNA, 7, 8, 10–11, 12, 13, 15, 17–18, 19, 20, 27, 28
  selenocysteine incorporation into eukaryotes, 69–73, 77, 78, 79
  selenophosphate as selenium donor for, 33–34
Selenodiglutathione, 181–182, 205, 209–210, 287, 294
Selenoenzymes, 2
  bacterial, 115–121
  cancer prevention and, 205, 208–209
  as peroxynitrite reductases, 259
Selenomethionine, 8, 9, 115–116, 302, 316
  cancer prevention and, 210, 211
  immune function and, 262
  toxicity of, 294
Selenophosphate, 12, 33–42, 181
  biosynthesis of, 35–37
  identification as selenium donor, 33–34
Selenophosphate synthetase (SPS), 27, 28, 33, 34, 35–39, 69, 75, 101, 102, 116t
  delivery of selenium to, 39–41
  functions, 105
  structure, 100
Selenophosphate synthetase 1 (SPS1), 105
Selenophosphate synthetase 2 (SPS2), 103t, 104, 105, 112
Selenoprotein A (SelA), 1, 99, 100, 102t
Selenoprotein B (SelB), 7, 12–16, 17, 18–19, 34, 66, 71, 72, 100, 102t
  domain structure and interaction with SECIS, 13–16
  functions, 70, 73, 77
Selenoprotein N (SelN), 51, 103t, 104, 106, 109
Selenoprotein P (SelP), 64, 74–75, 76, 103, 109, 123–134
  brain uptake, 132
  cloning and sequencing, 124–126
  diquat model, 130–132
  expression, 126–127

functions, 105–106, 130
heparin binding and, 127–128, 129–130
immune function and, 262
male reproduction and, 275
primary structure, glycosylation, and isoforms, 127–129
purification and characterization in plasma, 123–124
regulation of expression, 86–87, 88, 89, 93, 94
selenium content, 129
Selenoprotein Pa (SelPa), 108
Selenoprotein Pb (SelPb), 103
Selenoprotein R (SelR/SelX), 51, 103t, 104, 106
Selenoproteins, 2, 99–113
    archaeal, 101, 102t
    bacterial, 100–101, 102t
    cancer prevention and, 285, 286, 287
    eukaryotic, 102–109
    evolution of insertion, 107–108
    of glutathione system, 157–174
    15 kDa, see Sep15
    male reproduction and, 275–277
    prokaryotic, 100–102
    regulation of expression, 81–95
    selenium deficiency and biosynthesis of, 229–231
    selenium regulation of translation, 87–90
    of thioredoxin system, 179–187
    transcript abundance, 93
Selenoprotein T (SelT), 51, 103, 106, 108
Selenoprotein T2 (SelT2), 103
Selenoprotein W (SelW), 103, 106, 108, 137–145
    amino acid sequences, 139–140
    background, 137–138
    metabolic function, 145
    purification and physical characteristics, 138
    regulation of expression, 89
    structure and antibody production, 141–142
    tissue distribution, 142
Selenoprotein W2 (SelW2), 103
Selenoprotein Z (SelZ), 51
Selenoproteomes, 108–109
Selenosis, 304
Selenotrisulfide bonds, 212
Selenylsulfide, 185–186, 212
Sep15, 103t, 104, 106, 147–154, 315
    $A^{1125}/G^{1125}$ polymorphism and, 152
    amino acid sequences, 149, 150f
    pattern of expression, 151
    UGTR and, 147, 151–152, 154
Seryl-tRNA, 13, 20, 27, 28, 30, 70, 72
Seryl-tRNA synthetase, 11–12, 69, 108
Sjogren's syndrome, 186

Skin, 267
Skin cancer, 206, 267, 270
Sperm
    differentiation, 157
    function and morphology, 274–275
    maturation, 277–280
Spermatogenesis, 274
Sperm mitochondria-associated cysteine-rich protein (SMCP), 275, 277, 278–279
*Sps1* gene, 28
*Sps2* gene, 28
Squamous cell carcinoma, 206
Stimulating protein-1 (SP1), 126
Stomach cancer, 206, 207
Stroke, 173
Synthetic selenium compounds, 211–212

**T**
TGA codons, 2, 33, 99, 107, 110, 315
    bacterial selenoenzymes and, 115, 119–120, 121
    Sep15 and, 148
    thioredoxin reductase and, 183
Thioredoxin (TRX), 179, 182, 186, 210
Thioredoxin reductase (TR), 179
    gene, 16–17
    immune function and, 259, 260
    isoenzymes of, 186
    medical aspects of selenium in, 186–187
    structure and mechanism of mammalian, 180–181, 183–186
    substrate specificity, 182
Thioredoxin reductase-1 (TR1), 103t
    properties, 105
    regulation of expression, 81, 87, 88, 89, 93, 94
Thioredoxin reductase-2 (TR2), 103t, 105
Thioredoxin reductase-3 (TR3), 103t, 105
Thioredoxin reductases (system), 102
    cancer prevention and, 205, 207, 209, 270
    deiodinases and, 195
    general properties, 180–181
    male reproduction and, 275, 279
    overview of specific types, 104–105
    selenium reduction by, 181–182
    selenoproteins of, 179–187
Threshold hypothesis, 260
Thromboxanes, 259, 260
Thyroid cancer, 206
Thyroid gland, 189, 190, 191, 221, 222–223
Thyroid hormones, 189, 196, 197–198, 208, 220, 221–222
Thyrotropin (TSH), 86, 192, 196, 197, 222–223, 227

Thyroxine (T4), 193, 196, 197, 198, 221, 222, 223, 227
Tolerable Upper Intake Level (UL), 306, 307, 308–309
Transforming growth factor-$\beta$, 262
3,5,3'-Triiodothyronine (T3), 193, 195, 196, 197, 198, 208, 221, 222, 227
Trimethylselenide, 295
Trimethylselenonium, 210
Triphenylselenonium, 288
Triphenylselenonium chloride (TPSe), 212
tRNA, 229–230
   cysteyl-, 8, 12
   phosphoseryl-, 27–28
   selenocysteine, see Selenocysteine tRNA
   selenocysteyl-, see Selenocysteyl-tRNA
   seryl-, see Seryl-tRNA; Seryl-tRNA synthetase
*Trypanosoma cruzi*, 263
*Trypanosoma musculi*, 268
Tumor necrosis factor-alpha (TNF$\alpha$), 169, 248, 252, 261, 262

## U

UDP-glucose:glycoprotein glucosyltransferase (UGTR), 106, 147, 151–152, 154
UGA codons, 7, 10, 13, 20, 23, 27, 28, 30, 33, 34, 51, 55, 56, 63–64, 69, 72, 73, 75, 76, 77–78, 107, 108
   bacterial selenoenzymes and, 115, 120
   deiodinases and, 193, 195
   mechanism of decoding with selenocysteine, 17–19
   redthrough vs. termination, 16–17
   SECIS elements and, 45–47
   selenoprotein expression and, 81, 88, 89–90, 92
   selenoprotein P and, 123, 124, 127, 128, 129, 133
   selenoprotein W and, 139
   Sep15 and, 152

## V

Vascular adhesion molecule-1, 261
*vhuU* gene, 19
Viral suppression, 1, 2, 235–245, 316
Vitamin C, 305
Vitamin E, 205, 207, 219, 263, 268, 300, 305

## W

White muscle disease, 137, 138, 143, 145, 300
World Health Organization (WHO), 225, 300, 304–305

## X

Xanthine dehydrogenase, 101, 115, 116t, 117, 118–119